Project Evaluation *in* Regional Perspective

A WORLD BANK RESEARCH PUBLICATION

Project Evaluation
in
Regional Perspective

A Study of an
Irrigation Project
in
Northwest Malaysia

Clive Bell • Peter Hazell • Roger Slade

with contributions by
Shantayanan Devarajan

Published for The World Bank
THE JOHNS HOPKINS UNIVERSITY PRESS
Baltimore and London

The Johns Hopkins University Press
Baltimore, Maryland 21218, U.S.A.

Editor **Jane H. Carroll**
Maps **Yung Koo and Larry A. Bowring**
Figures **Pensri Kimpitak**
Book design **Brian J. Svikhart**
Binding design **Joyce C. Eisen**

Library of Congress Cataloging in Publication Data

Bell, Clive, 1943–
Project evaluation in regional perspective.
Bibliography: p.
Includes index.
1. Irrigation—Economic aspects—Malaysia.
2. Irrigation—Malaysia—Data processing.
3. Rice—Economic aspects—Malaysia.
I. Hazell, P. B. R. II. Slade, Roger, 1941–
III. World Bank. IV. Title.
HD1741.M3B44 338.9595'1 81-48173
ISBN 0-8018-2802-3 AACR2

Contents

Units of Measure

ALL DOLLAR VALUES in the text are Malaysian dollars ($). The exchange rates for 1966–79 were:

Malaysian dollars per U.S. dollar

1966	1967	1968	1969	1970	1971	1972
3.08	3.06	3.06	3.06	3.06	3.06	2.82

1973	1974	1975	1976	1977	1978	1979
2.44	2.41	2.40	2.54	2.46	2.32	2.19

Unit	*Metric equivalent*	*Imperial equivalent*
Acre	0.40 hectare	1.00
Relong	0.29 hectare	0.71 acre
Ton	1.02 tonnes	1.00
Picul	60.48 kilograms	133.33 pounds
Gantang	2.54 kilograms	5.60 pounds
Pound	0.45 kilogram	1.00
Mile	1.61 kilometers	1.00
Inch	25.40 millimeters	1.00

Preface

THIS BOOK ORIGINATED in ideas formulated in the early 1970s by staff of the Food and Agriculture Organization (FAO) and the World Bank. The initial idea was that successful projects may generate indirect benefits through consumption expenditures by the project's beneficiaries and that those benefits will be the greater, the larger is the share of incremental income spent on locally produced goods and services. To pursue this idea, a study of the Muda River Irrigation Project in Malaysia was undertaken by the FAO/World Bank Cooperative Program. The study began early in 1972, with funding largely from the World Bank, staff from the FAO, and the support of the Malaysian government. After two years of fieldwork in Malaysia, analysis commenced in Rome in 1974. Earlier, in a separate study concerning the dynamics of change in rural areas, the World Bank's Development Research Center had evolved ideas similar to those mentioned above. Consequently, given the greater resources available in the Bank, analytical work on the Muda data was transferred from the FAO to the World Bank in 1975. Intensive research continued in Washington until 1979 and resulted in this book.

In this study we have attempted the ambitious task of analyzing and quantifying changes in a fairly complex regional economy. This has required the application of a wide range of skills and close collaboration sustained over several years. As a result, although each of us has made a particular contribution, the book is a truly joint piece of work. Roger Slade, while working for the FAO/World Bank Cooperative Program, designed and executed the initial field studies and during his two years in Malaysia acquired a detailed knowledge of the region's economy, which was an indispensable element in the subsequent analytical stages. Clive Bell and Peter Hazell also gained first-hand knowledge through field visits before building the models that have been used in the analysis. Hazell was primarily responsible for the formulation of the farm sector model, while Bell led the development of the regional model. The fitting of the models to the data and the interpretation of the results were lengthy and laborious tasks which we did together. Shantayanan Devarajan collaborated with Bell in the social cost-benefit analysis reported in chapter 8 (which is solely their responsibility) and contributed to the estimation of the social accounting matrix.

As with all products resulting from a sustained period of research, ideas and techniques enter as if by osmosis, and after all is complete the mandatory disentangling of the contributions of others is susceptible to the notable sin of omission. Set out below are the acknowledgments that still remain clear to us, but we hope that any whom we may have omitted will be consoled by the thought that their contributions are in no way diminished by the absence of specific mention.

We acknowledge our debt to R. A. (Robbie) Bishop, then deputy director of the FAO/World Bank Cooperative Program in Rome, whose seminal ideas and unflagging energy in large measure defined the questions we have tried to answer and launched the initial field studies.

In the World Bank, John Duloy and Roger Norton (now at the University of New Mexico) sharpened the focus of the original research design, offered advice and guidance on technical problems, and provided solid administrative support throughout. At every stage of the drafting, Duloy was also a meticulous and perceptive critic. A particular word of thanks also goes to Graham Donaldson, who was instrumental in bringing about the marriage of the originally separate initiatives in Rome and Washington.

Our debt is no less great to several officials in the government of Malaysia, whose open approach and active support significantly helped turn the original proposals into concrete investigations. Special mention must be made of Dato Tamin bin Yeop, then general manager of the Muda Agricultural Development Authority (MADA) and S. Jegatheesan, the principal economist. Both gave freely of their time, helped guide the original field studies, and subsequently made available to us other important data. S. Selvadurai (then senior economist in the Ministry of Agriculture) provided additional and valuable assistance during the early years. The fieldwork and initial analysis were, in large measure, the product of the hard work and attention to detail provided by P. Haridas and Wong Hin Soon, who were the principals in the total staff of forty persons recruited through MADA for the FAO/World Bank field surveys. Ramesh Chander (then director of the Department of Statistics) kindly provided us with data and special tabulations on several occasions, while C. L. Robless (then head of the Economic Planning Unit) was constant in his encouragement of our endeavors.

From 1975 onward, we benefited greatly from the advice, comments, and criticism provided by numerous colleagues in the World Bank. Notable among these were Bela Balassa, Charles Blitzer, Wilfred Candler, Randolph Harris, Gerald O'Mara, Graham Pyatt, Anandarup Ray, and T. N. Srinivasan, while the vast computational burden was willingly and ably borne at various times by S. K. Chan, Jansen Chang, Cynthia Hwa, R. Inman, Gyu Lee and Y. K. Wen. Grateful thanks are also due Ann Van Aken, who typed the manuscript, framed the bibliography, and removed many of our more obvious blunders.

Finally, we would like to record our thanks to the International Food Policy Research Institute, whose director, John Mellor, graciously consented to Peter Hazell's continued involvement in the research and writing of this book after he left the World Bank in 1979.

Some parts of the book initially appeared elsewhere in somewhat different form. Chapters 6 and 7 include a substantially expanded and modified version of "Measuring the Indirect Effects of an Agricultural Investment Project on Its Surrounding Region," by Clive Bell and Peter Hazell, *American Journal of Agricultural Economics*, vol. 62, no. 1 (February 1980). Chapter 8 is a revision of "Towards a Synthesis of Semi-Input-Output and Little-Mirrlees: A Social Cost-Benefit Analysis with Multiplier Effects of an Irrigation Project in Northwest Malaysia," by Clive Bell and Shantayanan Devarajan, *Pakistan Development Review*, vol. 18, no. 2 (Summer 1979), pp. 165-85. The article formed part of a "Symposium on Shadow Pricing" to which the Summer 1979 issue of the *Pakistan Development Review* was wholly devoted. The appendix to chapter 8 is a slightly revised version of "Semi-Input-Output and Shadow Prices: A Critical Note," by Clive Bell and Shantayanan Devarajan, *Oxford Bulletin of Economics and Statistics*, vol. 42, no. 3 (August 1980), pp. 251-56. Permission to include the above material in this book is gratefully acknowledged.

Project Evaluation
in
Regional Perspective

1

Introduction

IN AREAS blessed with large and successful projects, casual observation of the bustle and energy in local markets and enterprises suggests that in addition to directly generating output and incomes, such projects may have powerful spread effects on their surrounding regions. Whatever the hazards of predicting them, the direct effects of a project are usually easy to identify and not too difficult to measure. A plant or scheme is constructed; workers are engaged; materials and services are purchased; and the resulting output is sold, perhaps at a profit. If, after allowance is made for the value of these resources in other uses, nothing else in the economy changes, the story is complete, even though the telling of it in every detail may require much time and effort.

But what if the workers thus hired earn more income than in their old pursuits, or the owners make a higher return on this investment than the going rate elsewhere and so have abnormally high profits to dispose of? What if the increase in supply owing to the project lowers the price of its output and thus increases the profitability of new investments in the sectors that use it as an input? Each of these results will normally cause an increase in economic activity in the rest of the economy, and this increase may, in turn, produce a further impulse of its own. Then the total effects of the project on incomes and the net demands for goods and services in the economy will be different from its direct effects. The differences between them, which are usually called the project's indirect or "downstream" effects, provide the main theme for this book.

The downstream effects of a private project are of no interest to the entrepreneur who undertakes it; but when such effects arise from a project in the public sector, they should be of keen interest to the people who evaluate and finance such projects. There are a number of obvious questions: How large are the downstream effects and who reaps the benefits or losses? Will they be concentrated largely in the immediate geographical region in which the project is located? What will be their impact on government revenues? These are not new questions. Yet there have not been many attempts to provide quantitative estimates of the downstream effects of particular projects. Thus, although this book places a strong emphasis on methods, it is based on a case study of a project that was completed in the mid-1970s, namely, the Muda Irrigation Project in a relatively poor agricultural region of northwest Malaysia. The region, its economy, its farming system, and the project are described in some detail in chapter 2.

The choice of an agricultural project was not accidental. When peasant farmers get a sustained and appreciable rise in incomes, their prosperity is often reflected in buoyant and vigorous activity in the local economy. Distributive trades and transport benefit from farmers' operations as well as their demands for nonfarm consumer goods. A fair part of the extra income goes for entertainment and eating out; jobbing engineering and repair workshops spring up; agroprocessing industries increase their throughput; and construction experiences a temporary boom as local activities expand to meet additional demand and households make outlays to improve old houses or build new ones. On the face of it, the downstream effects of agricultural projects appear to be both large and concentrated quite heavily on the local economy. Hence, while we certainly do not claim that downstream effects are always large, the main theme of this book makes agricultural projects promising cases to study.

There are also more narrowly professional reasons for our choice. Here, it may help to start by defining more carefully what we mean by "downstream effects." That projects in developing countries will usually generate substantial indirect effects, because of the character of their market structures, was first argued fully by Scitovsky (1954). He referred to such effects as "pecuniary external economies," that is, effects on the profits of one firm arising out of the choice of inputs and outputs by other firms. These effects stem in part from linkages between sectors of production: "backward" linkages in the form of demands for intermediate and capital goods (in our example, fertilizers, tractors, fuel, and the like) and "forward" linkages in the form of sales to others (farmers produce more paddy, which is sold to rice mills for processing and may make existing or additional investment in rice mills more profitable).[1] Downstream effects also arise from expenditures by households out of the extra income generated by the project, which might be called the consumption linkages associated with the project. Interestingly, Hirschman (1959), the great advocate of the importance of linkages in promoting development, took the position that the linkages associated with agriculture are weak, with the corollary that public investment funds should be steered to other sectors of the economy. Recently, however, both Mellor (1976) and Johnston and Kilby (1975) have renewed

1. The general case, in which extra investment in the purchasing sectors is triggered by a fall in the price of the good purchased, has been analyzed extensively by Chenery (1959).

the debate concerning the importance of these linkages for economic growth in developing countries. Mellor particularly emphasizes the crucial role of the linkages associated with final expenditures out of increases in agricultural incomes. We have more to say about this issue in later chapters.

Another professional issue, which arises naturally from our institutional setting, is the relation of the theme of the book and our choice of case study to the theory and practice of social cost-benefit analysis. We have made clear already that the evaluation of a project's social profitability should include its downstream effects. It is then natural to ask how, if at all, they are allowed for in the standard literature on social cost-benefit analysis. We return to this question at length in chapter 8.

In this case study, however, there is a further problem. It is widely agreed that the treatment of agricultural projects in, for example, the well-known work of Little and Mirrlees (1974) is rather sketchy, and that the evaluation of agricultural projects presents some rather special problems. Most important, such projects are often large enough to affect the levels of wage rates and rents, as well as farmers' techniques of production. Since the project may well change the distribution of incomes among the direct beneficiaries, it is necessary to estimate its direct effects with some care, not only because the distribution of *direct* benefits is central to an evaluation of the project's social profitability, but also because the size and composition of the downstream effects may depend on the character of that distribution. Hence, there are compelling reasons to analyze the project's direct effects in some detail. Accordingly, this is the first task that we tackle, and to do so we build a formal model of the farming sector. Only then do we turn to the downstream effects of the project and integrate them into a formal scheme of social cost-benefit analysis.

The Direct Effects of the Project

The Muda area is a paddy monoculture, for no other crop can be grown on its heavy clay soils.[2] Irrigation has made it possible to grow two crops a year instead of one. Thus, if farmers could carry out the same schedule of operations twice in the twelve-month calendar, with labor and other inputs available at constant costs, predicting the direct effects of the project would be trivial. The paddy farming economy would, in effect, be simply replicated; in an annual accounting all inputs, outputs, costs, and revenues would double without affecting the distribution of income among farmers, landlords, and farm workers.

2. Consistent with common usage, paddy is the term used to describe rice, whether growing or harvested, prior to the milling process. Farmers grow paddy, millers produce rice.

In fact, no such simple replication was possible. Growing two crops a year makes for a more hurried rhythm of cultivation; as one crop is harvested, the land is quickly tilled again and then planted to the next. Thus, any tightness in the supply of labor will manifest itself in a rise in wage rates. This is exactly what happened. But if wage rates rise, it may become cheaper to prepare the land with tractors than with buffaloes, and with a fixed supply of land, rents will be affected too. Since all these variables are simultaneously determined, a formal model is needed to estimate the effects of the project upon them.

Moreover, the arrival of irrigation was not the only change in the external environment of Muda's farmers. New varieties of paddy, which could also be grown under rain-fed conditions, were introduced at the same time, and there were sharp movements in the price of paddy relative to the prices of inputs such as fertilizer and fuel, which are determined independently of what happens in the Muda project area. Both changes would have occurred even if the project had not been undertaken. Hence, to disentangle the effects of the project from those caused by autonomous changes in the external environment, we need to simulate how the paddy farming economy would have looked if the project had not been undertaken and other autonomous changes had not occurred. A formal model of the paddy farming sector is certainly required for this task, and we have chosen mathematical (linear) programming as the basis for it.

Such a model is built, tuned, and tested in chapter 3; it is then put to work disentangling the contributions of the various sources of change from one another in chapter 4. To anticipate the main results for 1967–74, if exogenous changes in prices are set aside, the effects of irrigation alone on a set of key variables—output, incomes, wages, rents, and technique of production—are much larger than those of new varieties alone, except that irrigation triggers complete mechanization of plowing, whereas new varieties preserve the initial dominance of buffaloes. When the big changes in exogenous prices are taken into account, however, they account for just over half the overall change in aggregate nominal incomes, wage rates, and land rents and also encourage the use of tractors. Moreover, the distribution of income among farmers, landlords, and landless farm workers is indeed sensitive to variations in external conditions—the presence or absence of the project, the introduction of new varieties, and shifts in exogenously determined prices.

The Indirect Effects of the Project

As already noted, the indirect effects of a project take many forms. Some firms increase their output in response to the extra demand for inputs as other firms expand, or in response to cheaper supplies of inputs; households spend some of their additional income and save what remains; firms make new investment plans; the government gains (or loses) revenues, depending on the structure of taxation; and so on. To have any hope of tracing all these repercussions, we need a framework within which all the transactions in the economy can be ordered and accounted for in a consistent way.

Fortunately, just such a framework is ready to hand: the so-called Social Accounting Matrix, or SAM.[3] A SAM is simply a system of accounts in which (by convention) the various incomings of each and every account are represented by an ordered row and the corresponding outgoings by a similarly ordered column. Thus, the whole is a square tableau, each cell of which contains the transaction between a particular pair of accounts in the form of an outlay of one and an income of the other. This array provides a picture of all the transactions in the economy at a moment in time, and in so doing, it also reveals something of the structure of the economy.

So much for the general principles; but there are practical decisions to be made. It would make no sense to lay out such transactions in minute detail. What is needed is a set of accounts that will serve the purposes of our analysis without degenerating into a rococo scheme of disaggregation. As we are keenly interested in the distribution of income, we define several classes of households. Given the nature of the project, it is obvious to distinguish between agricultural and nonagricultural households, farm households within and outside the command of the irrigation scheme, and, perhaps, farms of different size. The other institutions are firms and government, and they have their place in this schema. For production activities, we need enough detail to distinguish between sectors producing goods that are readily tradable and those producing mainly or wholly to satisfy domestic demand; and sectors using very different technologies or employing factors owned mostly by one class of institution. This, of course, raises another question, which is critical to the distinction between traded and nontraded goods: Where should the line be drawn between local activities and those located outside the region? The answer is to be found in chapter 2, where the geographical boundary of the regional economy is defined.

These considerations led us to construct a regional SAM with forty-five accounts in all, the design and building of which are described in chapter 5 and appendix C. This snapshot picture of the regional economy in 1972 provides some valuable insights of its own as well as some pointers on how to approach the analysis in subsequent chapters. Two findings stand out.

3. For an elementary introduction to the nature and habitat of these creatures, see King (1981).

First, the linkages connecting production activities appear to be weak, whereas households spend a large fraction of their incomes on locally produced goods. Second, the outflow of private capital from the region is very large, which suggests that the chances of self-sustaining growth are poor.

Yet valuable though it is, this picture says rather little about how the regional economy actually works, and still less about how it would respond to changes from the outside, such as the construction of an irrigation system or movements in the prices of goods in which it trades. To explain such responses, a model of the economy is needed, and that entails, inevitably, making specific assumptions about individual and institutional behavior. Here, the SAM provides a basis for estimating the parameters of a quantitative model. If, as they should be, the categories of institutions and activities that make up the SAM have been defined in the light of what the model is to be used for, the SAM is an essential part of the model's foundations; but a SAM is not, in itself, a model of the economy.

Clearly, we are dealing with a small, open economy, which is heavily dependent on primary commodities. Goods can be traded with the rest of the world at given prices, and though labor has only limited mobility, capital can flow freely in or out, depending on the opportunities for profitable investment. Apart from short-term adjustments, under these conditions the output of tradable goods will be determined, not by the level of domestic demand, but by the level of installed capacity in the sectors producing them. Conversely, the output of nontraded goods will be determined by the level of domestic demand from industries and households alike. Similarly, if the prices of tradable goods should change, the prices of nontradables are likely to change too, and with them the level and distribution of both nominal incomes and domestic demand. In all this, the downstream effects of a project are likely to make themselves felt largely as changes in the output of nontradables.

In principle, of course, we are talking of a general equilibrium system with a set of exogenous conditions, in which prices, quantities, and incomes are mutually and simultaneously determined. Fortunately, the salient features of the economy discussed above can be captured in a relatively simple and robust breed of model, namely, the so-called semi-input-output method of Tinbergen (1966). In chapter 6 we develop two variants of a semi-input-output model. One of them features prices of nontradables that are endogenously determined, given the prices of tradables; the other is a special case of the first in which all prices are fixed. With certain assumptions, which are not especially strong, most of their parameters can be estimated directly from the regional SAM. Moreover, both models have the virtue of being computationally tractable.

The next step is to use these models to disentangle the contributions of the various sources of change that were at work between 1967 and 1974. Apart from the Muda project, there were autonomous changes in production and expenditures, as well as changes in the prices of tradable goods. The autonomous changes took two forms: first, growth in the output of traded goods other than paddy, such as rubber, fish, and sawmilling products; and second, an increase in investment and current outlays by the government, the latter being especially large. For such an economy, the sectors producing traded goods and the government's expenditure decisions will normally act as the engines of growth. As it happened, the prices of tradables also changed markedly over the period: the prices of primary commodities, which the region exports, were relatively high in 1967 and 1974, and relatively low in 1972. Their effects, too, must be accounted for.

Such an accounting is attempted in chapter 7, to which chapter 4 is both companion and servant. Servant because to each set of circumstances of the economy, whether hypothetical or actual, there corresponds a particular configuration of the paddy farming economy. For example, in simulating the regional economy in 1974 in the hypothetical absence of the project and the presence of the prices of tradables that ruled in 1972, there is a set of results in chapter 4 that must be spliced into the semi-input-output models that generate the results in chapter 7. This link between the analyses of the project's direct and downstream effects is important, even though the models themselves are not linked in a formal way.

It turns out that the downstream effects on the region were large: for every dollar of value added generated directly by the project, another 80 cents or so were generated downstream, and, as hypothesized by Mellor (1976), they originated primarily from household expenditure linkages. Of course, these additional 80 cents were not "free." Fewer workers migrated to other parts of Malaysia, and savings were diverted from other uses to finance the balancing investments needed to increase the region's output of nontraded goods. As one would expect, most of the value added downstream accrued to nonfarm households. Yet in the absence of the project, the distribution of income between paddy farming households in the irrigation command area and the other households in the region would have worsened considerably, for the autonomous changes overwhelmingly favored townspeople. As it was, aside from the changes in the region's terms of trade, paddy farming households were doing better in 1974 than in 1967—both relatively and absolutely—and they had the project to thank for that. Moreover, the movements in the region's terms of trade proved to have a potent effect on the distribution of incomes between farm and nonfarm households. But sharp though these movements were, they had only a slight effect on regional output and value added in the aggregate.

This suggests that agricultural price policy has a valuable redistributive role to play in shaping the region's growth in the future, an idea we follow up in chapter 9.

Social Cost-Benefit Analysis

It is now natural to ask how the downstream effects of the project are to be incorporated into a full-blown social cost-benefit analysis. Until now, we have concerned ourselves only with those affecting the region, while recognizing that they are not "free" because labor and capital have to be withdrawn from uses elsewhere. Thus, we must begin by specifying all the relevant effects of the project on the national economy. First, there are goods which can be traded between the region and the rest of Malaysia, but are not normally traded internationally; good examples are paddy for milling and electric power. As these flows are treated in chapter 7, part of the response in supply must come from elsewhere in the national economy, with a corresponding diversion of labor and capital from other activities. Second, and this is crucial, there is the matter of government income. Not only does the government construct and run the irrigation system, but it also levies taxes on incomes, expenditures, and international trade. Since the total effects of the project will normally include changes in the levels of exports and imports of tradable goods, as well as private incomes and outlays, the change in government income must be traced in full.

The project is then evaluated as follows. For each year of the project's life, we take the changes in the incomes of all the institutions in the economy which are brought about by the project. Then, by weighting these changes to reflect the social value of savings relative to private consumption, and of extra consumption by the well-to-do relative to the poor, we obtain the change in the social value of national income. Finally, by discounting this stream of changes in the social value of national income at the appropriate rate of interest over the project's entire lifetime, we arrive at the project's net present social value.

To do all this, we have extended the fixed-price model in chapter 7 to include the production activities that are affected elsewhere in Malaysia and a simple specification of the government's fiscal system. Such a model is developed in chapter 8. It is, in effect, a model of the entire economy, although only those elements that are affected by the project are specified. It turns out that the project yields a handsome social profit irrespective of whether the premium on government income relative to private income is high or zero, or the government's concern about the distribution of private incomes is strong or nonexistent.

At this point, it may be asked how our procedure for evaluating the project is related to the methods in the standard works on the subject.[4] The literature provides rules for deriving the shadow prices (that is, the social scarcities) of all goods and factors, which are applied solely to the *direct* inputs and outputs of a project. Do these shadow prices capture all downstream effects, too? In particular, what becomes of the 80 cents per dollar that was one of the highlights of the preceding section? The answer is, of course, that if both methods are used correctly, they should yield the same result, an equivalence that is established in the appendix to chapter 8. Since practitioners frequently use short-cuts, this issue is not quite closed, and we return to it in chapter 10.

In setting up the model in chapter 8, we tried to stay as close as possible to the spirit of Little and Mirrlees (1974), whose celebrated work has been immensely influential in this field. Although we do not pursue the issue in detail in this book, we have shown elsewhere (Bell and Devarajan, 1980*b*) that the shadow prices corresponding to the formulation in chapter 8 are intimately related to those in Little and Mirrlees. Thus, chapter 8 has a certain pedagogic value. It reveals that Tinbergen's semi-input-output method is an important part of the foundations of the standard methods of social cost-benefit analysis, something that has always been understood by those versed in the latter, but not always perceived more widely.

The Prospects for the Region

Thus far, we have devoted our attention to the changes that were brought about by the project when it reached maturity in the mid-1970s. To close our analysis, in chapter 9 we look forward, however speculatively, to the changes that are likely to occur over the next decade or so and how public policy might influence the course of events.

Although the project has done a great deal to raise the incomes of Muda's paddy farmers, the rest of the Malaysian economy has also grown rapidly, with every prospect of continuing to do so over the longer run. As the impetus provided by the project is now fully spent, other measures are needed if the gap between paddy farmers' incomes and those earned by households elsewhere in Malaysia is not to widen, and if heavy migration from the region is not to follow. Apart from general increases in government expenditures, which are hardly precise interventions, two specific kinds of public policies could help stave off these developments.

4. Little and Mirrlees (1974); Squire and van der Tak (1975); UNIDO (1972).

The first is to improve farm technology, especially the irrigation system and the varieties of paddy available to farmers. Such a program of agricultural intensification is already under way, though it will not be complete until about 1990. The second is to raise the support price for paddy, while subsidizing the prices of farm inputs. This is a rather delicate decision politically, but an increase of 25 percent in the support price relative to other prices should not arouse trenchant opposition if phased in gently. Using the models in chapters 3 and 6, we conclude, somewhat tentatively, that the two policies combined will just about hold the position over the next decade. Thereafter, it seems likely that heavy out-migration will ensue, with an upward pressure on the size of farm holdings. In turn, this may encourage the emergence of part-time farmers, with associated changes in the fabric of the social and economic organization of the paddy farming sector.

Concluding Remarks

In this chapter we sought to lay out our aims, our plan of campaign, and why we have undertaken it. Although there are certainly conclusions to be drawn that should be of interest to those who make policy, they are better discussed after the framework has been assembled and the analysis has been completed. For without all the qualifications and reservations which attend the analysis, it is hard to discuss the implications for policy without feeling that one is shadow boxing. Similarly, some technical issues and procedures deserve careful scrutiny after all has been done. This, too, has been deferred until chapter 10. In the interim, there is much terrain to be covered.

2

The Region and the Project

FOR THE MOST PART, this chapter is a descriptive prelude to the methodology and analysis that follow. It outlines some of the changes which the analysis must address and provides empirical support for the assumptions on which the analysis is based. It also sets the analysis in its context, which even those who are mainly interested in methodology will appreciate; for without an account of the region and the business of its people, the analysis would be dry indeed. Yet we have attempted to go beyond mere description, which would be equally tedious.

We first sketch the main features of the region's physical geography, population, settlement pattern, and economic structure. This introduces the criteria used to define the region's boundaries, and we discuss how closely they match the requirements of the semi-input-output framework used in later chapters. A brief account of the economic and social organization of paddy farming in the region provides an essential preliminary to a description of the project and its effects on the economy and technology of the paddy farming system. In turn, this description gives much of the empirical background for chapters 3 and 4.

The Regional Setting

Our region, which embraces the Muda irrigation project, lies in the northwest of peninsular Malaysia and comprises the state of Perlis and the greater part of the four most northerly administrative districts of the state of Kedah (see map 1). The core of the region is the northern Kedah plain, a landscape quite distinct from the rest of peninsular Malaysia. Stretching inland from the coast for some twelve miles, the plain makes up the greater part of an unusually flat terrain that extends north and south for about forty miles and sprawls along the banks of the Kedah River, with the town of Alor Setar at its center. To the south, the plain ends abruptly in the forested slopes of Kedah Peak (nearly 4,000 feet high). To the east, it terminates in the low rolling foothills of the ranges of inner Malaysia, from which short

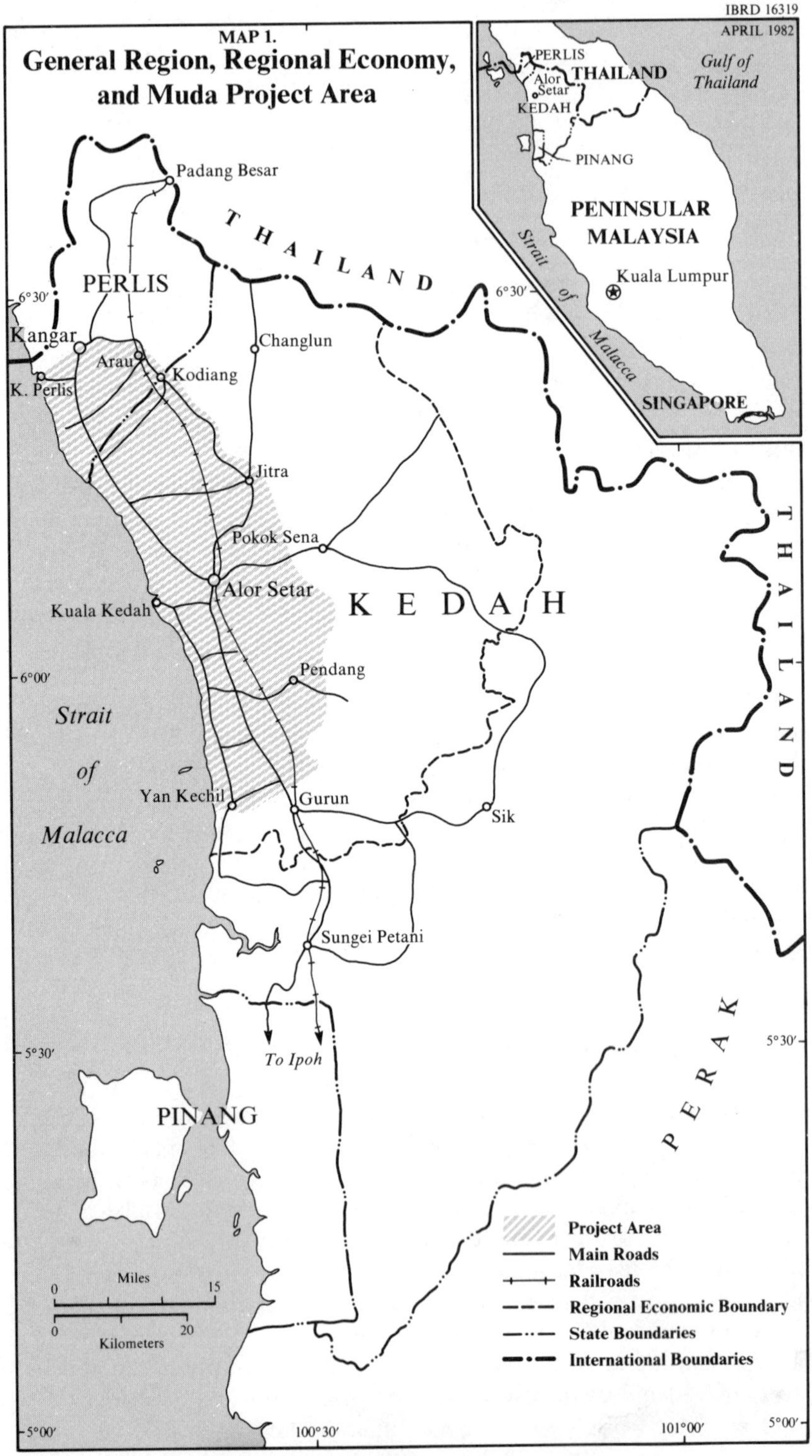
IBRD 16319
APRIL 1982
MAP 1.
General Region, Regional Economy, and Muda Project Area
Padang Besar
THAILAND
PERLIS
Kangar
Arau
Kodiang
K. Perlis
Changlun
Jitra
Pokok Sena
Alor Setar
Kuala Kedah
KEDAH
Pendang
Strait of Malacca
Yan Kechil
Gurun
Sik
Sungei Petani
To Ipoh
PINANG
PERAK
THAILAND
Gulf of Thailand
Alor Setar
KEDAH
PINANG
PENINSULAR MALAYSIA
Kuala Lumpur
Strait of Malacca
SINGAPORE
6°30′
6°00′
5°30′
5°00′
100°30′
101°00′
Miles
0
15
0
20
Kilometers
Project Area
Main Roads
Railroads
Regional Economic Boundary
State Boundaries
International Boundaries

streams discharge to the plain from small catchment basins. To the north, the plain continues beyond Kedah into Perlis, where it is disrupted by monolithic and sometimes spectacular limestone outcrops. Most of the plain is less than ten feet above sea level. Spring tides average 4.5 feet above mean sea level, and only the coastal bund (embankment) and smaller bunding around the fields prevent regular flooding by the sea for several miles inland.

The climate is tropical, with two monsoon seasons. Rain does not fall uniformly over the area, and the climatic records show wide variations between rainfall stations in daily, monthly, and annual amounts; within the project area the mean annual rainfall varies between 70 and 130 inches. In any month, total rainfall is unpredictable. Often one or two storms of high intensity bring a deluge lasting only a few hours, and even in the monsoon season several days can pass without rain. It is hot and humid throughout the year, with the daily temperature usually ranging between 73° and 85° F and relative humidity fluctuating between 70 and 95 percent. There are only slight monthly variations in day length.

The estimates of land use in the region (see table 2-1) are based on a 1966 aerial survey. Although these estimates are somewhat dated and have not escaped criticism, they do provide a satisfactory guide to the general pattern of land use. The importance of rice cultivation is evident. For the most part, it is confined to the coastal plain, which embraces most of the project area. In northern Perlis the valley bottoms are also under paddy cultivation, and there is a series of independent small schemes, some benefiting from controlled drainage or irrigation facilities. Outside the plain, land is put to other uses. Since 1970, scattered pockets of tobacco cultivation have appeared, together with two sugar plantations covering 20,000 acres. But these are small exceptions in a landscape dominated by

Table 2-1. *General Land Use of the Region, 1966*

Land use	*Percentage of area*
Forest (including unproductive secondary jungle)	34.0
Paddy	28.6
Scrub (mainly unproductive)	13.1
Rubber	9.9
Mixed horticulture (mainly kampong land[a])	5.8
Other productive	1.8
Urban	0.5
Other unproductive (swamp, cliffs, quarries)	6.3
Total	100.0

a. Village land occupied by houses, fruit trees, and gardens.

Source: Extracts from "District Land Use Acreage Abstracts," unpublished work sheets prepared for Wong (1966).

rubber plantations and smallholdings divided by stretches of secondary jungle. In the extreme east, in the foothills of the mountains, logging, charcoal burning, and upland paddy cultivation are widespread. The large areas with little, if any, productivity are concentrated mainly in the upland part of the region.

The morphology of the plain is largely the result of prolonged sedimentation. Nearly everywhere, the soil consists of very fine tenacious clay, which ranges from black to grey because of a high humus content. The extreme flatness and the confused nature of the natural drainage pattern combine with this heavy clay morphology to render most of the soil acidic in varying degree. These conditions are suitable for growing rice, but the acidity largely precludes the cultivation of other crops. The plain has been settled and reclaimed mainly since about 1880, and the habitations form a rectilinear pattern beside drains and roads, away from both the seaward and the landward edges of the plain.

While few reliable data are available for the years before 1925 (Kedah was dominated by Siam until 1909), two early estimates of the population make it possible to infer the acreage under paddy. The first was made in 1850 or thereabouts by local chiefs, who reckoned the population to be 6,500 men (Logan, 1851), from which we can infer that the total population was about 35,000. From our knowledge of the cultivation techniques used (hand hoes and buffaloes), we deduce that about 26,000 acres were under cultivation. According to the second estimate, in 1911 there was a population of about 145,000, which implies a cultivated area of about 106,000 acres (Dobby, 1951). This suggests that during the last half of the nineteenth century the cultivated area was expanding a little faster than 2 percent a year.

The statistics available for the years since 1931 are fairly complete. Between 1931 and 1944 the area under paddy increased at an annual average rate of about 3,000 acres, from a base of 160,000 acres. Part of the expansion in the 1920s and 1930s was due to three more or less separate drainage schemes, which, to a large extent, have determined the basic pattern of the present Muda scheme. (At the time the project was appraised in 1965, the World Bank estimated that 130,000 acres were receiving the limited benefits of improved drainage.) After World War II the rate of reclamation was faster, as land which had become derelict during the Japanese occupation was brought back into cultivation. By 1949 the area under cultivation had reached 220,000 acres. Between 1950 and 1975, the acreage increased by a further 20,000 to 30,000 acres, mainly under the aegis of government schemes and development efforts.

The population of the plain has remained mostly Malay despite the long period of political domination by Siam (Thailand) that ended in 1909. Thai, Indian, and Chinese Malaysians account for about 31 percent of the entire population—a sizable proportion, but well below the national average of 49 percent. The non-Malays are found mainly in the towns, especially Alor Setar. There is, however, a small and probably declining group of Chinese farmers (the law prohibiting non-Malays from owning paddy land came into effect later in Kedah than elsewhere in Malaysia). There are also isolated pockets of Thai farmers, and Siamese is widely spoken over much of the northern part of the plain. These two minorities have diversified the settlement pattern, since both prefer houses built on the ground in their fields, in contrast to the lines of houses on stilts in the Malay *kampongs* (villages).

A dense network of roads now covers the whole of the plain, but this, too, is a recent development. In 1950 it was still possible to write:

> The pattern of transport results from a fusion of two different stages. Modern road transport running centrally to the southern part of the Plain and on the east of the northern part, carries only a fraction of the gross movement taking place. There is really no major flow of goods or people beyond the Plain northward or southward, though the latter has the greater movement. Roads inland through the eastern hills are all dead-end and without large flows of traffic. The railway, keeping nearer to the eastern fringe, the traditional land-route across the Plain, carries less traffic than the road and, when through routing to Siam was possible, never moved commodities or people in any great volume: it has yielded significance to the road, but there are only some 85 miles of road to serve the 800 square miles and quarter million population of the Plain. Much of the padi movement from the countryside is by water, chiefly on the drains and canals whose influence on transport relate to their significance in settlement distribution. Today, the importance of these waterways is greatest as latitudinal lines of movement with junction and transhipment sites where they meet the road, the railway and the longitudinal waterlines. (Dobby, 1951, p. 314.)

By contrast, there are now some 800 miles of metaled roads (excluding 400 miles of farm roads), and the north to south movement is large. The railway, however, suffers from single-track operations, manually operated level crossings, outmoded rolling stock, high maintenance costs, and a proliferation of small stations. As a bulk carrier, it is dwarfed by the road network.

Defining the Region

The region's boundaries were defined in part by physical geography, but the primary consideration was quite different. In 1971, while the project was under construction, the Food and Agriculture Organization (FAO) and

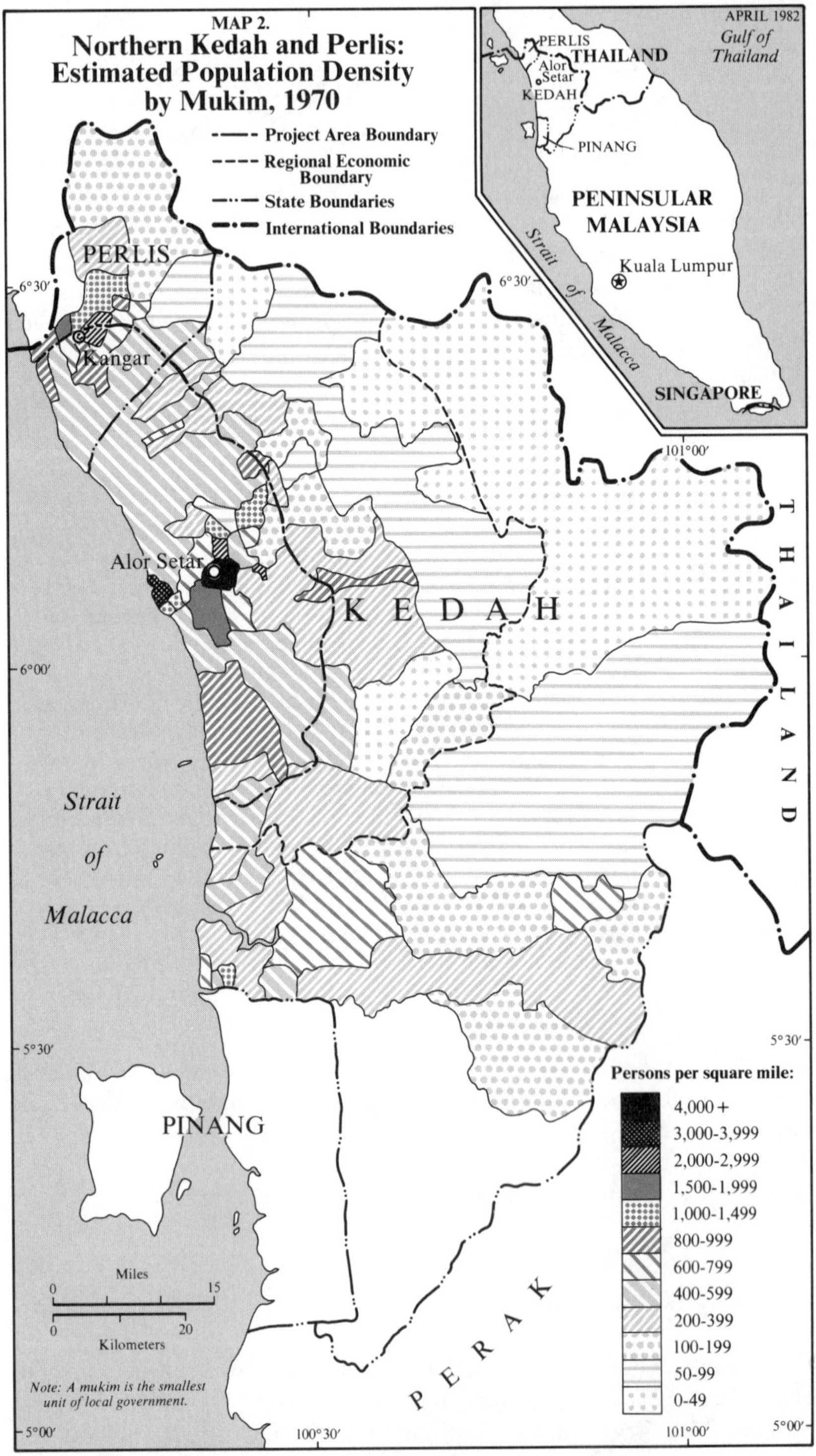
MAP 2.
Northern Kedah and Perlis: Estimated Population Density by Mukim, 1970
Project Area Boundary
Regional Economic Boundary
State Boundaries
International Boundaries
APRIL 1982
Gulf of Thailand
PERLIS
THAILAND
Alor Setar
KEDAH
PINANG
PENINSULAR MALAYSIA
Strait of Malacca
Kuala Lumpur
SINGAPORE
6°30′
101°00′
PERLIS
Kangar
Alor Setar
K E D A H
T H A I L A N D
6°00′
Strait of Malacca
5°30′
Persons per square mile:
4,000 +
3,000-3,999
2,000-2,999
1,500-1,999
1,000-1,499
800-999
600-799
400-599
200-399
100-199
50-99
0-49
PINANG
Miles
0
15
0
20
Kilometers
Note: A mukim is the smallest unit of local government.
P E R A K
5°00′
100°30′
101°00′
5°00′

the World Bank decided to undertake a set of surveys to quantify many of the economic changes that were taking place (FAO, 1975; see also appendix A). Such surveys have since become a common feature of the Bank's loans, but this was, at the time, a novel and pioneering initiative. More important, the Muda surveys were designed to measure the indirect effects of the project on the local economy, especially those arising from household expenditures out of extra income, as well as the direct benefits accruing to the farming households living within the project area. Thus, the scope of enquiry was wider than that of many recent evaluations of other projects, which have tended to focus on short-term monitoring or the measurement of the project's direct benefits. A key consideration in designing the Muda surveys was to arrive at a suitable definition of the region.

Spatially, the secondary effects of a project tend to be more diffused than its direct effects. And the size of the surrounding region in which most of these secondary effects are felt depends on the geographical location of the sources of supply that meet the total (direct and indirect) demand thus generated. For example, the linkages arising from households' expenditure decisions will extend to all places where consumers do their spending and the goods they purchase are produced. At the time, it was decided to confine the surveys to the places of purchase. In essence, this required drawing an "economic boundary" which would encompass most of the day-to-day transactions of the region's households. As argued below, there are strong grounds for adopting this procedure, which amounts to defining the region in such a way that most of what are normally treated as nontradable goods in the context of international trade are produced within the region's geographical boundaries. In this connection, the strongly defined features of the area's physical geography—the Straits of Malacca to the west, the Thai border to the north, rubber plantations and Kedah Peak to the south, and the mountainous jungles of inner Malaya to the east—made the task of drawing such an economic boundary somewhat easier.

The first step was to construct a map of the population density in northern Kedah and Perlis on the basis of the population census of 1970 (see map 2). This enabled the most populous areas to be identified and suggested the economic nuclei of the region. Where people conducted their daily business and the extent to which the region satisfied both the material and spiritual needs of its inhabitants were also studied. These enquiries revealed that daily necessities were invariably obtained within two miles of the house. Higher level services, such as western medical attention or government and banking services, were almost exclusively provided in Kangar and Alor Setar, the state capitals of Perlis and Kedah, respectively. In a graduated fashion, all subsidiary towns in the region, with the exception of two in the south (Gurun and Bedong), looked inward to Alor Setar. In short, Alor Setar was (and no doubt remains) the principal "central place." In 1972, for example, it accounted for 41 percent of retail business activity.

These studies, reported in detail in FAO (1975), provided the principal data used to determine the boundary of the regional economy. The boundary was also influenced by the networks of other nontradables, in particular the extent and density of postal services, the distribution of domestic water and electricity, the service areas of the telephone system, the frequency and extent of public transport, and the density of traffic flows. The precise boundary for the region was defined by establishing the watershed of activity for each of these networks, superimposing one on another, and then relating the whole to the salient features of the local typography, land use, population density, and the urban hierarchy. Finally, the actual boundary used for fieldwork and the interpretation of the data from secondary sources was slightly adjusted to coincide with the smallest units (*mukims*) of local government.

It must now be asked whether this definition of the region's boundaries squares with our choice of semi-input-output analysis as the framework for estimating the project's indirect effects. In principle, they should be consistent with one another, for the essence of semi-input-output lies in the distinction between tradable and nontradable goods. Within national frontiers this is not a difficult distinction to draw, but in a regional context the differences are blurred. For example, if the region's demand for transport services should increase, trucks and buses can be brought in from other regions, perhaps at the cost of extra congestion on the roads. Similarly, migrant workers may be drawn in if the region's labor market becomes sufficiently tight. Thus, a balance must be struck between extending the region's boundaries to the point where they include virtually all activities producing nontradables and preserving the integrity of the region's labor markets.

As already noted, the boundaries were drawn without an explicitly formulated methodological framework for measuring the project's downstream effects. Since the boundaries had to be accepted as given, the question of whether they were suitable could be answered only by proceeding with the business of estimating the relevant flows within our chosen framework. Together, the a priori considerations discussed above and the survey data themselves gave much assurance that all would be well, and as the findings reported in subsequent chapters unfolded, it turned out that our confidence was not misplaced. This suggests that one can use the above method to settle on the boundaries of a region without giving too many hostages to fortune in the analysis that follows; but it is obviously more sensible to settle at the outset on the categories and activities which will appear in the analysis.

Table 2-2. *Regional Population, 1972*
(thousands of persons)

Area	*Muda paddy farming*	*Other farming*	*Landless paddy workers*	*Non-farming*	*Total*
Urban[a]	16.2	13.2	1.2	78.6	109.2
Rural	271.8	118.6	20.1	167.4	577.9
Total	288.0	131.8	21.3	246.0	687.1

a. Inhabitants of towns containing 5,000 or more persons.
Source: FAO (1975).

Population and Economic Structure

In 1972 the region's population was 687,000, 16 percent of whom were living in urban areas, and the great majority in Alor Setar itself (table 2-2). About 45 percent depended on paddy farming as the principal source of income, and almost two-thirds on farming of one sort or another. Population growth during the intercensal period 1957–70 was quite modest, the quinquennial growth rates for the urban and rural subpopulations being 4.6 and 3.7 percent, respectively. Since the crude natural rate of increase was about 2 percent a year, out-migration was correspondingly heavy.

The simple urban structure consists of a hierarchy with Alor Setar (population 110,000) at its apex, followed by some thirty smaller towns ranging in population from 9,000 downward. Forming the base structure is the rural community of interlinked kampongs. The large difference in size between Alor Setar and the smaller towns is reflected in the distribution of services in the region, summarized in table 2-3.

That people were leaving the region in large numbers in the 1960s is certainly consistent with the low incomes and strong interregional disparities of that period. Hoerr (1970) estimated that the region was one of the poorest areas in West Malaysia. In 1965 regional gross domestic product (GDP) per capita was about $525, compared with $850 for the whole country and $1,085 for the four most prosperous western states.[1] Then, as now, the region was mainly a producer of primary products, notably rice, rubber, and fish, with their associated processing industries and a supporting infrastructure of services and utilities. Other manufacturing activities were, and still remain, relatively unimportant.

1. All dollar values in the book are Malaysian dollars. For exchange rates per U.S. dollar for 1966–79, see the Units of Measurement in the front of the book.

Table 2-3. *Percentage Share of Alor Setar in Selected Regional Services, 1972*

Service	*Percent*	*Service*	*Percent*
Hospital beds	65	Tailors	36
General practitioners	85	Hotels	83
Dentists	58	Banks	64
Telephones	70	Petrol stations	41
Supermarkets	100	Secondary school pupils	45
Jewelers	22	Primary school pupils	33
Pawnshops	20	Cinema seats	65

Source: FAO (1975).

In the early 1970s rice production provided a livelihood in whole or in part for some 60,000 to 65,000 farming families (50,000 of whom lived on the plain), about 5,000 landless workers, and many others engaged in occupations which depended on, or serviced, the farming community. In the 1960s the region accounted for 27 percent of the nation's output of rice; by 1974 this had risen to about 45 percent, largely as a consequence of the Muda scheme.

Rubber, fishing, and forestry are rather weakly integrated into the region's productive structure. Indeed, the rubber estates, which are much smaller in the aggregate than rubber smallholdings, are virtually economic enclaves. Following old and established traditions, rubber estates house, educate, and feed their laborers, who live on the estates; there is little off-estate processing, and decisions on the running of the estates are centrally controlled by the large estate and trading houses in Kuala Lumpur. Similarly, the forestry and logging industries are essentially controlled by firms based outside the region. Some of the products of the local sawmills go to the construction industry, and then mostly for rural housing; the rest is exported, much of it abroad. Logging generates few local jobs directly, since logging contractors tend to bring in their own teams of skilled workers to man their highly mechanized operations.

The fishing industry, which produces about 12 percent of the nation's marketed fresh fish through the ports of Kuala Perlis and Kuala Kedah, has rather similar characteristics inasmuch as its equipment is purchased, and most of its production sold, outside the region. It does, however, have certain direct production linkages within the region, particularly in Alor Setar, where there is a small group of ship chandlers and ice factories. More important, the fishing industry was the principal means of livelihood for some 7,000 to 8,000 families in the early 1970s (Ministry of Agriculture and Fisheries, 1971), although mechanization has gone at a smart pace since. There was also some overlap between the fishing industry and rice cultivation in the single-cropping era. It was common for rice farmers to engage in part-time fishing when they were not cultivating their fields. With the advent of double-cropping, however, part-time fishing by paddy farmers has declined.

In the late 1960s the region's industrial base was very small; annual value added in manufacturing industry per capita was $15, compared with $93 for West Malaysia as a whole (Department of Statistics, 1968*a*). Typically, a manufacturing establishment in the region had a turnover of less than $20,000 a year and was controlled by an individual proprietor who employed fewer than four paid employees. As noted already, rubber processing, rice milling, and sawmilling dominated the manufacturing sector, and that dominance continued, in large part because of the surge in paddy output that followed the introduction of irrigation. As is often the case, these major industries have provided relatively few jobs in manufacturing, most of the employment in this sector being in numerous small firms. Alor Setar has also remained the main manufacturing center. In 1968 it accounted for 52 percent of all establishments and 68 percent of all gross sales (Department of Statistics, 1968*a*).

The largest formal employer is the government: the local arms of federal government departments, the two state governments, and quasi-government organizations. Informal employment in the service industries appears to be even greater, however, with jobs usually going to members of the extended family. One survey (Department of Statistics, 1968*b*) reported that among the 176 small businesses studied in Alor Setar in 1968, there were 157 working proprietors and partners, employing 97 unpaid family workers and 116 paid full- or part-time employees. This small-scale sector—retail and wholesale trade, restaurants and coffee shops, and small-scale manufacturing—is made up of a multitude of Chinese family businesses.

The Economic and Social Organization of Paddy Production

Muda's paddy farmers, like those in other parts of Malaysia, are peasants. There are no estates, nor are there capitalistic enterprises cultivating hundreds of acres. There are a few Chinese households whose entrepreneurial energies are devoted, in part, to 50 or at most 100 acres, and most of this land is rented from Malays. The peasantry shows a measure of social and economic differentiation (Ali, 1975), but compared with South Asia, say, such differentiation is not especially marked. The distribution of farm size by operational holdings is not strikingly uneven, and although a high proportion of the land cultivated is rented from others, most rental contracts appear to involve kin. In such cases, a newly married couple may pay a nominal rent to a well-off parent or uncle, whereas a farmer in his prime is likely to pay the going market rate on land rented from his widowed mother. Absentee landlordism is significant; but although it is a source of inequality, it does not bear down upon the peasantry with the force of its counterpart in eastern India, for example. Thus, while tenancy is extensive, it is not oppressive in the large, much of it being a reflection of kinship relations linked by the domestic life cycle (Ali, 1975; FAO, 1975; Jegatheesan, 1977).

The household structure is generally nuclear (Masuo and Yoshihiro, 1966, 1967; Ali, 1975), but it is not uncommon for newly married couples to live with the husband's or wife's parents, or even alternate between them, while working on the parents' farms or for others. Similarly, unmarried children can work off their parents' farm if they find the opportunity. As for the household's economic decisions, the studies cited suggest that the household head, who is normally a male in his late thirties or forties, disposes of the household's resources as he sees fit in the light of his responsibility to ensure the household's welfare. Income from the farm in the form of sales of produce and services, and any earnings he makes from other activities, accrue directly to the household head. Income earned by or accruing to his wife is also passed on to him (if there is no actual physical transfer, he will retain at least a joint right to determine when and on what it is spent). The same holds for income earned by minors. In addition, the household head receives a payment in cash or kind or labor from those who earn outside income, and in return he provides them with food and shelter.

As we have seen, tenancy and kinship are usually closely intertwined, as are tenancy and the domestic life cycle. The pure tenant is often a young man who has just started a family, while someone who is both owner and tenant usually has a holding larger than the average in his locality, is a successful cultivator in his prime, and has also scaled to the peak of the domestic cycle. The Chayanovian parallel (Chayanov, 1966) in all this should not be overdrawn because paddy land is very scarce, rather than being there for the taking at the extensive margin. Nevertheless, in other respects the parallel is close enough to have some suggestive value in the Malaysian context.

These, then, are the qualitative features of the paddy farming system, the organization of which has changed little since 1960. A more detailed quantitative account of the changes that occurred between the mid-1960s and mid-1970s, in which the Muda project played a large part, follows a description of the project itself.

The Muda Scheme

One of the largest agricultural development projects ever undertaken in Malaysia, the Muda scheme was designed to bring 255,000 acres of paddy land (farmed by some 51,000 farm families) under double-cropping where previously only a single rain-fed crop was possible. When the scheme was planned in the early 1960s, imports of rice were large and the Malaysian balance of payments was under serious strain. The Muda scheme and Kemubu, its smaller sister project on the east coast of the peninsula, were primarily attempts to achieve national self-sufficiency in rice production. Although considerations of welfare and equity had little influence on the project design, they have, as is the fashion, become increasingly important in the evaluation of the project's impact.

Construction work began in 1966, and the first release of irrigation water took place in the off-season (March–August) of 1970. By 1974 the scheme was fully operational. The civil engineering works consist of: the Muda and Pedu dams and their reservoirs, with a storage capacity of about 800,000 acre-feet; a five-mile tunnel connecting the two reservoirs; river improvements; diversion works; two main canals and distribution systems; drains, saltwater barriers, and tidal gates; and the equipment and facilities for the operation and maintenance of these structures. By 1974 total outlays on the capital account, including interest charges, had reached $270 million, of which about $100 million was financed by a loan from the World Bank to cover the project's direct foreign exchange costs.

The basic design contains some inherent operational problems. In particular, only 23 percent of the project's total demand for water is met by controlled flows through the reticulation system, the balance being supplied by rainfall and natural rivers. A computerized water control system was installed in the early 1970s, but it did not appear at first to be fully suited to the physical system, which relies on field-to-field flooding over distances of up to two miles. This, in conjunction with the irregularities of the microtopography, conspires to produce a ragged pattern of water distribution, with delays of up to thirty days between the arrival of water at plots on one side of an irrigation block and those on the other. This results in a pattern of widely staggered planting dates, which in turn creates problems for individual farms in harvesting, transporting, and drying. In addition, such a spread of cropping activity provides favorable conditions for the build-up and carry-over of pests and diseases. Improvements in the computerized control system have been made, and follow-up investments to improve the reticulation system are also under way. Their effects are discussed at some length in chapter 9.

It was originally envisaged that the Drainage and Irrigation Department of the Malaysian Ministry of Agriculture would execute the project. But in 1968 the government appointed a project coordinator who concurrently held the post of state agricultural officer for Kedah. Subsequently, the Muda Agricultural Development Authority (MADA) was established as a statutory corporation, by regulation in 1970, and by act in 1972. Thus MADA represents an interesting initiative to combine the authority of government with the flexibility and initiative of private enterprise. From its inception, the project received strong political backing, and MADA's chief executive had direct access to the highest reaches of government.

MADA's technically competent staff carried out careful preparatory work before the introduction of double-cropping, the most important aspect of which was the development of a series of pilot demonstration schemes. These were designed to publicize the forthcoming changes in agricultural practices and verse farmers in a new, more demanding agricultural calendar and improved methods of farm and water management. Steps were taken to ensure adequate supplies of high-yielding, short-duration varieties of rice, and the first Farmers' Associations were founded to act as keystones for improved extension advice and credit disbursement. In addition, MADA has the authority to promote and coordinate regional development through productive ventures which rely on or complement the greater productivity of the farming system. It has found relatively few opportunities to exploit, however, and has thus turned back to an intensification of the engineering system in an effort to make the region still more prosperous. Moreover, MADA's real scope to pursue regional ventures is limited by the power of the purse, for it has no revenue of its own, but is dependent on federal budgetary support. Thus, there is some indirect control of its programs and staffing levels by both the Ministry of Agriculture and Rural Development and the Ministry of Finance. Nevertheless, MADA enjoys considerable independence, largely as a result of its considerable success and strong political support.

The basic development philosophy embodied in the Muda scheme has been that the physical infrastructure is only a part of an overall package of innovations. The core of this package is a system of twenty-seven Farmers' Associations (FAs) housed in strategically located Agricultural Development Centers (ADCs), which, in addition to their designated functions, have become informal meeting places for the farming community. The ADCs provide the physical facilities for activities and services, such as marketing advice, machinery repair, water management information, and veterinary services, to implement and control double-cropping. Each ADC is situated in a farm locality and serves about 2,000 farmers, none of whom lives more than three or four miles away.

The purpose of the FAs, which have voluntary membership, has been well described in MADA (1972, pp. 10–13), and two short quotations will suffice:

> The FA provides a package of services including extension, supply, credit and marketing. Each of these functions supports and is dependent upon the others to a high degree. Therefore, the strengthening of the farm locality through the FA is viewed as a single implementation device incorporating all four functions. The FAs serve as an administrative base for motivating and organizing farmers into dynamic communities for the achievement of socio-economic development. The development of the FA is emphasized and accelerated so that it may become a self-sustaining unit as soon as possible . . . The extension work of an FA includes demonstration plots on land easily accessible to the farmer, pilot projects, courses at the Rural Training Center, water management, marketing and extension services. Throughout these and other services the emphasis in extension work is on motivating farmers to adopt the scheduling and new techniques required by double-cropping. The FA approach in extension work is social and economic rather than purely technical. Training programs are geared toward making the farmer aware of the importance of his organization and the importance of the individual farmer in the organization. The long term benefits of the Scheme and the farmers' role in the achievement of the Scheme's objectives are also strongly emphasized.

Credit is supplied to the farmer, via Bank Pertanian (the National Agricultural Bank), through Local Credit Centers (LCCs), which include the FAS.

> Loans are provided in the form of coupons to farmers who have been approved by LCC. The coupons may be exchanged for only the inputs included in the package and only by registered stockists and contractors. As a condition of loan the farmer will take all the recommended inputs in the recommended amounts and follow extension advice on their proper application. Visits are scheduled to the farm to assist farmers in proper utilization. The coupon system is arranged to enforce proper timing of application of the various inputs by a date of negotiation limit placed in the coupon. Such a system coordinates the credit scheme with supply and with extension.

The project has brought about a great increase in the output of paddy and far-reaching changes in the technology used to produce it, and paddy farming households have been much better off as a result. In turn, these changes spurred economic activity in the regional economy.

Table 2-4. *Area of Paddy Cultivation, 1967–74*
(thousands of relongs of cultivated paddy)

	Muda scheme			*Non-Muda*			
Year	*Main season*[a]	*Off-season*[b]	*Annual total*	*Main season*	*Off-season*	*Annual total*	*Regional annual total*
1967	334.2	—	334.2	104.2	12.2	116.2	450.4
1968	338.6	—	338.6	102.9	18.9	121.8	460.4
1969	338.1	—	338.1	103.9	11.6	115.5	453.6
1970	329.6	114.6	444.2	101.0	11.2	112.2	556.4
1971	327.0	157.6	484.6	99.7	11.2	108.9	593.5
1972	324.4	234.7	559.1	99.7	14.0	113.7	672.8
1973	327.4	296.8	624.2	99.8	14.0	113.8	738.0
1974	327.8	320.3	648.1	99.8	14.0	113.8	761.9

— Not applicable.

a. Revised in the light of the latest estimates of the Muda area as presented in MADA (1976). See also FAO (1975, table 64), and Ministry of Agriculture and Cooperatives (1967).

b. Unrevised because off-season acreages were estimated independently by MADA.

Changes in Output and Technology in the Paddy Sector, 1966–74

The gross cropped area within the scheme's command expanded from 334,000 relongs in 1967 to 648,000 relongs in 1974 (table 2-4). In 1974 the cropping intensity reached 1.98, which is somewhat above the long-run average of 1.80 envisaged at the design stage. This expansion of the gross cropped area was achieved without major engineering problems, and the irrigation system operated with a water-use efficiency of about 30 to 40 percent.[2] The net physical area devoted to paddy cultivation declined by about 6,000 relongs, a reflection of the net losses to the project's engineering works and urban encroachment, some of which was due to the project's effects on the regional economy.

Production grew at a faster rate than the cropped area because yields were also increasing with the introduction of improved varieties and the greater use of fertilizers and pesticides. By 1974 output had reached 11.1 million piculs from a base of 4.6 million piculs in 1967, an increase of 141 percent (table 2-5).

Before the project, farmers grew traditional varieties of paddy. Derived from Malay and Siamese native material, these were long-stemmed, leafy, photosensitive, highly susceptible to lodging (the breaking of the plants' stems in adverse weather), and had a growing period of 160 to 180 days. During the mid-1960s several improved varieties were released, the most important being Malinjah and Mahsuri,[3] both of which promised much higher yields, but proved to be susceptible to rice blast and neck rot. In 1965, when the project was appraised, the vital importance of such varieties was not appreciated, but it was soon realized that if farmers did not adopt varieties that mature quickly, double-cropping would probably not be feasible. Problems with the new varieties thus caused considerable consternation, and continued work on plant breeding led to the release in 1968 of Bahagia, a precursor of IR5. This variety had good resistance to blast and other pests, a growing period of 135 to 140 days, and good yield potential. It proved to be generally satisfactory and was the main short-term variety when double-cropping was first introduced. Subsequent

2. Efficiency could be substantially increased by improving in-field water control, which would permit a more timely pattern of water distribution, eliminate the staggered planting of the off-season crop, and make harvesting times more uniform. As chapter 3 makes clear, these scheduling problems have a profound effect on the operation of the farm economy. Plans are now well in hand to improve this situation; they are discussed at some length in chapter 9.

3. A cross between Pebifoon (Taiwan) and Siam 29, and Taichung 65 and Moyang Ebos 80, respectively.

Table 2-5. *Changes in Production and Yield, 1967–74*

Year	*Yield (gantangs per relong)*		*Gross production (thousands of gantangs)*			*Regional production (annual)*
	Main season	*Off-season*	*Main season*	*Off-season*	*Annual*	
			Muda scheme			
1967	333	—	111,289	—	111,289	
1968	318	—	107,674	—	107,674	
1969	325	—	109,883	—	109,883	
1970	363	375	119,645	42,975	162,620	
1971	368	393	120,336	61,937	182,273	
1972	384	398	124,570	93,411	217,981	
1973	285	402	126,049	114,268	240,317	
1974[a]	403	413	132,103	132,103	264,387	
			Non-Muda			
1967	296	358	30,843	4,368	35,211	146,500
1968	283	373	29,121	7,050	36,171	143,845
1969	291	335	30,235	3,886	34,121	144,004
1970	289	336	29,189	3,763	32,952	195,672
1971	342	356	34,097	3,987	38,084	220,357
1972	334	279	33,300	3,906	37,206	255,187
1973	329	292	32,834	4,088	36,922	277,239

— Not applicable.

Note: Yields are from crop-cutting experiments scaled down by 0.88, this being the ratio of experimental yields in 1967 and 1973 to those reported in Ministry of Agriculture and Cooperatives (1967) and FAO (1975). The result gives yields gross of deductions for moisture, chaff, and other waste.

a. Non-Muda figures are not available for 1974.

breeding work and trials led to the introduction in 1973 of Jaya (international designation, C4-63), which matures in 120 days, has a short stem, and produces high yields of good quality rice. Subsequent work has produced further improvements in the varieties available, and farmers have not been slow to adopt them. By 1972, about 90 percent of the farmers were using improved varieties during the main season, and 95 percent during the off-season (FAO, 1975).

The introduction of these new varieties was accompanied by changes in agricultural practices, especially in the use of fertilizers and the methods of land preparation. In 1967, when traditional varieties of paddy were dominant, inputs of nitrogen were very low, averaging 15 pounds of nitrogen per relong, excluding the nitrogen content of traditional organic fertilizers, the input of which was small (Ministry of Agriculture and Cooperatives, 1967, table 21). By 1972–73 the average dose had risen to about 30 pounds per relong (FAO, 1975), which, in the light of the experimental data discussed in chapter 3, was about optimal for the high-yielding varieties then under cultivation. The use of pesticides also increased with the project, and by 1972 nearly 50 percent of the farmers were using insecticides or herbicides, although average levels of expenditure were probably too low to have any significant impact on pest populations or the incidence of disease.

Traditionally, the land was prepared for cultivation by buffaloes. In 1967 it was estimated that only 5.2 percent of the area under paddy was prepared exclusively by tractors, although a further 33.3 percent of the area was prepared by a mixture of hand, buffalo, and tractor methods (Ministry of Agriculture and Cooperatives, 1967). By the mid-1970s, however, the stock of tractors had grown dramatically (see table 2-6), and virtually all

Table 2-6. *Regional Tractor Sales and Stocks*

Item	*1966*	*1967*	*1968*	*1969*	*1970*	*1971*	*1972*
Sales							
Four-wheel tractors[a]	23	22	38	65	105	60	62
Two-wheel tractors[b]	n.a.	105	330	591	580	558	662
Average brake horsepower for four-wheel tractors[b]	54.4	56.1	61.9	65.9	66.4	66.5	n.a.
Stocks							
Four-wheel tractors[c]	263	n.a.	n.a.	n.a.	n.a.	n.a.	506
Two-wheel tractors[c]	250	n.a.	n.a.	n.a.	n.a.	n.a.	2,271

n.a. Not available.

a. FAO (1975, table 67).

b. Jegatheesan (1972).

c. A working life of ten and six years is assumed for four-wheel and two-wheel tractors, respectively.

tillage was done by machine. Over the same period, the stock of buffaloes declined steeply, in response to the stricter scheduling of farm operations under double-cropping and the disappearance of free-range grazing on fallow fields in the off-season. In 1967 the total buffalo stock was 42,800 (Ministry of Agriculture and Cooperatives, 1967), and by 1973 it had fallen to 22,500 (FAO, 1975); in both cases about half would have been mature work animals. In 1972 more than 70 percent of farmers owned no work animals (FAO, 1975).

This rapid switch from traditional to mechanical methods of tillage has some interesting features. Perceiving a potential scheduling problem early on, MADA began to encourage the use of four-wheel tractors. Yet the response came, not from paddy farmers (who could hardly bear the risks of such large investments and still less finance them), but from agents outside the farming community who independently offered largely uniform contracts to till at competitive rates across the project area. At the same time, farmers began to purchase the far less costly two-wheel tractors.[4] Although the cost of land preparation by these machines is substantially more than that by their larger and more powerful counterparts under test conditions (MADA, 1970*a*), they do offer the advantages of timeliness and security of supply, as well as better mobility in difficult field conditions. There are also the less tangible benefits of status and prestige for their owners. Moreover, as the project approached maturity, there were further developments that made two-wheel tractors economically more attractive. As early as 1974, the widespread use of large four-wheel tractors was giving rise to serious problems. In some areas, deep cultivation began to break down the plow pan, while in others, the soil has an inherently low load-bearing capacity. Similarly, large tractors tend to damage bunds and levees, thereby giving rise to acrimonious disputes. As a result, the composition of the draft-power stock began to shift in favor of pedestrian tractors during the postproject period (see chapter 9).

Changes in Farm Structure and Incomes

The typical paddy farming household changed very little, if at all, between 1966 and 1973 (see table 2-7). It consisted of five or six persons, nearly half of them children under sixteen. Most adults were economically active, and almost all of them worked on the household's paddy farm. Children under sixteen years were occasionally given tasks on the farm,

4. The FAO/World Bank survey data reveal that an insignificant number (0.1 percent) of paddy farmers owned four-wheel tractors in 1972, but about 4 percent owned pedestrian tractors.

Table 2-7. *Details of the Average Paddy Farming Household*

Characteristic	*1966*	*1973*
Family size (persons)	5.5	5.6
Economically active members	3.0	3.0
Literacy rate of household heads (percent)	52.3	68.1
Total landholding (relongs)	6.5	6.5
Paddy	5.7	5.6
Other crops	0.3	0.2
House lot	0.5	0.7
Paddy area owned	2.7	3.1
Paddy area leased in	3.1	2.9
Paddy area leased out	0.1	0.4
Paddy area operated	5.7	5.6

Note: The net area leased in declined from 3.0 to 2.5 relongs per paddy farming household. This is land rented from non–paddy farming households, including kin and absentee landlords.

Sources: 1966: Ministry of Agriculture and Cooperatives (1967); 1973: FAO (1975).

but their contribution to total labor inputs was very small. Virtually all households depended wholly on the farm for their livelihood, and relatively few heads of households had secondary occupations. There were some gains in literacy over this period, although very few farmers were educated beyond primary school; indeed, most of those who were literate received little or no formal education. The situation changed quickly with the expansion of the educational system, however, and in 1973 nearly 90 percent of all households contained at least one male other than the household head who was not only literate, but usually better educated than his father (FAO, 1975).

The area of the average holding did not change over the period. In 1973 almost half the area under paddy was leased, about 60 percent of it from close kin, some of whom were not themselves engaged in paddy farming. The remainder (20 percent of the total area) was leased from households primarily engaged in other economic sectors. Many of the latter are absentees, and the remnants of the aristocracy are reputed to figure prominently among them. This evidence points to a very active market in land. Indeed, it appears that there was even some recontracting during the course of a single season: some farms varied in size by as much as 200 percent during one or other of the two seasons in 1972–73 (FAO, 1975).

In part, the extensiveness of tenancy is because Malay paddy farmers are reluctant to permit holdings to be endlessly partitioned (Wincott, 1976). Traditionally, widows, sons, and daughters inherit equal shares of the family land. This traditional pattern is not universally observed, however, and the Islamic rule, under which sons receive twice as much as daughters and widows one-sixth, is also followed, perhaps under some prodding by

Table 2-8. *Distribution of Farms by Size of Operated Area in 1966*

Size class (relongs)	*Percentage of farms*	*Cumulative percentage of farms*	*Percentage of area*	*Cumulative percentage of area*
0–0.9	0.89	0.89	0.07	0.07
1–1.9	9.05	9.94	2.36	2.43
2–2.9	14.38	24.32	6.34	8.77
3–3.9	13.77	38.09	8.55	17.32
4–4.9	12.84	50.93	10.28	27.60
5–5.9	12.20	63.13	11.96	39.56
6–6.9	6.38	69.51	7.40	49.96
7–7.9	6.10	75.61	0.82	47.78
8–8.9	5.57	81.18	8.47	56.25
9–9.9	3.51	84.69	5.97	62.22
10 +	15.31	100.00	37.78	100.00
Gini coefficient 0.354				

Sources: Summary tabulation worksheets for Ministry of Agriculture and Cooperatives (1967), reproduced in Jegatheesan (1977).

the administration. Wincott suggests that there is a preference among richer farmers for the latter system. In any event, these patterns of inheritance almost certainly account for the large share of rented land that is leased from kin. Moreover, when holdings are fragmented, farmers try to lease out awkwardly located parcels and replace them with rented plots which are more conveniently situated.

Despite an active rental market, there was still an appreciable spread in the distribution of operational holdings. The data in tables 2-8 and 2-9

Table 2-9. *Distribution of Farms by Size of Operated Area in 1972*

Size class (relongs)	*Percentage of farms*	*Cumulative percentage of farms*	*Percentage of area*	*Cumulative percentage of area*
0.0–1.0	4.37	4.37	0.66	0.66
1.1–2.0	14.59	18.96	4.77	5.43
2.1–3.0	14.34	33.30	7.62	13.05
3.1–4.0	12.22	45.52	8.57	21.62
4.1–5.0	15.96	61.48	14.40	36.02
5.1–6.0	9.48	70.96	10.10	46.12
6.1–7.0	7.86	78.82	9.90	56.02
7.1–8.0	6.11	84.93	8.99	65.01
8.1–9.0	2.87	87.80	4.69	69.70
9.1–10.0	4.49	92.29	8.42	78.12
10.1 +	7.71	100.00	21.88	100.00
Gini coefficient: 0.360				

Source: FAO (1975).

reveal no change in overall inequality, but the proportion of holdings below one relong increased fivefold over the period covered by the two surveys. By 1972–73 there was a small but significant group of households which were, in effect, landless. There is no clear explanation for this shift at the lower end of the distribution, and it may be simply statistical, inasmuch as the methods and qualities of the surveys differed. It is possible, however, that although those with the smallest holdings retained a small parcel of land to meet their own needs for paddy, they found it more profitable to move into the expanding market for hired labor—perhaps because the new system of cultivation favors farms that exceed some minimum size.

A breakdown of farms by the tenurial status of the operation (table 2-10) reveals marked differences that stem, at least in part, from the fact that a large portion of the total pool of land leased in is rented from kin. In particular, more than 70 percent of all rental contracts are between kin (Affifuddin, 1973*b*). Some of this land is probably inherited, with actual transfer taking place before the death of its owner. This does not, however, explain why some farmers do not participate in the land market. Various explanations have been put forward, the life-cycle hypothesis being one of the better known (University Sains, 1974), but none is entirely convincing.

What is clear is that land rents have adjusted to new levels under double-cropping (table 2-11), real rents being 30 percent greater on average in 1973 than in 1967. There were, of course, great variations around the mean levels in both years. In 1967, 10 percent of rental contracts were at rates of less than $39 per relong, while a further 29 percent were in excess of $80. Comparable variations were observed in 1973.

There is no wholly reliable evidence concerning the effect of kinship between landlord and tenant on the levels of rents. Wincott (1976) suggests that rental agreements between kin are likely to provide the tenant with

Table 2-10. *Farm Size by Tenurial Status*

	1966		*1973*	
Status	*Percentage of farmers*	*Mean farm size (relongs)*	*Percentage of farmers*	*Mean farm size (relongs)*
Owners	44.1	5.1	41.6	4.8[a]
Tenants	40.9	5.3	39.8	5.1
Owner-tenants	13.5	9.0	18.1	8.1[a]
Others	1.4	n.a.	0.5	n.a.

n.a. Not available.

a. In 1973 owners and owner-tenants leased out, on average, 0.6 relong. This is not, however, a general phenomenon among all households. Of the 478 sampled farmers in these two groups, only 56 actually leased out land.

Source: FAO (1975).

Table 2-11. *Annual Rent per Relong*
(Malaysian dollars)

Rent	*1967*	*1967 in 1973 prices*[a]	*1973*
Cash	65.5	86.6	99.7
Kind	50.5	66.8	102.9
All	58.1	76.8	99.9

a. Inflated by 1.32, which is the relevant consumer price index for 1973 with 1967 as base (table 7-1).

Sources: Ministry of Agriculture and Cooperatives (1967), and FAO (1975).

various noneconomic advantages. Be that as it may, the 1972–73 survey revealed no differences between average rents on tenancies involving kin and those not involving kinfolk. There are, however, some grounds for believing that rentals in tenancy contracts between kin adjust more slowly to market forces. In particular, if the rental market does not clear because rentals are not sufficiently flexible upward, then kin may be able to claim first place in the queue for land to be leased out. As a farmer put it to one of us: "If land is rented to outsiders when kinsmen are in need, then heads will get broken."

The labor market is also highly active in the Muda area, the suppliers being farm households, which are the most important source of farm labor, landless farm worker households, and migrant workers. There was a small decrease in the number of farm operators between 1966 and 1973, but a clear increase in the number of farm workers, a result perhaps, as we have already speculated, of the growing overall pressure on the land forcing previously marginal farmers into the ranks of the landless.[5] The evidence on labor supplied by migrant workers is sketchy. The records of the number of migrant workers reaching the region in 1967 from Thailand and Kelantan, which is on the northeast coast of Malaysia, can be treated with some confidence. Supplies from these sources were sharply curtailed between 1967 and 1973. The Kemubu irrigation project generated much on-farm employment in Kelantan, and migrant Thai workers were prohibited from entering Malaysia in 1970. The evidence from FAO (1975) indicates that this shortfall has been made good from elsewhere, but it provides no clue as to the source. It is safe to conclude, however, that some Thai migrants were able to make their way in, if only because the border is fairly open. Also, Goldman and Squire (1977) report that in 1976 increasing numbers of migrant laborers were being drawn into the project area from

5. Many of these landless households do operate minute holdings (0.14 relong on average), but even this land is invariably rented on a seasonal basis.

Table 2-12. *The Use of Family Labor*
(hours per year)

Family labor	*Single-cropping households, 1969–70*	*Double-cropping households, 1972–73*[a]
Own farm	645[b]	949
Wage employment		
Paddy	172[c]	312
Other	63[c]	30
Other productive activity[d]	990[e]	855
Total	1,870	2,146

a. FAO (1975).

b. From Doering (1972). His data are presented per relong, which we have scaled by the mean farm size of 5.61 relongs in 1972–73.

c. From the FAO (1975) data on single croppers in 1972–73, adjusted for family size and the absence in 1969 of double-cropping.

d. Defined as time spent attending to livestock and poultry, working on the house lot and in the house, fishing, and producing handicrafts and other crops (mostly rubber), but the normal domestic duties of women are excluded.

e. Unadjusted figures for single-cropping households in 1972–73.

other parts of Kedah, especially Baling and Sik, which are well to the south and southwest of the region.

In the absence of reliable data for the preproject period, it is difficult to arrive at firm estimates of how much additional family labor was called forth by the demands of double-cropping. We have approached this question by using the evidence from a small survey undertaken in 1969–70 (Doering, 1972 and 1973) in combination with some bold inferences from the FAO data for single-cropping households in 1972–73 (table 2-12). These calculations suggest that, on average, the amount of work put in by the family on its own paddy farm over a year increased by 47 percent, and work in paddy cultivation of all kinds increased by 54 percent. As double-cropping provides another season of activity, this estimate is broadly plausible. On the same basis, the use of hired labor increased more strongly still (table 2-13),

Table 2-13. *Total Labor Use in Paddy Farming*

	Single-cropping households, 1969–70		*Double-cropping households, 1972–73*	
Labor	*Hours annually per relong*	*Percent*	*Hours annually per relong*	*Percent*
Family	115	58	169	51
Hired	83	42	160	49
Total	198	100	329	100

Sources: Single-cropping: Doering (1972); double-cropping: FAO (1975).

which accords with intuition about the likely effects of strong peaks in the demand for labor under double-cropping.

There were also changes in wage rates during this period, though how much they stemmed from changes in the supply of and demand for labor in the project area is not entirely clear, since real wages were rising elsewhere in Malaysia. Leaving causes aside, the evidence on changes in wage rates is sketchy. The figures in table 2-14 from Goldman and Squire (1977), although much qualified by those authors, represent the best information available. Since the tasks in question are performed with a standard technique, it appears that real wage rates rose by about 50 percent between 1970 and 1974.

As we have seen, labor inputs per unit of land did not double with the introduction of double-cropping. Farmers adopted labor-saving techniques, and labor inputs per relong per crop were 17 percent lower after the introduction of double-cropping than before. This decline was due largely to the switch from buffaloes and hand hoes to tractors: for each relong, inputs of labor for land preparation fell by half between 1970 and 1973. Inputs of labor for transplanting also declined, which could imply that planting densities also fell. Goldman and Squire (1977) speculate that this did happen; there is some evidence in Yamashita, Jegatheesan, and Wong (1976) that planting densities have fallen; and the substitution of capital and land for labor in the face of an increase in the wage-rental ratio is the standard explanation for such a change in farm practice. Nevertheless, there is no really firm evidence from the preproject period to support the assertion that planting densities declined.

Estimates of what happened to farm incomes over this period are again plagued by the frailty of the evidence for the earlier years. FAO (1975) provides detailed data on incomes for 1972–73 and estimates household income for farms with double-cropping at just over $1,900 (table 2-15). For the preproject era, Jegatheesan (1977) has made a gallant attempt to

Table 2-14. *Contract Rates for Hired Labor by Task*
(Malaysian dollars per relong)

	Transplanting		*Harvesting and threshing*	
Year	*Current prices*	*Constant prices*[a]	*Current prices*	*Constant prices*[a]
1967	5.7	5.7	24.0	24.0
1970 (off-season)	11.9	11.0	24.7	22.9
1972 (off-season)	20.9	19.0	42.0	38.1
1973 (off-season)	24.7	18.5	49.3	37.0
1974 (off-season)	30.0	18.6	57.8	35.8

a. 1967 prices, using the cost of living index for landless households (table 7-1).
Source: Goldman and Squire (1977).

assemble and interpret the data available for 1966–67, and we use his estimates. Unfortunately, he does not describe in sufficient detail how he allocated farm income and expenditures to arrive at his estimate of net annual disposable household income. As we have no basis for changing his

Table 2-15. *Details of Average Incomings and Outgoings for Paddy Farmers, 1972–73*
(Malaysian dollars)

1. *Household revenues*	
Value of paddy production	2,278
Revenue from livestock	72
Other farm revenue	106
Wage income from paddy cultivation	67
Other wage earnings	83
Miscellaneous income	162
Total	2,768
2. *Current farm expenditures*	
Seed	14
Fertilizer	114
Pesticides	6
Other inputs	40
Hired labor	292
Hired buffalo	4
Hired machinery	57
Livestock costs	3
Machinery operating expenses	50
Insurance	1
Taxes	131
Land rent	111
Other expenses	9
Total	832
3. *Household income* (1 – 2)	1,936
4. *Net purchase of assets*	
Purchase of vehicles and farm machinery	96
Purchase of livestock	23
Purchase of land	15
Farm debt repayment	103
Other debt repayment	233
Less:	
Loans for farming (modern)	24
Loans for farming (traditional)	117
Loans for household expenditures	99
Total	230
5. *Household consumption expenditure*[a] (3 – 4)	1,706

a. Excludes changes in financial holdings (such as cash and bank deposits), house values, and farm stocks.

Source: FAO (1975).

Table 2-16. *Average Income from Paddy Farming, 1967–75*
(current Malaysian dollars)

Paddy income[a]	*1966–67*	*1972–73*	*1974–75*
Value of paddy production	1,026	2,278	4,961
Wage income	135	150	310
Gross income from paddy farming	1,161	2,428	5,271
Less variable production costs	417	700	1,861
Net income from paddy farming	744	1,728	3,410

a. Excludes nonfarm income and sales of other farm produce.
Sources: 1966–67 and 1974–75: Jegatheesan (1977); 1972–73: FAO (1975).

estimate for 1966–67, we have therefore tried to reorder the more detailed data for 1972–73 in the same way (table 2-16). In current prices, the gross value of paddy production increased by 122 percent and net income from paddy farming by 132 percent.

We have also included Jegatheesan's estimates for 1974–75 based on data from Yamashita, Jegatheesan, and Wong (1976). Although this sample of households is not statistically representative of the project area, it provides the only other detailed postproject estimate of farm incomes. On the basis of these estimates, by 1975 nominal net paddy farm income had virtually doubled over its level in 1972–73. Thus, between the pre- and postproject periods, our best estimate is that the gross value of paddy output rose by 380 percent and income from paddy farming by 360 percent, both in nominal terms. Earnings from employment in paddy cultivation merely doubled.

Conclusions

The outline of the story is clear. The Muda project matured quite quickly, bringing about a large increase in both the output of paddy and the incomes of paddy farmers, with attendant changes in technology and the rhythm of cultivation. It is now time to analyze the workings of the paddy farming sector, not only to gain a deeper understanding of its response to the components of the project, but also to provide a basis for predicting the effects of other public policies. In the next two chapters, we will build and use a model of the sector designed for that purpose.

3

A Model of the Muda Paddy Economy

IN CONTRASTING the situations that prevailed before and after the project, we relied in chapter 2 on sketchy and incomplete data. This was especially true of the preproject situation, for which little information exists on such basic variables as household incomes; even the data relating to the post-project period are unreliable for several key variables and incomplete with respect to seasonal wages and the cost of draft power. Further, in the absence of any theory about how the paddy farming economy works, we were unable to explain the changes that took place. Hence, no attempt was made to separate the changes induced by the project from other autonomous changes, such as increases in yields arising from the introduction of improved varieties of seed and changes in the price of paddy relative to farm inputs.

In chapters 3 and 4 we try to overcome these weaknesses. In particular, we develop an analytical framework based on a formal mathematical model of the paddy farming sector, which solves for equilibrium values of paddy production activities and determines the corresponding quantities and prices of land, labor, and draft power traded in the local markets. In this chapter the model is constructed and validated for 1973 using the FAO/World Bank farm survey data (FAO, 1975). Then in chapter 4 we supply the model with the available historical evidence and use it to provide pictures of the paddy farming sector in 1967 and 1974.

At the outset, it should be noted that the model simulates steady states, in which all adjustments are complete. Hence, these pictures of the economy may not be exact reflections of the actual situations that prevailed in those years, a matter treated in some detail below. Subsequently, we analyze the direct effects of the project by controlling for autonomous changes and solving the model for different hypothetical situations.

Basic Modeling Considerations

To meet our objectives, the paddy sector model must be based on solid microeconomic foundations. This requires a detailed treatment of not only important paddy production decisions, but also their impact on the regional labor, draft-power, and land rental markets. Our first modeling effort, perhaps overambitiously, attempted to define a number (eventually four) of representative farms linked at the regional level through the operation of the principal factor markets. Experimentation with this model encountered severe problems of model degeneracy and slow convergence. Although slow convergence was partly overcome by using column selection rules different from those currently available in commercial linear programming systems (Bisschop and Meeraus, 1977), these and other problems led us to represent the Muda project area by a single aggregate farm. Such an approach clearly raises the problem of aggregation bias; but, as we argue below, this can be expected to be small. Furthermore, such an aggregate model cannot capture interfarm trading in the factor markets, and the total farm income it generates can be allocated to different classes of household only on an ad hoc basis after the model has been solved. Nevertheless, the markets for land rented from nonfarming households, labor supplied by the landless and migrants, and four-wheel tractor services can be satisfactorily incorporated into the model, the solutions to which yield the payments to these factors.

The farming system in the project is essentially a monoculture operating within a system of very active factor markets.[1] If these factor markets are perfect and the production function is homogeneous of degree one, and if the influence of managerial skill and luck is ignored, then a household's income will depend only on the parametrically given set of rentals for factors and its endowments of land, labor, and draft power. It follows, therefore, that any attempt to base a model on representative farms under these conditions must lead to model degeneracy, since an infinite number of alternative interfarm trading arrangements will be equally good for attaining the optimal factor-use ratios for each farm. A farm with a high land-man ratio could, for example, either hire labor from a farm with a low land-man ratio, or lease out land to that farm. In either case, identical land-labor ratios can be obtained, and both the total income and its distribution between the two farms will remain unchanged. The distribution of income may, however, be affected by interfarm trading in factors if, first, the production function is not homogeneous of degree one, so that profits or losses remain to be allocated after all factors have been paid their marginal products; or if, second, market imperfections are sufficient either to impede the attainment of equal factor-use ratios on all farms or to cause

1. Most farm households do have small food gardens (about 0.7 relong on average) and keep a number of small livestock, mostly goats and poultry. These activities are mostly confined to land unsuited to paddy production, and their outputs are primarily for family consumption.

the prices of factors to be different from the values of their marginal products.

The available evidence suggests that the land rental market in Muda is distorted in that not all farmers have equal access to rental land: when there is excess demand for leases, the kinfolk of the landlords seem to get the land in preference to non-kin. Provided this allocation is not correlated with farmers' managerial skill, this distortion appears insufficient to affect aggregate paddy income in a significant way because other factor markets seem to permit offsetting and compensatory transactions. In other words, aggregation bias will be small even if our model is not founded on a number of representative farms.[2] For example, farmers who have restricted access to rented land may make compensatory sales (where the transactions cost of such sales are equal) in the labor and draft-power markets. Hence, when the lump sum transfers implicit in the rationing of leases are netted out, they attain much the same income levels as if they had access to as much land as they wished. This proposition was confirmed through experiments conducted with the early version of the model based on four representative farms. First, an initial, perfectly competitive solution was obtained. Then, land access constraints were incorporated, as well as somewhat arbitrary, but small, differential transactions costs between hiring and selling activities. The resultant aggregate use and pricing of factors, as well as total income and its distribution among the four farm groups, were little different from those when the model was specified with perfect factor markets (table 3-1).

If the factor markets other than that for land are reasonably competitive, the land-labor utilization ratio should be approximately the same on all farms. Moreover, if our assumptions are correct, we should expect to find that the observed land-labor utilization ratios are invariant with respect to farm household resource endowments. Accordingly, we formulate the following linear expression:

$$Y = \alpha + \beta_1(OA) + \beta_2(NAW),$$

where Y = the ratio of total farm labor used (in hours) to cultivated area

2. Aggregation bias has been defined precisely by Day (1963*a*). Let x_k denote the linear programming solution vector of the kth individual farm model, and let x denote the solution vector of the aggregate farm model. Then aggregation bias exists if $\Sigma_k x_k \neq x$. Maintaining our assumption that all farms face identical prices and technology, Day has established that x will be an unbiased estimate of $\Sigma_k x_k$ if all farms have identical ratios of resource endowments. This condition is required to hold only for nontraded resources, since the prices of traded factors are given exogenously to all farms. In Muda, as we have argued, all resources are freely traded except land, for which Day's condition must be satisfied, since endowments of a single resource are always proportional.

Table 3-1. *Income Distribution and Land Rents under Alternative Land Market Formulations*

Income and land rents	*Competitive formulation*[a]	*With land constraints*[b]
Household net income		
Pure tenants	1.0	1.05
Small farms owning land	1.0	0.90
Large farms owning land	1.0	1.02
Large farms owning land and tractors	1.0	1.01
Average	1.0	1.00
Land rents	1.0	1.03

a. Results indexed to 1.0.

b. Land areas constrained to observed values in 1973.

measured in relongs; OA = relongs of paddy land owned; and NAW = number of adult family members employed on the paddy farm.

Using data from the FAO/World Bank household survey, we obtained the following estimates (t values are given in parentheses):

$$Y = \underset{(25.00)}{426.6} - \underset{(4.47)}{7.69}\,(OA) - \underset{(0.55)}{3.13}\,(NAW).$$

$$R^2 = 0.025 \qquad F = 10.932 \qquad \bar{Y} = 393.7$$

While these results show β_1 to be statistically significant, Y has only a very small partial elasticity (0.061 percent) with respect to OA, and the R^2 for the estimated equation is very small (2.5 percent). We may therefore safely assume that factor endowments are of little consequence in explaining farmers' land-labor utilization ratios. This finding certainly contrasts with received wisdom on land-labor ratios in other parts of the world, especially South Asia, but then Muda has rather well-developed and competitive markets for inputs (with the exception of land).

A further methodological consideration was our choice of modeling technique. The obvious importance of seasonality in determining resource allocation patterns, and hence factor prices, in the Muda project area called for a model in which the important decision variables could be time-dated with sufficient precision to depict seasonal bottlenecks in resource supplies. In addition, since there had been changes in technology over time and we wished to analyze the determinants of these changes, we needed a model able to accommodate considerable microdetail and explicitly portray technological options. These considerations led us to select a mathematical programming approach and to depart from the more standard production function methodology used by Barnum and Squire (1979) in their analysis of labor demand in the Muda paddy economy.

Having opted for a mathematical programming model, we had then to choose between a static equilibrium model and a dynamic model. The impact of a project is best measured with a model that tracks history year by year and incorporates specific assumptions about the way in which farmers form their yearly price and yield expectations, the speed with which the factor markets adjust to changes in production and resource use, and the dynamics of investment decisions in draft-power resources. Such requirements can be met, in part, by incorporating flexibility (or adjustment) constraints into the model and then using recursive programming (Day, 1963*b*; Day and Singh, 1977); but the parameters specifying these constraints are inherently difficult to measure, especially in the absence of reasonable time series data.

Hence, we have chosen to use a static equilibrium model, which, given prices, technology, and fixed resource endowments, can be used to depict the paddy economy in different steady states. Decisions regarding investment in buffaloes and tractors are treated endogenously, so that the equilibrium stocks of these resources are obtained from the model. The model does not, therefore, correspond to any specific calendar year, but to the steady state corresponding to the exogenous variables and parameters that characterize the year in question. This approach clearly makes the validation of the model against any specific year difficult. Nevertheless, the model permits an examination of the economic forces acting within the paddy economy and, when its results are compared with observed events, provides some indication of the speed at which the paddy economy is moving from one steady state to another. A model of this kind also permits the conduct of policy experiments on a comparative static basis. For example, we can examine the postproject equilibrium which would have occurred had prices not changed from preproject levels, or derive an equilibrium in which postproject prices and resource endowments prevailed, but in which the project had not been constructed. And, as we shall see, it is through experimentation of this kind that we are able to tackle the problem of first estimating and then decomposing the direct benefits of the project.

Specification of the TIGER Model

The Muda paddy model, henceforward called TIGER,[3] is an annual equilibrium model which treats the scheme area as a single aggregate farm.

3. The name TIGER was originally chosen as being totally devoid of any significance for the Muda area. Subsequent research has established, however, that the tiger is an important animistic symbol in local Malay folklore, as well as being the brand name of a particularly good brew of local beer.

It is concerned with the following decisions: the total area and seasonal scheduling of paddy production during the year; the choice of technology, namely, fertilizer treatment and use of draft power from buffaloes or pedestrian or four-wheel tractors for land preparation—choices that can differ by soil type and between the main-season and off-season crops; seasonal use of farm family labor; seasonal hiring of landless laborers and seasonal migrant workers, and the determination of seasonal wages; the total area of land rented from nonfarm landlords and the determination of rents; and the optimal regional stocks of buffaloes and pedestrian and four-wheel tractors and the seasonal use of these resources.

The model does not include all production decisions. Livestock activities (other than tending to buffaloes) and food gardening are taken as exogenously given, with resources being set aside for these activities. Similarly, allowance is made for employment other than in paddy farming by household members. According to the FAO/World Bank survey data, none of these activities is very important, and in total they amount to only 7 percent of total family farm labor.

The important resource constraints in the model are: the regional area of paddy land, defined by two soil types (acid and nonacid) and the areas available for single- and double-cropping; the seasonal amounts of farm family labor available for paddy farm work; and the seasonal supplies of landless laborers and migrant workers, which are defined in the model by upward sloping labor supply functions.

An early version of the model also included constraints on the seasonal use of farm credit, but the farm enterprise cash flow was found to be such that farm credit was never taken up. Hence, these constraints have been omitted from the present model. This does not imply that farmers have no need for credit, but only that in an equilibrium state the average farmer's seasonal cash flow is sufficient to obviate borrowing for paddy production. In practice, individual farmers may still seek credit; for example, they may have to roll over historically accumulated debt, or purchase tractors or land, or they may suffer a severe crop failure or the ravages of natural pestilence. Similarly, important family events may require heavy cash outlays which are met by borrowing. Indeed, the data in table 2-15 show that the average farmer has significant debt. TIGER, however, deals with equilibrium situations and does not take such considerations into account.

The model is specified using 1973 (the FAO/World Bank survey year) as the base year. Most of the technical coefficients are derived from the FAO/World Bank survey data, although some use has been made of other surveys undertaken by the Malaysian Ministry of Agriculture and MADA. The technical reports published by MADA have been particularly useful, especially those dealing with labor utilization, mechanization, and the response of paddy varieties to fertilizer inputs. Last, but not least, discussions with MADA officials and the farmers themselves provided an additional and invaluable source of material.

Nursery Scheduling Decisions

The technology decisions involved in paddy farming are relatively simple. A crop of paddy is raised twice each year, and each must be planted in accordance with the tight seasonal scheduling of the irrigation calendar. The tightness of this calendar clearly has a profound bearing on the seasonal distribution of the demand for labor and draft power.

To facilitate the analysis, we have divided the year into twenty-six two-week periods (table 3-2). On this basis, the key operations of nursery

Table 3-2. *Definition of Time Periods in TIGER*

Fortnight ending		*Fortnight (t period)*	*Four weeks (j period)*	*Season (s period)*[a]
December	9	1	1	
	23	2	1	
January	6	3	2	
	20	4	2	
February	3	5	3	
	17	6	3	
March	3	7	4	1
	17	8	4	1
	31	9	5	1
April	14	10	5	1
	28	11	6	1
May	12	12	6	1
	26	13	7	
June	9	14	7	
	23	15	8	
July	7	16	8	
	21	17	9	
August	4	18	9	2
	18	19	10	2
September	1	20	10	2
	15	21	11	2
	29	22	11	2
October	13	23	12	2
	27	24	12	2
November	10	25		
	25	26		

a. Defined on the basis of the permitted nursery dates: 1 = off-season; 2 = main season.

seeding, transplanting, and harvesting for a typical variety with a twenty-week growing period are timed as follows:

	Fortnight performed
Nursery	t
Transplant	$t + 2$
Harvest	$t + 10$

For convenience, we assume that each operation is performed at the midpoint of the corresponding fortnight.

The nursery date is the key decision variable, since this determines the dates of all subsequent operations. The range of nursery dates available to the farmer is constrained by the date on which his irrigation water arrives for the off-season crop and by the rainfall pattern during the main season. This irrigation date is assumed to be the midpoint of the ninth fortnight.[4] In this event, the feasible nursery dates permitted in TIGER and their corresponding harvest dates, are:

	Nursery period		*Harvest period*	
	Fortnight	*Date*	*Fortnight*	*Date*
Off-season crop	7–12	February 18–May 12	17–22	July 8–September 29
Main-season crop	18–24	July 22–October 27	2–8	December 10–March 17

The earliest date for nursery seeding in the off-season is constrained by the need to puddle the fields with irrigation water before transplanting. The latest date is based on the assumption that serious water control problems would arise if transplanting were not completed within eight weeks of the irrigation date.[5] The options for the main-season crop are based on average rainfall distribution data for 1960–72 (see appendix B). These data suggest that paddy planted before July 22 (beginning of period 18) is likely to receive excessive amounts of water at the critical flowering and maturation stages. In addition, the soil is likely to be too dry for land preparation and transplanting before period 20 in most years. Nursery dates after October 27 (end of period 24) are judged infeasible because, on

4. Our year is defined as beginning on November 26, 1972, and ending on November 25, 1973, to retain compatibility with the FAO/World Bank survey year. The midpoint of period 9 is therefore March 24, which is quite close to the average irrigation date in 1973.

5. In reality, the irrigation date varies within the Muda project, since the water is released by sections over a four- to six-week period. The distance of a field from the nearest irrigation canal also affects the arrival date for irrigation water. These details cannot be incorporated into our aggregate farm model, and rather than unduly restrict the model, we have specified the nursery date options on a more flexible basis than might be encountered by the individual farm. Our choice of nursery date options encompasses more than 90 percent of the planting dates observed in the FAO/World Bank survey.

Table 3-3. *Main-season Yield Response to Nitrogen and Form in Which Applied*

Nitrogen level (pounds per relong)	Paddy yield after deductions[a] (gantangs per relong)		Fertilizer (percent)	
	Acid soils	Nonacid soils	Basal	Urea
0	248	251	0	0
15	290	322	100	0
25	297	350	80	20
30	295	357	73	27
35	290	362	67	33

a. Deductions for chaff and other waste account for 17 percent of gross yields in the data source used (see appendix B).

average, insufficient rainfall is received to meet the crop's requirements for water.

Choice of Fertilizer Treatment

FAO (1975), using data from Soo (1972), showed that two main soil types can be identified (acid and nonacid) which significantly affect yields. Consequently, separate paddy growing activities are included in the model for both the acid and nonacid soils. For each soil type an excellent approximation to nitrogen response functions was obtained by offering five fertilization levels in the model.[6] These nitrogen levels and their associated yields by soil type are presented in table 3-3 for the main-season crop. These data were derived from statistical analysis of a cross-section of farm data collected in 1974–75 by MADA (1976; see appendix B in this volume). They relate to a modern variety of seed with a twenty-week growing period from nursery to harvest. In the light of varietal improvements between 1973 and 1975, the estimated yield response function was scaled down by 5 percent for the purposes of the 1973 model.[7] A similar analysis of the farm data for the 1974–75 off-season crop proved less fruitful (appendix B) and forced us to settle on the simple expedient of increasing the main-season yields in table 3-3 by 7 percent for the off-season crop. This percentage reflects the average yield differential observed between the two seasons in the FAO/World Bank survey.

6. Local research (Samy and others, 1970) indicates that paddy is not very responsive to phosphorus or potassium, so that we consider only the nitrogen content of fertilizer.

7. Average paddy yields increased by 25 percent over the 1969–74 period. This is equivalent to an average annual growth rate of 4.5 percent, but probably most of the yield increase occurred before 1973, so that a 5 percent adjustment for two years is not unreasonable.

The form in which the nitrogen is applied is also given in table 3-3. Basal has a 16 percent nitrogen content, while urea contains 46 percent nitrogen. Basal is typically applied at the time of transplanting, and urea is applied as a topdressing two weeks later. The timing of these applications is built into the model, while the associated labor requirements are estimated to be 0.4 man-day per relong at all permitted levels of application.

By affecting yields, the amount of nitrogen used also affects labor requirements for harvesting operations. From the FAO/World Bank data we calculate threshing, transporting, drying, and winnowing to be 1.30 man-days per picul of paddy. We also assume that all harvesting is done manually, as the FAO data reveal no cases of mechanized harvesting. (Although this assumption is certainly realistic for 1973, it was untenable for the late 1970s. It is relaxed in chapter 9 when we examine the consequences of increased mechanization and other recent developments.)

Method of Field Preparation

In preparing the fields for transplanting, three mechanization options are offered. All involve labor, but they differ in their source of draft power, which may be provided by buffaloes, pedestrian tractors, or four-wheel tractors. These options and their resource requirements are shown in table 3-4, in which the resource requirements provide for two plowings (or rotivations) of the land, and are based on an average performance of one acre a day for pedestrian tractors and 5.7 acres a day for four-wheel tractors

Table 3-4. *Techniques for Field Preparation*

		Method of field preparation (days per relong)		
Requirement	*Period*[a]	*Buffaloes*	*Pedestrian tractors*	*Four-wheel tractors*
Labor	t	3.54	3.38	3.27
	$t+1$	3.10	—	—
	$t+2$	2.67	3.59	2.40
Buffaloes	t	0.30	—	—
	$t+1$	2.00	—	—
	$t+2$	1.00	—	—
Pedestrian tractors	t	—	0.14	—
	$t+2$	—	1.42	—
Four-wheel tractors	t	—	—	0.03
	$t+2$	—	—	0.25

— Not applicable.

a. t denotes nursery date. Entries in rows labeled t represent resource requirements for preparation of the nursery. All other figures are resource requirements for the preparation of fields before transplanting.

(MADA, 1970*a*). Tractors are, of course, faster than buffaloes, and mechanized methods also avoid the need for fields to stand two weeks between plowings.[8] This saving in time offers an important advantage, given the stringencies of the irrigation calendar, and it is reflected in the model by allowing field preparation with tractors to be conducted in the same two-week period as transplanting. With buffaloes, however, land preparation must be undertaken in the fortnight before transplanting, and labor and land must be made available at that time.

As with field preparation, various sorts of draft power can be used for nursery preparation. The nursery area is very small (about 4 percent of the transplanted area), thus we simply assume the technique used is the same as that selected for field preparation.[9]

The Draft-power Investment Choices

The choice of technique for land preparation implies decisions about the optimal stocks of buffaloes and tractors at the aggregate level. These stocks are endogenous to the model, and they are determined through the incorporation of investment activities. Buffaloes and pedestrian and four-wheel tractors can be purchased at fixed annualized capital costs (which include resources for feeding and maintenance in the case of buffaloes), and each unit purchased is then available for work in all fortnightly periods in the model. A variable cost is levied on the use of tractors to cover expenses for fuel, oil, and repairs. Hence, the total annual cost of a tractor depends on the utilization rate, and this is endogenous to the model. Pedestrian and four-wheel tractors are limited to working a maximum of eleven days a fortnight to allow for "down time" and vehicle transport between fields. Buffaloes are limited to working a maximum of nine days a fortnight.

Treatment of the Paddy Market

As Malaysia is not self-sufficient in rice, increases in domestic paddy production simply substitute for rice imports and have a negligible effect on the domestic price. The government also maintains a minimum price, which often determines the price that farmers receive. These considerations suggest that the Muda area faces a highly elastic demand for its paddy output, even though it now produces about 45 percent of Malaysia's rice

8. With buffalo-drawn plows, fields are allowed to stand for about two weeks between cultivations to allow time for the clods to soften and break down and for weeds and other vegetative matter to be cleared. Mechanical rotivation produces a much finer tilth on the first pass and shreds all vegetative matter.

9. This is a slight departure from reality, since in practice most nursery plots are prepared by hand.

requirements. Hence, the model is specified with a fixed output price. On the assumption that all farmers face the same price each year and that they maximize profits,[10] maximizing the producer surplus from paddy production in the region will lead to a perfectly competitive solution in the paddy market.

Treatment of the Land Market

It has already been argued that different allocations of land among farm households will have little effect on aggregate paddy production or on aggregate farm income. An immediate corollary is that the value of the marginal product of land can be satisfactorily estimated in an aggregate farm model in which the total amount of land available sets a single, overall land constraint. Indeed, all that we need to know about the land market is: how much, if any, paddy land in the Muda scheme area will be left idle each season; and, if no land is left idle, what the value of its marginal product will be. The latter should, of course, determine the rental value of land in a competitive market. Actual land rents, however, appear to have adjusted only slowly to the introduction of double-cropping; thus, the value of the marginal product of land obtained from the dual values in the model solution must be regarded as a pointer to the level of land rents after adjustments are complete.

The model contains a set of aggregate constraints that limit total land use each fortnight to be equal to or less than the available project area. The two basic soil types (acid and nonacid) in the project area (FAO, 1975) are treated separately in the model to allow for different levels of yield. We do not, however, differentiate between land owned by farm and nonfarm households. Therefore, in maximizing the model's objective function, the producer surplus from paddy production for the project area includes the income accruing to nonfarm households through rental payments for their land.

Treatment of the Draft-power Markets

As buffaloes and pedestrian tractors are owned only by farm households, we need specify only aggregate constraints on their use. Supplies of these resources are generated through the investment activities described above. Hence, the constraints, which are defined fortnightly, are simple

10. Yield and price variations do create some risk, but if farmers are unable to diversify into alternative crops, there is little they can do to reduce their income risk. Optimal fertilizer levels may be affected, but Roumasset (1976) has shown that this need not be a significant consideration with irrigated rice.

balance rows which equate the supply and demand of each resource. This treatment implicitly assumes that service charges for all interfarm transactions in draft power equal the full average cost of each technique, which in turn depends on the utilization rates for buffaloes and pedestrian tractors and is therefore endogenous to the model.

Four-wheel tractors are owned by nonfarm households, but both their rental and utilization rates depend primarily on the demand for their services within the project area. To endogenize these values in the model, the treatment of four-wheel tractors is identical to that of buffaloes and pedestrian tractors. This specification implies that four-wheel tractor services are charged at their average cost, and that profits, other than any rents incorporated in the overhead costs, are zero.

Thus, our treatment of the draft-power markets assumes perfect competition, so that all farmers have equal access to draft-power services, regardless of whether they own a buffalo or pedestrian tractor.

Treatment of the Labor Market

We differentiate between farm family labor and day labor, the latter being supplied by local landless workers and seasonal migrants from elsewhere in Malaysia and Thailand.

Since TIGER is an aggregate farm model, we ignore interfarm trading in farm family labor and treat family labor as if there were one large family through the incorporation of aggregate fortnightly constraints on the available supplies. Labor requirements for farm overhead work and the care of livestock (other than buffaloes) and food gardens are deducted from the available supply of family labor. Allowance is also made for employment other than in paddy farming, which is determined exogenously to the model. The use of farm family labor is charged a fixed reservation wage, which places a lower bound on the marginal product of family labor (Duloy and Norton, 1973). We thus assume that family members place some value on leisure time, so that they do not work to the point at which the marginal product of their labor is zero. This does not, however, prevent the marginal value product of family labor from rising above the reservation wage when the market wage is higher.

Our treatment of the day labor provided by landless workers and seasonal migrants is rather different. We assume that these workers have alternative employment opportunities and that Muda's farmers must compete for their services. More specifically, we depict each month's supply of this labor by an upward sloping supply function. The wage rate in each month then depends on the amount of labor hired.

Monthly employment and wage data for 1973 from the FAO/World Bank survey were used to estimate econometrically a four-week day labor supply function (appendix B). This function is linear and of the form $W_t = \alpha + \beta L_t$, where W_t is the wage in period t (Malaysian dollars per day), L_t is the amount of labor hired in period t (days), and α and β are parameters which are the same for each period. The total wage bill in the model is, therefore, $\Sigma_{t=1}^{13} W_t L_t$.

Some care is needed when incorporating the wage bill and labor supply into the model. If we were to substitute $\alpha + \beta L_t$ for W_t and subtract $\Sigma_t(\alpha + \beta L_t)L_t$ directly from total paddy income in the objective function, the solution so obtained would correspond to monopsonistic behavior in which the marginal value product of labor is equated to its marginal cost. To obtain a competitive market solution, in which the average cost of labor is equated to its marginal value product, we must modify the objective function so as to maximize the sum of the producer and consumer surpluses over the monthly labor markets (Hazell, 1979).

These possibilities are portrayed in figure 3-1, in which the demand curve for labor is the marginal value product curve implicitly embedded in TIGER, and the labor supply function is the average cost function, $W = \alpha + \beta L$, corresponding to the marginal factor cost function

Figure 3-1. *Alternative Labor Market Equilibria*

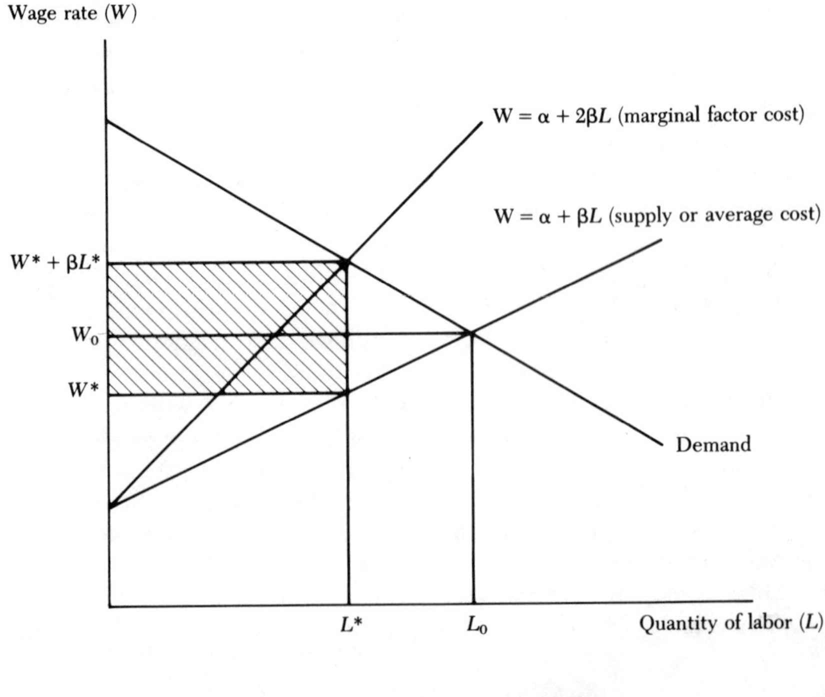

$W = \alpha + 2\beta L$. If the model's objective function were maximized net of the wage bill, this would yield a solution in which L^* of labor is hired, the marginal value product of labor is equal to $W^* + \beta L^*$, and the actual wage paid is only W^*. Monopsonistic profits, which are equal to the shaded rectangle, would accrue to farm households. By maximizing a modified objective function corresponding to the area between the labor demand and supply schedules, the competitive market solution (L_0 units of labor hired at wage W_0) can be obtained. The modified objective function which yields the competitive solution specifies that $(\Sigma_t \alpha L_t + 0.5\ \beta L_t^2)$ be subtracted from total paddy income instead of the total wage bill $(\Sigma_t \alpha L_t + \beta L_t^2)$. The former is the sum of the areas under the labor supply schedules, while the latter is the sum of the areas under the marginal factor cost curves.

An Algebraic Statement of the Model

With the aid of the full set of symbols set out in table 3-5, we provide below a complete statement of the equations of the model. The only fixed resources in the model are land (LND_i) and farm family labor (LAB_t). All other resources can be purchased in unlimited quantities, either at parametrically fixed prices (in the case of tractors and fertilizers) or at increasing average costs (in the case of day labor). Most of the constraints are therefore balance or tie rows, which simply force in the required resource purchases.

The equations of the system may be written as follows:

Objective function (maximand)

(3.1) MAX (CAPS)

$$= \Sigma_s\, P_s \cdot PDYPD_s - c_u \cdot BUYU$$

$$- c_b \cdot BUYB - c\Sigma_i \Sigma_f \Sigma_m \Sigma_\tau X_{fm\tau i}$$

$$- r \cdot TOTFLB - [\alpha \sum_j TOTDLB_j + 0.5\beta \sum_j (TOTDLB_j)^2]$$

$$- (c_2 \cdot TOT2WT + c_4 \cdot TOT4WT) - (K_2 \cdot 2WINV + K_4 \cdot 4WINV)$$

$$- K_B \cdot BFINV$$

CAPS = [Gross value of paddy output] − [Cost of urea fertilizer]

− [Cost of basal fertilizer] − [Miscellaneous production costs]

$$- \begin{bmatrix} \text{Reservation value} \\ \text{of family labor} \end{bmatrix} - \begin{bmatrix} \text{Sum of areas} \\ \text{under day labor} \\ \text{supply schedules} \end{bmatrix}$$

$$- \begin{bmatrix} \text{Operating cost} \\ \text{of two- and four-} \\ \text{wheel tractors} \end{bmatrix} - \begin{bmatrix} \text{Annual fixed cost} \\ \text{for two- and four-} \\ \text{wheel tractors} \end{bmatrix}$$

$$- \begin{bmatrix} \text{Annual fixed} \\ \text{cost for} \\ \text{buffaloes} \end{bmatrix}$$

If an additional $0.5 \cdot \beta \Sigma_j (TOTDLB_j)^2$ were subtracted from the objective function (3.1) and the reservation value of family labor ($\tau \cdot TOTFLB$) were added back in, *CAPS* would measure the total net income accruing from paddy production in the Muda project area, which is the aggregate return to farm households and nonfarm landlords. The total wage bill for hired labor is not, however, deducted because we wish to obtain a competitive solution in the labor market, and *CAPS* must measure the sum of the producers' and consumers' surpluses over the thirteen four-week labor markets, net of the reservation value of family labor.

Equation (3.1) is maximized subject to the following constraints:

Paddy production balances — *Number of equations*

(3.2) $\Sigma_i \Sigma_\tau \Sigma_m \Sigma_f \, q_{fti\tau} \cdot X_{fm\tau i} - 23.8 \, HV_t = 0$, all t — 26

$$\begin{bmatrix} \text{Quantity of paddy} \\ \text{produced by various} \\ \text{techniques} \\ \text{(in gantangs)} \end{bmatrix} - \begin{bmatrix} \text{Quantity of} \\ \text{paddy} \\ \text{harvested} \\ \text{(in piculs)} \end{bmatrix} = 0, \text{ in each fortnight}$$

(3.3) $\Sigma_{t\varepsilon s} HV_t - 0.16 \, \Sigma_{\tau \varepsilon s} \Sigma_m \Sigma_f \Sigma_i X_{fm\tau i}$

$- PDYPD_s = 0$, all s — 2

$$\begin{bmatrix} \text{Seasonal quantity} \\ \text{of paddy harvested} \end{bmatrix} - \begin{bmatrix} \text{Seasonal seed} \\ \text{requirement} \end{bmatrix}$$

$$- \begin{bmatrix} \text{Seasonal production of paddy} \\ \text{net of seed requirement} \\ \text{(that is, disposable production)} \end{bmatrix} = 0, \text{ each season}$$

Fortnightly land resource constraints — *Number of equations*

(3.4) $\Sigma_\tau \Sigma_m \Sigma_f \, a_{mt\tau} \, X_{fm\tau i} \leqslant LND_i$, all t, i — 104

$$\left[\begin{array}{c}\text{Total area of land used for}\\ \text{paddy production in various}\\ \text{techniques, each fortnight}\end{array}\right] \leqslant \left[\begin{array}{c}\text{Available}\\ \text{land}\\ \text{area}\end{array}\right] \text{for all soil types}$$

Fortnightly family labor constraints

(3.5) $\Sigma_i \Sigma_\tau \Sigma_m \Sigma_f \, g_{fmt\tau} \, X_{fm\tau i} + h \cdot HV_t + a_{Bt} \cdot BFINV - HDLB_t \leqslant LAB_t$, all t — 26

$$\left[\begin{array}{c}\text{Labor requirement}\\ \text{for paddy}\\ \text{production in various}\\ \text{techniques in period } t\end{array}\right] + \left[\begin{array}{c}\text{Labor}\\ \text{requirements for}\\ \text{paddy harvesting}\\ \text{in period } t\end{array}\right]$$

$$+ \left[\begin{array}{c}\text{Labor for buffalo}\\ \text{maintenance}\\ \text{in period } t\end{array}\right] - \left[\begin{array}{c}\text{Hired day}\\ \text{labor}\\ \text{in period } t\end{array}\right] \leqslant \left[\begin{array}{c}\text{Available}\\ \text{family labor}\\ \text{in period } t\end{array}\right]$$

Fortnightly draft-power constraints

(3.6) $\Sigma_i \Sigma_\tau \Sigma_m \Sigma_f \, u_{mt\tau} \, X_{fm\tau i} - e_2 \cdot 2WINV \leqslant 0$, all t — 26

$$\left[\begin{array}{c}\text{Two-wheel tractor}\\ \text{requirement for}\\ \text{paddy production}\\ \text{in various techniques}\\ \text{in period } t\end{array}\right] - \left[\begin{array}{c}\text{Available}\\ \text{two-wheel}\\ \text{tractor stock}\\ \text{in period } t\end{array}\right] \leqslant 0$$

(3.7) $\Sigma_i \Sigma_\tau \Sigma_m \Sigma_f \, v_{mt\tau} \, X_{fm\tau i} - e_4 \cdot 4WINV \leqslant 0$, all t — 26

$$\left[\begin{array}{c}\text{Two-wheel tractor}\\ \text{requirement for}\\ \text{paddy production}\\ \text{in various techniques}\\ \text{in period } t\end{array}\right] - \left[\begin{array}{c}\text{Available}\\ \text{four-wheel}\\ \text{tractor stock}\\ \text{in period } t\end{array}\right] \leqslant 0$$

(3.8) $\Sigma_i \Sigma_\tau \Sigma_m \Sigma_f \, w_{mt\tau} \, X_{fm\tau i} - e_B \cdot BFINV \leqslant 0$, all t — 26

$$\left[\begin{array}{c}\text{Buffalo requirement}\\ \text{for paddy production}\\ \text{in various techniques}\\ \text{in period } t\end{array}\right] - \left[\begin{array}{c}\text{Available}\\ \text{buffalo stock}\\ \text{in period } t\end{array}\right] \leqslant 0$$

Accounting row for use of family labor — *Number of equations*

(3.9) $\Sigma_t \, (\Sigma_i \Sigma_\tau \Sigma_m \Sigma_f \, g_{fmt\tau} \, X_{fm\tau i}) + h\Sigma_t HV_t + (\Sigma_t \, a_{Bt}) \cdot BFINV - AGDLB - TOTFLB = 0$ — 1

$$\left[\begin{array}{c}\text{Annual total labor}\\ \text{requirement for}\\ \text{paddy production}\\ \text{in various}\\ \text{techniques}\end{array}\right] + \left[\begin{array}{c}\text{Annual total labor}\\ \text{requirement for}\\ \text{paddy harvesting}\end{array}\right]$$

$$+ \left[\begin{array}{c}\text{Annual total}\\ \text{labor require-}\\ \text{ment for}\\ \text{buffalo}\\ \text{maintenance}\end{array}\right] - \left[\begin{array}{c}\text{Total}\\ \text{annual use}\\ \text{of hired}\\ \text{day}\\ \text{labor}\end{array}\right] - \left[\begin{array}{c}\text{Total}\\ \text{annual}\\ \text{use of}\\ \text{family}\\ \text{labor}\end{array}\right] = 0$$

Accounting rows for use of draft power

(3.10) $\Sigma_t(\Sigma_i \Sigma_\tau \Sigma_m \Sigma_f \, u_{mt\tau} \, X_{fm\tau i}) - TOT2WT = 0$ — 1

$$\left[\begin{array}{c}\text{Annual total}\\ \text{two-wheel tractor}\\ \text{requirements for}\\ \text{paddy production}\\ \text{in various techniques}\end{array}\right] - \left[\begin{array}{c}\text{Total annual use}\\ \text{of two-wheel}\\ \text{tractors}\end{array}\right] = 0$$

(3.11) $\Sigma_t(\Sigma_i \Sigma_\tau \Sigma_m \Sigma_f \, v_{mt\tau} \, X_{fm\tau i}) - TOT4WT = 0$ — 1

$$\left[\begin{array}{c}\text{Annual total}\\ \text{four-wheel tractor}\\ \text{requirements for}\\ \text{paddy production in}\\ \text{various techniques}\end{array}\right] - \left[\begin{array}{c}\text{Total annual}\\ \text{use of four-}\\ \text{wheel tractors}\end{array}\right] = 0$$

(3.12) $\Sigma_t(\Sigma_i \Sigma_\tau \Sigma_m \Sigma_f \, w_{mt\tau} \, X_{fm\tau i}) - TOTBUF = 0$ — 1

$$\left[\begin{array}{c}\text{Annual total}\\ \text{buffalo requirements}\\ \text{for paddy production}\\ \text{in various techniques}\end{array}\right] - \left[\begin{array}{c}\text{Total annual}\\ \text{use of}\\ \text{buffaloes}\end{array}\right] = 0$$

Fertilizer balances *Number of equations*

(3.13) $\Sigma_i \Sigma_\tau \Sigma_m \Sigma_f \, \theta_f \, X_{fm\tau i} \; - \; BUYU \; = \; 0$ 1

$$\begin{bmatrix}\text{Total demand for}\\ \text{urea fertilizer}\end{bmatrix} - \begin{bmatrix}\text{Amount of urea}\\ \text{fertilizer}\\ \text{purchased}\end{bmatrix} = 0$$

(3.14) $\Sigma_i \Sigma_\tau \Sigma_m \Sigma_f \, \phi_f \, X_{fm\tau i} \; - \; BUYB \; = \; 0$ 1

$$\begin{bmatrix}\text{Total demand for}\\ \text{basal fertilizer}\end{bmatrix} - \begin{bmatrix}\text{Amount of basal}\\ \text{fertilizer}\\ \text{purchased}\end{bmatrix} = 0$$

Accounting rows for hired day labor

(3.15) $\Sigma_{t\varepsilon j} \, HDLB_t \; - \; TOTDLB_j \; = \; 0, \text{ all } j$ 13

$$\begin{bmatrix}\text{Term to sum hired}\\ \text{day labor in two-}\\ \text{week } (t) \text{ totals}\\ \text{to four-week } (j)\\ \text{totals}\end{bmatrix} - \begin{bmatrix}\text{Total use of}\\ \text{hired day labor}\\ \text{in period } j\end{bmatrix} = 0$$

(3.16) $\Sigma_j \, TOTDLB_j \; - \; AGDLB \; = \; 0$

$$\begin{bmatrix}\text{Sum of four-}\\ \text{week totals}\\ \text{of hired day}\\ \text{labor use}\end{bmatrix} - \begin{bmatrix}\text{Total annual}\\ \text{use of hired}\\ \text{day labor}\end{bmatrix} = 0$$

As formulated, the model is a quadratic programming problem because the objective function *CAPS* is quadratic in $TOTDLB_j$. Since the objective function is concave in $TOTDLB_j$, however, it can be approximated to a linear programming form using a technique described by Duloy and Norton (1975).

The method is applicable to the term $Q_j = 0.5\beta\Sigma_j\,(TOTDLB_j)^2$ and entails defining new activities, $y_{kj} \geqslant 0$, $k = 1$ to K, which are subject to a convex combination constraint $\Sigma_k y_{kj} \leqslant 1$. Then with each y_{kj} there is associated a value of $TOTDLB_j$, say, L_{kj}, such that $TOTDLB_j = \Sigma_k L_{kj} \cdot y_{kj}$. A corresponding value of $Q_{kj} = 0.5\beta \cdot L_{kj}^2$ is also assigned, and Q_j is then equal to $\Sigma_k Q_{kj} y_{kj}$. TIGER is then modified so that $\Sigma_j \Sigma_k Q_{kj} y_{kj}$ is subtracted from (3.1) instead of $0.5\beta\Sigma_j\,(TOTDLB_j)^2$, and the following equations are added:

(3.17) $$\sum_k y_{kj} \cdot L_{kj} - TOTDLB_j = 0, \text{ all } j$$

(3.18) $$\sum_k y_{kj} \leqslant 1, \text{ all } j$$

Table 3-5. *Notation and Symbols for TIGER*

Symbol	Description	Unit of measure	Subscripts	Permitted values
I. Activities				
$X_{fm\tau i}$	Paddy growing	Relongs	f = level of applied nitrogen in pounds per relong	0, 15, 25, 30, 35
			m = type of mechanization for land preparation	A = buffaloes B = pedestrian tractors C = four-wheel tractors
			τ = nursery date (fortnight)	7 to 12 (off-season) 18 to 24 (main season)
			i = soil type	PD = acid, double-cropped PS = acid, single-cropped GD = nonacid, double-cropped GS = nonacid, single-cropped
HV_t	Paddy harvested	Piculs	t = period (fortnight)	1 to 26
$PDYPD_s$	Total paddy production net of seed requirement	Piculs	s = season	1 = off-season 2 = main season
BUYB	Purchase of basal fertilizer	44-lb. bags		
BUYU	Purchase of urea fertilizer	44-lb. bags		
TOTBUF	Total annual use of buffaloes	Days		
TOT2WT	Total annual use of two-wheel tractors	Days		
TOT4WT	Total annual use of four-wheel tractors	Days		
BFINV	Stock of buffaloes	Animals		
2WINV	Stock of two-wheel tractors	Tractors		
4WINV	Stock of four-wheel tractors	Tractors		
$HDLB_t$	Hiring of day labor	Days		
$TOTDLB_j$	Total use of day labor	Days	j = four-week period	1 to 13
AGDLB	Total annual use of day labor	Days		
TOTFLB	Total annual use of family labor	Days		
II. Costs and prices				
P	Paddy	$/picul		

[cut off]	[Cost of basal fertilizer]	[$/bag]	
c	Miscellaneous production costs[a]	$/relong	
r	Reservation wage for family labor	$/day	
c_2	Operating cost of two-wheel tractors	$/day	
c_4	Operating cost of four-wheel tractors	$/day	
K_B	Annual cost for care and replacement per buffalo	$/animal	
K_2	Annualized fixed cost per two-wheel tractor	$/tractor	
K_4	Annualized fixed cost per four-wheel tractor	$/tractor	
III. Coefficients			
α	Intercept of four-week day labor supply functions	$/day	
β	Slope of four-week day labor supply functions	$/thousand days	
$a_{nt\tau}$	Land requirement	Relongs	
$q_{fti\tau}$	Yield of paddy	Gantangs/relong	
$g_{mt\tau}$	Nonharvest labor requirement	Days/relong	
h	Harvest labor requirement	Days/picul	
$u_{nt\tau}$	Two-wheel tractor requirement	Days/relong	
$v_{nt\tau}$	Four-wheel tractor requirement	Days/relong	
$w_{mt\tau}$	Buffalo requirement	Days/relong	
ϕ_f	Input of basal fertilizer	Bags/relong	
θ_f	Input of urea fertilizer	Bags/relong	
a_{Bt}	Labor requirement for buffalo care	Days	
e_B	Fortnightly buffalo availability	Days/buffalo	
e_2	Fortnightly two-wheel tractor availability	Days/tractor	
e_4	Fortnightly four-wheel tractor availability	Days/tractor	
LAB_t	Total supply farm family labor	Thousands of days	Fixed
LND_i	Total supply of land	Thousands of relongs	Fixed

a. Details in appendix B.

An excellent degree of approximation was obtained by using 100 activities ($K = 100$) for each $TOTDLB_j$.

For programming models such as TIGER, tableaux are a help in understanding the structure of such models. In a somewhat compressed form, we have set out such a tableau in table 3-6.

Validation of the Model for 1973

Thus far, we have dealt with TIGER in quite general terms because it can be calibrated to reflect different years or policy experiments. We turn now to the question of actually testing and validating the model, and for this purpose we use 1973 as the base period. This, of course, is the year in which the FAO/World Bank survey was undertaken, and hence for which there is a rich data base drawn on a basis representative of the project. The year is an unfortunate choice only in that the irrigation system had not quite attained full coverage, and farmers and local factor markets were still adjusting to the new double-cropping system. It is unlikely that an equilibrium model can replicate reality with great accuracy under these circumstances, but something in the right neighborhood is to be hoped for. It is in this spirit that we proceed, both in this chapter and the next, to draw comparisons between the model's results and the empirical evidence for particular years.

Numerical Specification of the Model

The principal technology coefficients of the 1973 model have already been discussed, and supplementary information is set out in appendix B. It thus remains to specify the regional land and labor constraints, and the prices exogenous to the model.

The land constraints are based on the 1973 figures in table 2-4. A small area is deducted to allow for the land farmed by landless workers and part-time farmers, and the remaining area is allocated to acid and nonacid soils (25 percent and 75 percent, respectively).

To derive the available supply of family labor, it is necessary to deduct all labor requirements for overhead and nonpaddy work, and for wage employment in nonfarming activities, since they are treated exogenously in TIGER. The derivation proceeds as follows: First, we take the number of adult household members working on the average family farm, as reported in the FAO/World Bank survey. These figures are then converted to man-days of supply per four-week period. An average work week of 4.5 days is assumed for men and 3.5 days for women, which allows for weekends, local holidays, sickness, farm inspection and marketing activities, and, in the case of women, domestic chores. Then, on a four-week basis,[11] we deduct all the exogenous labor requirements mentioned above, using the values reported in the FAO/World Bank survey for 1973. The resultant net labor supplies are multiplied by the estimated number of farms in the Muda project area that year (51,571), and then converted from a four-week to a fortnightly basis by dividing by two. Pertinent data are to be found in table 3-7.

Since the prices of day labor, draft power, and land are endogenous to the model, the key prices that remain to be specified are those for paddy and fertilizer, the variable and fixed costs for draft-power investment, and the reservation wage for family labor. One problem in validating a model to a specific year is that farmers might have had different price expectations from the prices actually realized. As different production decisions might have been made if farmers had had better foresight, we need to be sure that the prices assumed for 1973 are reasonably in accord with what farmers might have anticipated that year.

For fertilizer prices, we simply take the observed prices for 1973. These are $5.28 per 44-pound bag for basal and $6.54 per 44-pound bag for urea. In real terms, these prices had changed very little since the late 1960s, and there is no reason to believe that farmers' expectations would have changed when making their 1973 decisions.

The price of paddy presents a more difficult problem, since it began to increase quite rapidly toward the end of the 1973 off-season, after being virtually constant in the preceding five years. This increase had its origin in abrupt price movements in the world market, and it is therefore unlikely that Muda farmers anticipated the increase in making their decisions in 1973. Accordingly, since our first task is to validate the model for 1973, we assume that farmers acted that year on the basis of the previous year's price. Our estimates of farm-gate prices, which are defined as gross of deductions (see the definition of yields in table 3-3), are $15.05 per picul for the off-season crop, and $15.35 per picul for the main-season crop.

The variable and overhead costs for tractors are based on those in a study by MADA (1970*a*) and are inflated by the import price indices for mineral fuels and machinery (Malaysia, 1974). For two-wheel tractors, operating costs per day (c_2) are $6.26, and overhead costs per year (K_2) are $610; for four-wheel tractors, $c_4 = \$33.08$ and $K_4 = \$2{,}074$.

The annual overhead cost for a buffalo (K_B) comprises animal feed, veterinary services, and medicine, together with a portion (one-eighth) of the annual cost of rearing a replacement animal, adjusted for the value of a culled buffalo. On the basis of the FAO/World Bank data, the estimate of K_B

11. The relevant FAO/World Bank survey data are available only by four-week periods.

Table 3-6. *Schematic Layout of Linear Programming Tableau for TIGER Model*

Equation number	*Row description*	*Model activities* $X_{fm\tau i}$	HV_t	$PDYPD_s$	*BUYB*	*BUYU*	*TOT4WT*	*TOT2WT*	*TOTBUF*	*4WINV*
3.1	Objective function	$-c$		P_s	$-c_b$	$-c_u$	$-c_4$	$-c_2$		$-K_4$
3.2	Paddy production counts in fortnight t	$q_{fti\tau}$	-23.8							
3.3	Paddy production net seed in season s	-0.16	1	-1						
	Constraints in period t									
3.4	Land	$a_{mt\tau}$								
3.5	Labor	$g_{fmt\tau}$	h							
3.6	Pedestrian tractors	$u_{mt\tau}$								
3.7	Four-wheel tractors	$v_{mt\tau}$								$-e_4$
3.8	Buffaloes	$w_{mt\tau}$								
	Total usage									
3.9	Family labor	$\Sigma_t\, g_{fmt\tau}$	h							
3.10	Pedestrian tractors	$\Sigma_t\, u_{mt\tau}$						-1		
3.11	Four-wheel tractors	$\Sigma_t\, v_{mt\tau}$					-1			
3.12	Buffaloes	$\Sigma_t\, w_{mt\tau}$							-1	
3.13	Urea	θ_f				-1				
3.14	Basal	ϕ_f			-1					
3.15	Day labor counts in period j									
3.16	Total use of day labor									
3.17	Identity rows between $TOTDLB_j$ and approximating activities									
3.18	Convex combination constraints on y_{kj}									
	Number of activities	780	26	2	1	1	1	1	1	1

Equation number	Row description	Model activities							Nature of constraint	Number of equations
		2WINV	BFINV	$HDLB_t$	$TOTDLB_j$	AGDLB	TOTFLB	y_{kj}		
3.1	Objective function	$-K_2$	$-K_B$		$-\alpha$		$-r$	$-Q_{kj}$	Maximize	1
3.2	Paddy production counts in fortnight t								$= 0$	26
3.3	Paddy production net seed in season s								$= 0$	2
	Constraints in period t									
3.4	Land								$\leq LND_i$	104
3.5	Labor		a_{Bt}	-1					$\leq LAB_t$	26
3.6	Pedestrian tractors	$-e_2$							≤ 0	26
3.7	Four-wheel tractors								≤ 0	26
3.8	Buffaloes		$-e_B$						≤ 0	26
	Total usage									
3.9	Family labor		$\Sigma_t a_{Bt}$			-1	-1		$= 0$	1
3.10	Pedestrian tractors								$= 0$	1
3.11	Four-wheel tractors								$= 0$	1
3.12	Buffaloes								$= 0$	1
3.13	Urea								$= 0$	1
3.14	Basal								$= 0$	1
3.15	Day labor counts in period j			1	-1				$= 0$	13
3.16	Total use of day labor				1	-1			$= 0$	1
3.17	Identity rows between $TOTDLB_j$ and approximating activities				-1			L_{kj}	$= 0$	13
3.18	Convex combination constraints on y_{kj}							1	≤ 1	13
	Number of activities	1	1	26	13	1	1	1,300	2,157[a]	283[b]

a. Total activities b. Total equations

Table 3-7. *Supply of Farm Family Labor*
(thousands of days)

Period j[a]	*Date ending*	*Total labor supply*	− *Overhead requirements*	= *Net labor supply*
1	December 23	2,335.13	660.62	1,674.5
2	January 20	2,335.13	659.59	1,675.5
3	February 17	2,335.13	500.24	1,834.9
4	March 17	2,335.13	651.86	1,683.3
5	April 14	2,335.13	566.25	1,768.9
6	May 12	2,335.13	477.03	1,858.1
7	June 9	2,335.13	628.65	1,706.4
8	July 7	2,335.13	881.35	1,453.8
9	August 4	2,335.13	802.96	1,532.1
10	September 1	2,335.13	471.87	1,863.3
11	September 29	2,335.13	494.57	1,840.6
12	October 27	2,335.13	514.68	1,820.5
13	November 25	2,335.13	604.41	1,730.7
Total		30,356.69	7,914.08	22,442.6

a. Four-week periods.

is so small that it is set at zero in the model. Buffaloes do require labor for general care and maintenance, however, and this is charged in the model at four man-days per fortnight, or 104 days a year. This requirement includes time needed for tethering the animals and collecting forage; buffaloes are essentially scavengers in the Muda area, and only negligible amounts of land are devoted to growing forage crops.

The reservation wage for family labor is not directly observable, but a plausible value is the intercept coefficient of the four-week day labor supply functions, this being the lowest wage at which any day laborer would work. This wage is $1.53 per day (see appendix B), but we also report solutions of the model in which the reservation wage was set at other levels.

Results from the 1973 Model

Table 3-8 presents the results for different values of selected parameters, together with the values of each variable calculated from the FAO/World Bank sample survey data. Solution C results from the set of parameter values described in the preceding section. In particular, this solution corresponds to a reservation wage (r) for family labor of $1.53 per day, the same value as the intercept of the day labor supply functions.

A comparison of solution C with the FAO/World Bank values reveals some noticeable discrepancies. First, the estimate of aggregate net income obtained from the survey data is $80.6 million, compared with $107.8 million in solution C. At the same time, the estimates of total paddy production are very similar at 8.7 million and 8.9 million piculs, respectively. Since the total expenditure on day labor and land rents is also lower in the survey data than in the model's solution, the problem can be traced to a major discrepancy in the estimates of the price of paddy. According to the FAO/World Bank survey, the imputed paddy price received by farmers is about $12 per picul—appreciably lower than the average price assumed in the model of $15.20 per picul, which is based on the 1972 farm-gate prices recorded by MADA. (MADA recorded even higher prices than this in 1973, when a shortage of rice in world markets pushed local prices to $25 per picul in the 1973 off-season.) The imputed survey price of $12 per picul is consistent neither with actual prices, nor with regionally consistent estimates of value added from the paddy sector.[12] We conclude that the FAO/World Bank survey data underestimate total income, and that the model is a more reliable estimator on this score. By way of comparison, however, consider solution G, in which the paddy price is reduced by $3 per picul. This leads to a very good match between the model's and the survey's estimates of total income.

Second, the land rents in solution C are about twice as large as those reported in the FAO/World Bank survey. This is probably because the land rental market had only begun to adjust to double-cropping in 1973. Rents averaged only $65 per relong annually in 1967 when all the land was single-cropped (table 2-11), but they increased rapidly as the project came on stream and were reported to be well over $200 per relong annually by 1977.

Third, the survey's estimates of the stocks of buffaloes and tractors in 1973 are far above the equilibrium stocks required in the model. Solution C uses only four-wheel tractors and buffaloes to meet draft-power requirements, whereas in reality farmers also used pedestrian tractors in 1973. When pedestrian tractors are converted to four-wheel tractor equivalents,[13] the estimated 1973 stock was 921 four-wheel tractor equivalents, which compares with only 600 tractors in solution C. At the same time, the survey's estimate of the 1973 buffalo stock is 16,700 animals, whereas solution C requires only 9,346 animals.

There is little reason to believe that the draft-power stocks were anywhere near their equilibrium levels in 1973. Draft-power needs were increasing rapidly with the expansion of double-cropping, and a major transition from the use of buffaloes to tractors was under way. The stock of four-wheel tractors had increased from about 260 in 1967 to 500 by 1973, and the stock of pedestrian tractors had increased from 250 to 2,400 over

12. See subsequent chapters on the estimated gross output of paddy for 1972 and 1974.

13. A four-wheel tractor is equivalent to 5.7 pedestrian tractors on the basis of estimated work rates (MADA, 1970*a*).

Table 3-8. *Results from TIGER, 1973*

Item	Model solutions							FAO/World Bank survey values (H)
	$r=0$ (A)	$r=1.25$ (B)	$r=1.53$ (C)	$r=1.75$ (D)	$r=1.53$ $\beta=-25\%$ (E)	$r=1.53$ $\beta=+25\%$ (F)	$r=1.53$ Paddy price $3 lower (G)	
Aggregate net income ($ thousands)[a]	107,830	108,177	107,817	105,581	109,055	106,770	81,082	80,600
Total paddy production (thousands of piculs)	8,964	8,943	8,948	8,954	8,942	8,954	8,947	8,700
Total usage								
Nitrogen (thousands of pounds)	18,661	17,754	17,761	17,768	17,753	17,766	17,759	15,600
Family labor (thousands of days)	20,936	20,652	20,218	18,670	20,520	19,980	20,331	16,923
Day labor (thousands of days)	5,269	4,661	4,448	5,055	5,336	3,916	4,494	6,308
Total labor (thousands of days)	26,205	25,313	24,666	23,725	25,856	23,896	24,825	23,231
Stock of four-wheel tractors	170	416	600	937	222	867	553	500
Stock of pedestrian tractors	0	0	0	0	0	0	0	2,400
Stock of buffaloes	16,039	12,401	9,346	5,207	14,257	6,123	10,082	16,700
Land rents ($/relong/year)								
Nonacid soil	237	234	234	232	242	228	141	101
Acid soil	180	176	176	172	182	171	102	90
Monthly wages ($/day)								
Nov. 26–Dec. 23	1.61	1.53	1.53	1.75	1.63	1.53	1.53	1.88
Dec. 24–Jan. 20	1.53	1.55	1.62	1.82	1.68	1.56	1.62	1.50
Jan. 21–Feb. 17	2.04	1.96	1.84	1.92	1.75	2.01	1.86	1.30[b]
Feb. 18–Mar. 17	2.54	2.52	2.46	2.34	2.24	2.57	2.48	2.19
Mar. 18–Apr. 14	1.60	1.53	1.63	1.75	1.70	1.54	1.61	1.20
Apr. 15–May 12	1.74	1.66	1.62	1.75	1.66	1.57	1.62	1.54
May 13–Jun. 9	1.53	1.55	1.62	1.75	1.55	1.59	1.61	1.80
Jun. 10–Jul. 7	1.53	1.53	1.53	1.75	1.53	1.53	1.53	1.60
Jul. 8–Aug. 4	2.76	2.80	2.86	2.78	2.57	3.11	2.84	1.45
Aug. 5–Sep. 1	3.00	2.98	2.92	2.76	2.74	3.11	2.92	1.85[b]
Sep. 2–Sep. 29	3.44	3.08	2.95	2.77	2.82	3.11	2.99	2.52
Sep. 30–Oct. 27	1.94	1.80	1.66	1.75	1.73	1.62	1.68	2.33
Oct. 28–Nov. 25	1.66	1.64	1.58	1.75	1.60	1.57	1.58	2.88
Average wage ($/day)	2.79	2.72	2.66	2.40	2.45	2.88	2.68	1.82[b]
Annual wage payments ($ thousands)	14,718	12,669	11,850	12,117	13,064	11,274	12,023	10,332[b]

a. Defined as the net return to farm households and nonfarm landlords. Wage payments to day laborers are not included.
b. The wages reported for January 21–February 17 and August 5–September 1 are questionably low (see appendix B).

the same period. By 1977 there were still about 500 (mostly small) four-wheel tractors, but the stock of pedestrian tractors had increased to more than 5,000. The region's stock of buffaloes also fell by more than 40 percent between 1967 and 1973, and continued to fall by about 10 percent a year thereafter. These rapid rates of change must inevitably have led to some carry-over of excess draft-power resources, both because owners of buffaloes would be reluctant to dispose of animals which still had some years of working life and because investments in four-wheel tractors are made by an entirely different group of households. Only the long-term influences of the market can serve to synchronize the decisions of farmers and tractor contractors in arriving at an optimal mix of investments, and this is unlikely to have happened by 1973.

Selected seasonal results from solution C are to be found in table 3-9, together with their counterparts from the FAO/World Bank survey. These results show that the model follows the main-season crop schedule quite closely, except that it advances the off-season crop by about four weeks. In reality, there was some slippage of the irrigation date that year because many farmers had not completed the harvesting of their main-season crop by the targeted irrigation date. In fact, this slippage has become an important problem in the Muda project. Progressive delays, combined with insufficient water in the reservoirs, caused the entire 1978 off-season crop to be forgone, with the hidden blessing that farmers gained an opportunity to catch up with the double-cropping calendar. For these reasons, the model's results should be viewed as the seasonal production pattern that should have been followed in 1973.

We have identified several discrepancies between the model's results and the FAO/World Bank survey data for 1973. But rigorous validation of the model is inherently difficult: first, because there is little reason to believe that the sector was close to an equilibrium state in 1973, and second, because some of the FAO/World Bank survey data are of questionable accuracy. The model's results presented here and those reported in chapter 4, where we undertake additional validation, are generally quite plausible; no outrageous answers violate our secondary data sources or our general understanding of the regional economy. Nevertheless, before proceeding to apply the model, we examine its sensitivity to changes in the values of key parameters.

Solutions A, B, and D in table 3-8 report the effects of changing the reservation wage (r) for family labor. In solution A this wage is set at zero; the values in solutions B and D are less drastic departures from C. The main changes from solution C are in the number of buffaloes and four-wheel tractors, with consequent adjustments in labor use and in the wage payments to day laborers. Total income, paddy production, and land rents

Table 3-9. *Further Results from TIGER for 1973*

Four-week period	Four-week distribution of paddy production (percent)		Four-week day labor hire (thousands of days)	
	Model solution C	FAO/World Bank survey	Model solution C	FAO/World Bank survey
Nov. 26–Dec. 23	5.2	2.6	0	62
Dec. 24–Jan. 20	12.6	7.1	75	212
Jan. 21–Feb. 17	17.0	17.3	236	1,330
Feb. 18–Mar. 17	15.6	14.1	701	505
Mar. 18–Apr. 14	0	0.8	75	106
Apr. 15–May 12	0	0	71	454
May 13–Jun. 9	0	0	70	208
Jun. 10–Jul. 7	0	0.6	0	13
Jul. 8–Aug. 4	18.7	8.0	996	385
Aug. 5–Sep. 1	18.4	21.7	1,037	1,117
Sep. 2–Sep. 29	12.5	20.6	1,065	654
Sep. 30–Oct. 27	0	5.7	90	498
Oct. 28–Nov. 25	0	1.3	33	135
Total	100.0	100.0	4,448	5,679

are hardly affected. On these grounds, there is little reason to change the reservation wage in the model from the initial value of $1.53 per day.

Solutions E and F in table 3-8 report the effects of changes in the slope coefficient of the four-week day labor supply functions. A reduction of 25 percent in this coefficient (solution E) reduces the average wage by 8 percent, and leads to a significant substitution of buffaloes for four-wheel tractors. But paddy production remains the same, and aggregate net income (which is defined net of wage payments) increases by only 1.1 percent. A 25 percent increase in the slope coefficient (solution F) has a symmetrically opposite effect. Aggregate net income declines by 1 percent, the average wage increases by 8 percent, and there is a strong switch to tractors and away from buffaloes.

4

The Sources of Change in the Paddy Farming Economy

THE MAIN REASON for constructing TIGER was to analyze the impact of the project on the paddy farming economy. To accomplish this, we must compare the values taken by relevant variables in the steady state characterized as "project maturity" with their counterparts in the hypothetical case that the project had not been undertaken. This set of variables includes not only the level of output, but also the incomes accruing to the various classes of agents, the use and timing of inputs, and the rentals at which their services are traded—all of which are generated by TIGER. Some of the conditions under which factors are supplied change exogenously with the passage of time, as do the prices of certain commodities and the technology available to farmers. Since these must be specified if TIGER is to do its job, it is necessary to settle on a particular year as representing the underlying conditions of project maturity.

We have chosen 1974 for this purpose. This was the first year in which the whole of the command area received irrigation, and the latest year for which we have sufficient data to estimate, without resort to unrestrained speculation, the set of exogenous parameters that drives TIGER. We hasten to add that we do not believe the paddy farming economy was in steady-state equilibrium in 1974. Indeed, inasmuch as farmers were learning new practices and a very different rhythm of cultivation from that dictated by rain-fed conditions, and relative prices had started to shift appreciably after a long period of comparative stability, there are strong grounds for supposing that the economy was not in a steady state in 1974. Hence, in using TIGER to assess the impact of the project, we are not attempting to complete a quantitative account of the economy as it actually was in that year, on the basis of the patchy evidence available. Rather, our aim is to simulate the steady-state equilibrium corresponding to the exogenous conditions that prevailed in 1974.

In view of the changes in the paddy farming economy over the 1967–74 period, it is natural to ask which exogenous forces were at work and what each contributed to the changes that occurred. In addition to the introduction of irrigation and exogenous changes in certain relative prices, improved varieties appeared that can be, and are, grown under rain-fed conditions. We shall try to disentangle these contributions to change from one another by using suitably formulated experiments with TIGER. Once again, it must be emphasized that the analysis is confined to comparative statics. In particular, the picture for 1967 that emerges from TIGER is the steady-state configuration of the economy under the conditions then prevailing. Even though the *Farm Economic Survey* data (Ministry of Agriculture and Cooperatives, 1967) provide a fairly complete picture of things as they actually were in that year, these data do not enter into the decomposition analysis except insofar as they provide a solid basis for the estimation of the exogenous parameters for TIGER.

It is important at the outset to be clear about the meaning of comparisons of the steady-state values of variables yielded by TIGER with those observed in the corresponding year. Any discrepancies can be interpreted either as shortcomings in TIGER as a predictive tool or as a measure of the difference between the actual and steady-state configurations of the economy. In the sections that follow, the comparisons provide further validation of TIGER only insofar as it can be plausibly argued that when the economy was observed, it was also in steady-state equilibrium. It so happens that such a case can be made for many variables in 1967; but the amount of prior calculation and guesswork needed to create the solution for 1967 is much greater than that needed for 1972, so that the real onus of validation lies on chapter 3.

A Scheme for Decomposition

Let Z be the vector of variables of interest, and let p denote the vector of exogenously given prices. Then the changes in the paddy farming economy between 1967 and 1974 are represented by $Z_{74}(p_{74}) - Z_{67}(p_{67})$. Any attempt to disentangle the contributions of irrigation, new varieties, and price changes from one another encounters a well-known problem of "path-dependence." In particular, the magnitude of the system's response to irrigation, with everything else held constant, depends on the prevailing level of prices; the same is true of the introduction of new varieties. Hence, when measuring their effects, one must choose between base year and final year prices.

If the project is being evaluated ex ante and the exogenous prices in the base year are also the best estimates of those that will rule when the project reaches maturity, then the project's effects should be estimated on that assumption. Conversely, estimating its effects at the prices that actually ruled when the project reached maturity provides an ex post evaluation. The difference between the ex ante and ex post estimates is then a measure of good or bad luck in an intrinsically uncertain world.

When situations in which there is no project are denoted by the superscript ^{o}, and those in which there are no new varieties in 1974 by an asterisk, the ex ante decomposition at base year prices is:

$$(4.1)\quad Z_{74}(p_{74}) - Z^o_{67}(p_{67}) \equiv [Z_{74}(p_{74}) - Z_{74}(p_{67})] + [Z_{74}(p_{67}) - Z^*_{74}(p_{67}) + Z^o_{67}(p_{67}) - Z^o_{74}(p_{67})] + [Z^*_{74}(p_{67}) - Z^o_{67}(p_{67})] + [Z^o_{74}(p_{67}) - Z^o_{67}(p_{67})]$$

$$[\text{Total change}] \equiv \begin{bmatrix}\text{Price and price in-}\\ \text{teraction effects}\end{bmatrix} + \begin{bmatrix}\text{Project-variety}\\ \text{interaction effect}\end{bmatrix} + [\text{Project effect}] + [\text{Variety effect}].$$

Similarly, the decomposition at final year prices is:

$$(4.2)\quad Z_{74}(p_{74}) - Z^o_{67}(p_{67}) \equiv [Z^o_{67}(p_{74}) - Z^o_{67}(p_{67})] + [Z_{74}(p_{74}) - Z^*_{74}(p_{74}) + Z^o_{67}(p_{74}) - Z^o_{74}(p_{74})] + [Z^*_{74}(p_{74}) - Z^o_{67}(p_{74})] + [Z^o_{74}(p_{74}) - Z^o_{67}(p_{74})]$$

$$[\text{Total change}] \equiv \begin{bmatrix}\text{Price and price in-}\\ \text{teraction effects}\end{bmatrix} + \begin{bmatrix}\text{Project-variety}\\ \text{interaction effect}\end{bmatrix} + [\text{Project effect}] + [\text{Variety effect}].$$

It may be helpful to formulate the above identities within the familiar setting of the price-quantity diagram for a single commodity. To illustrate the procedure, we use the net income of producers; but the principles apply, *mutatis mutandis*, to other variables. In figure 4-1, S^{o*} is the supply function for paddy when there are neither improved varieties nor irrigation. Hence, the income accruing to paddy producers in 1967 is given by the area A. For the decomposition we adopt base year prices, that is, equation (4.1). In the absence of the project, and hence of irrigation, the introduction of improved varieties would have shifted the supply function to S^o, so the "variety effect" on producers' incomes is given by the area B_1. Similarly, in the absence of new varieties, the introduction of the project would have shifted the supply function to S^*, so that this would have increased producers' incomes by the area $B_1 + B_2$. With the benefit of both new varieties and irrigation, the supply function shifts to S, and producers'

Figure 4-1. *Decomposition at Base Year Prices*

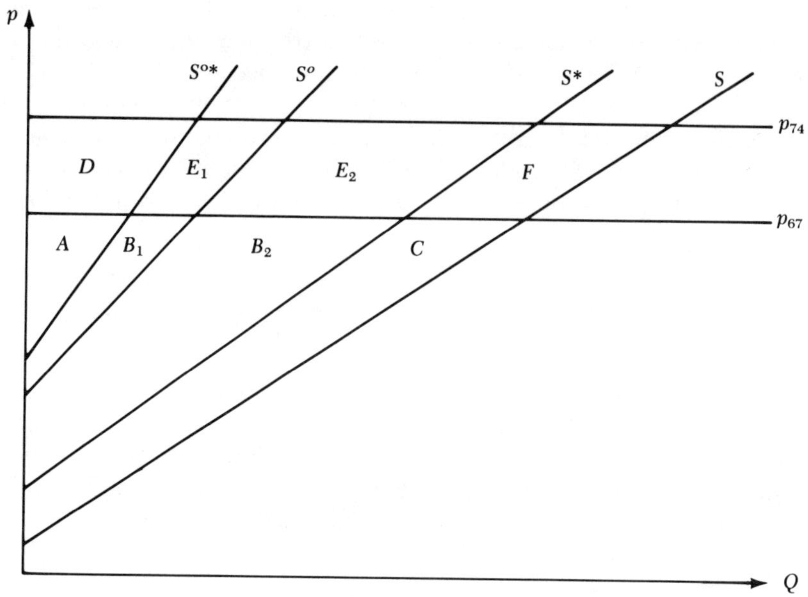

incomes rise by $B_1 + B_2 + C$. Hence, the effect of project-variety interactions on incomes is given by the area $C - B_1$. If the price of output were to rise in the interim, producers' incomes would increase by an amount given by the sum of the areas D, E_1, E_2, and F, which is the combined effect of a "pure" price effect and a string of price-variety-project interactions. If the prices of producers' inputs were to rise too, S would shift to the left, with a corresponding fall (G, say) in incomes. (This shift in S has not been shown in figure 4-1, which is already rather cluttered.) Then the effects of changes in prices and the associated interactions on producers' incomes is $D + E_1 + E_2 + F - G$. Ex post, this is a measure of the evaluator's good or bad luck—or incompetence at forecasting.

All the considerations that enter into the choice of year when examining the effects of prices apply with equal force to the treatment of the "pure" effects of irrigation and new varieties. As they are treated above, it has been assumed that the introduction of new varieties is not foreseen when the irrigation project is considered; conversely, the effects of the new varieties are evaluated as if irrigation were absent. In the present case, these were exactly the perceptions of the decisionmakers and their advisers (see chapter 2, the section "Changes in Output and Technology in the Paddy Sector"). Even if their error was not culpable, it is interesting to pursue the question of decomposition in the event that the conjunction of the introduction of irrigation and new varieties is actually foreseen.

This is readily done by using figure 4-1 as an expository device. Retaining base year prices for the decomposition, the starting point is the intersection of S, the supply function in the presence of both new varieties and the project, with p_{67}. Thus, the "pure" project effect on producers' incomes is now $C + B_2$, the "pure" variety effect is C, and the project-variety interaction effect is $B_1 - C$. As before, the combined effects of a rise in prices to p_{74} are the sum of D, E_1, E_2, and F. We will pursue both decompositions below.

Changes in the Paddy Farming Economy, 1967–74

From 1967 to 1974 there were exogenous changes in the prices of certain inputs as well as in the price of paddy. There were also changes in factor supplies, together with the introduction of irrigation and the availability of improved varieties of paddy. Each of these must be specified for the model experiments that go into the decompositions defined above.

Exogenously determined prices are laid out in table 4-1, along with their counterparts for 1973. The prices of paddy are taken directly from various surveys, as are the agrochemical and farm machinery costs for 1973. The costs of these inputs for 1967 and 1974 have been derived from those for 1973 by using suitable import price indices.

The adjustment of the labor supply parameters needs more detailed discussion. The parameters α and β characterize the alternatives open to

Table 4-1. *Exogenous Prices for Various Years*

Symbol	*Coefficient*	*1967*	*1973*	*1974*
α	Intercept of supply function for day labor ($/day)	1.15	1.53	1.85
β	Slope of supply function for day labor ($/million days)	1.00	1.33	1.61
r	Reservation wage of family labor ($/day)	1.15	1.53	1.85
c_u	Cost of urea ($/44 pounds)	4.97	6.54	10.00
c_b	Cost of basal fertilizer ($/44 pounds)	4.01	5.28	8.96
c	Miscellaneous paddy costs ($/relong)	3.80	5.00	8.50
P_1	Off-season paddy price ($/picul)	0.00	15.05	26.44
P_2	Main season paddy price ($/picul)	13.20	15.35	22.32
K_2	Annualized fixed cost of pedestrian tractor ($)	460.00	610.00	677.00
K_4	Annualized fixed cost of four-wheel tractor ($)	1,564.00	2,074.00	2,302.00
K_B	Annualized fixed cost of buffalo ($)[a]	0.00	0.00	0.00
c_2	Operating cost of pedestrian tractor ($/relong)	5.28	6.26	7.05
c_4	Operating cost of four-wheel tractor ($/relong)	27.87	33.08	37.25

a. As in chapter 3, the only effective fixed cost for buffaloes is labor for feeding and maintenance. This is built into TIGER at four days per animal for each two-week period. The cost of this labor is endogenously determined.

those hired as casual laborers by Muda's paddy farmers. But since there are no data from which α and β can be estimated for 1967 and 1974 as they have been for 1973, it is necessary to make some rather rough-and-ready assumptions. From what we know of conditions in the labor markets in both the region itself and the areas from which seasonal migrants were drawn during this period, it seems reasonable to suppose that the real supply price of labor did not change appreciably. This does not, of course, imply that the real wage rate was constant, for the employment of casual workers rose in the face of an upward-sloping supply function for such labor. Assuming that the real wage rate remained constant for each level of employment of casual workers, the values of α, β, and r for 1967 and 1974 are given by deflating or inflating their values in 1973 by an index of the cost of living for such workers and paddy farming households. Accordingly, the parameters for 1973 have been scaled by the consumer price indices for these groups (see table 7-1).

The endowments of land and family labor are set out in table 4-2; the net areas under cultivation in 1967 and 1974 are derived from table 2-4. Even if

Table 4-2. *Fixed Resource Supplies for Different Experiments*

Resource	1967	1973	1974: With project	1974: Without project
Land (thousands of relongs)				
Acid soil, single-cropped	83.6	6.5	—	82.5
Nonacid soil, single-cropped	250.6	19.6	—	247.5
Acid soil, double-cropped	—	72.8	81.0	—
Nonacid soil, double-cropped	—	218.5	243.0	—
Total land	334.2	317.4	324.0	330.0
Net family labor supply[a] (thousands of days)				
Fortnight				
1,2	860	837.3	842	901
3,4	795	837.8	843	834
5,6	845	917.5	923	886
7,8	824	841.7	847	864
9,10	707	884.5	890	742
11,12	741	929.1	934	777
13,14	754	853.2	858	791
15,16	779	726.9	731	818
17,18	795	766.1	771	834
19,20	879	931.7	937	922
21,22	925	920.3	926	970
23,24	826	910.3	915	866
25,26	862	865.4	870	905
Annual net family labor supply	21,180	22,443.6	22,574	22,220

a. Net of overhead requirements for farm maintenance, other farm work, and employment in nonfarming activities.

the project had not been undertaken, there would have been some losses of paddy land to urban growth, which we estimate to be 4,200 relongs, spread over acid and nonacid soils in the same ratio as their overall proportions in 1967.

The figures for family labor supply for 1973 require two adjustments. First, there were changes in the population of farms. From the *Farm Economic Survey*, we estimate that there were 53,254 farms and 500 landless families in 1967. Since there were about 4,500 landless households in 1973 (FAO, 1975), it appears that by 1974 some 4,000 of the households with marginal farms had joined the ranks of the landless. Meanwhile, the rural population had been growing, but only slowly because of heavy out-migration. If the number of farms had been growing at the same rate, this would have led to an additional 2,614 farms. In sum, this gives an estimate of 51,868 (53,254 − 4,000 + 2,614) farms for 1974, which is consistent with the FAO estimate of 51,571 farms for 1973. If the project had not been undertaken, then the economic pressures which created the additional landless workers probably would not have materialized. Hence, we postulate that there would then have been 55,868 (53,254 + 2,614) farms in 1974. Given these populations of farms, the initial estimates of family labor supplies for 1967 and 1974 were obtained from those for 1973 (table 3-7) by multiplying by the appropriate ratio of farm populations.[1]

The second type of adjustment to family labor supplies concerns changes in off-farm employment and the labor needed for farm maintenance and other jobs on the farm not connected with paddy cultivation. These "overhead" uses of labor are exogenous to the model and are subtracted from gross family labor supply to yield net labor supply for paddy cultivation (see table 3-7). We do not have any data on these exogenous labor requirements in 1967, but the 1973 FAO/World Bank survey did include a stratum of single-cropping farms which were not then receiving irrigation.[2] The data on overhead labor requirements for these farms are used instead of the requirements in table 3-7 when calculating net family labor requirements for 1967 and the hypothetical case of 1974 in the absence of the project.

Finally, there were changes in the technology for producing paddy. In 1967 the modern, quick-maturing varieties had not yet been introduced; it took about twenty-six weeks to grow the traditional varieties. The 1967 model is therefore specified to reflect this longer growing period, and the yields for different doses of nitrogen are specified as the values for the traditional varieties in table B-6 of appendix B. The yield response to

1. The average number of economically active household members was found to be the same in the 1967 *Farm Economic Survey* and the 1973 FAO/World Bank surveys (table 2-7), so no adjustment in the family work force seemed necessary.

2. See appendix A.

nitrogen for the 1974 model is the same as that for the 1973 model, which is consistent with the data in table 2-5.[3] The various labor requirements for cultivated paddy in both the 1967 model and that for 1974 in the absence of the project have been increased above those for 1973 to allow for additional weed cutting and clearing activities in land preparation. Two man-days a relong are added in the fortnight in which the land is prepared. The double-cropping calendar has eliminated these activities, since the land is no longer left idle in the dry season.

TIGER's estimates of the steady-state values of key variables in 1967 and 1974 are shown in solutions J and L in table 4-3. They indicate a remarkable increase of 330 percent in the aggregate nominal net income from paddy production between 1967 and 1974, which, after deflation by a cost of living index for paddy farming households,[4] implies an increase of about 175 percent in aggregate real income. Annual paddy production is 150 percent higher in 1974, and the total use of labor is two-thirds greater, with the creation of an additional 10 million man-days of work. The average nominal wage paid to casual laborers more than doubles, there is a switch from buffaloes to four-wheel tractors, and average land rents are five times higher. The average yield of paddy goes up from 272 to 352 gantangs a relong, and the average dose of nitrogen used each season rises from 21.5 to 30.5 pounds a relong.

The earnings of casual laborers rise proportionally more than aggregate net income from the cultivation of paddy. Despite the wholesale displacement of buffaloes by four-wheel tractors, casual laborers still have three times as much work, and their total nominal wage earnings increase more than sixfold, which is much greater than the proportional change in aggregate net income. The average real wage rate, however, is only a third higher in 1974 than in 1967. For households supplying casual workers, the proportional rise in their annual real incomes would therefore have been less than the increase in aggregate real producers' incomes, unless they were able to increase substantially the number of man-days worked each year or had other rapidly growing sources of income. With the advent of irrigation, local landless households would have had little difficulty in doubling, if not tripling, the number of man-days worked in growing paddy without cutting back on other lines of employment. In this case, the results

3. The average yields observed in the 1974–75 agroeconomic survey (Yamashita, Jegatheesan, and Wong, 1976) were 5 percent higher (see appendix C). This survey was not designed to be representative, however, and the average yields for the scheme area given in chapter 2 reflect both the use of some traditional varieties (about 8 percent of the area), as well as doses of nitrogen that were less than optimal on many farms.

4. Local cost of living indexes for agricultural households are given in table 7-1. On average, these increased by about 60 percent from 1967 to 1974.

Table 4-3. *Results from TIGER, 1967–74*
(nominal Malaysian dollars)

Item	Model solution[a]			
	1967		1974	
	1974 prices (I)	1967 prices (J)	Without project (K)	With project (L)
Aggregate net income ($ thousands)	71,442	43,526	89,958	187,240
Total paddy production (thousands of piculs)	3,750	3,826	4,751	9,584
Total usage				
Nitrogen (thousands of pounds)	6,065	7,212	9,747	19,740
Family labor (thousands of days)	12,794	12,878	13,644	19,586
Day labor (thousands of days)	1,393	1,562	1,649	4,566
Total labor (thousands of days)	14,187	14,440	15,293	24,152
Stock of four-wheel tractors	91	39	185	1,526
Stock of pedestrian tractors	0	0	0	0
Stock of buffaloes	15,064	15,736	13,806	0
Land rents ($/relong/year)				
Nonacid soil	143	85	194	453
Acid soil	109	64	147	349
Monthly wages ($/day)				
Nov. 26–Dec. 23	1.85	1.15	2.53	1.85
Dec. 24–Jan. 20	1.85	1.15	2.35	2.05
Jan. 21–Feb. 17	2.10	1.35	1.98	2.73
Feb. 18–Mar. 17	1.93	1.20	2.01	2.85
Mar. 18–Apr. 14	2.03	1.28	1.85	1.85
Apr. 15–May 12	1.85	1.16	1.85	1.93
May 13–Jun. 9	1.85	1.15	1.83	2.01
Jun. 10–Jul. 7	1.85	1.15	1.85	1.85
Jul. 8–Aug. 4	1.85	1.15	1.85	3.50
Aug. 5–Sep. 1	2.35	1.49	2.16	3.50
Sep. 2–Sep. 29	2.18	1.37	2.06	3.48
Sep. 30–Oct. 27	2.59	1.65	2.49	1.89
Oct. 28–Nov. 25	2.01	1.28	1.90	1.91
Average wage ($/day)	2.31	1.46	2.34	3.20
Annual wage payments ($ thousands)	3,221	2,285	3,855	14,615

— Not relevant.

a. The terms in equation (4.1) to which these solutions correspond are: J = $Z^{o}_{67}(p_{67})$; L = $Z_{74}(p_{74})$; M = $Z^{*}_{74}(p_{67})$; N = $Z_{74}(p_{67})$; O = $Z^{o}_{74}(p_{67})$. I and K are additional solutions required for determining the impact of the project.

Model solution[a]			*Impact of the project*[b]	
1974 with project		*1974 without project*	*1967 prices*	*1974 prices*
1967 prices 1967 varieties (M)	*1967 prices 1974 varieties (N)*	*1967 prices 1974 varieties (O)*	$\frac{N-O}{N-J} \times 100$	$\frac{L-K}{L-I} \times 100$
75,932	112,362	55,235	83.0	84.0
7,534	9,594	4,775	83.5	82.8
13,705	20,517	10,570	74.8	73.1
14,138	20,532	13,796	88.0	87.5
7,064	5,997	1,930	91.7	91.9
21,202	26,529	15,726	89.4	88.9
1,736	640	39	100.0	93.4
0	0	0	0	0
0	10,708	15,924	103.7	91.6
159	258	117	81.5	83.5
119	200	89	81.6	84.2
1.15	1.15	1.60	—	—
1.15	1.21	1.47	—	—
1.15	1.64	1.26	—	—
2.07	1.94	1.28	—	—
2.52	1.34	1.15	—	—
2.13	1.35	1.15	—	—
1.28	1.32	1.15	—	—
1.15	1.15	1.15	—	—
1.15	2.39	1.15	—	—
1.67	2.31	1.37	—	—
1.93	2.46	1.32	—	—
3.15	1.40	1.61	—	—
1.53	1.28	1.22	—	—
2.38	2.09	1.47	98.4	96.6
16,842	12,563	2,846	94.5	94.4

b. At 1967 prices the implied ratio is $[Z_{74}(p_{67}) - Z^o_{74}(p_{67})]/[Z_{74}(p_{67}) - Z^o_{67}(p_{67})]$, whereas at 1974 prices the ratio is $[Z_{74}(p_{74}) - Z^o_{74}(p_{74})]/[Z_{74}(p_{74}) - Z^o_{74}(p_{74})]$.

for the steady states suggest that they would fare as well as, if not better than, many paddy farming households. Extra employment for migrants, however, would almost certainly go to new entrants.

There is also a change in the share of aggregate net income accruing to landlords who do not themselves cultivate, who own 45 percent of the net area cultivated. Since the rent on land is equal to the value of its marginal product in the steady state, aggregate net income can be allocated between noncultivating landlords and farmers by taking the total imputed rent on all land and giving 45 percent thereof to the former. The results for 1967 and 1974 are shown in table 4-4. Payments to landlords increase proportionally more than the income accruing to cultivators, but the average income accruing to farm households from paddy production still increases fourfold in nominal terms, which implies that average real income increases two and a half times.

Since TIGER features but one farm, we cannot make detailed statements about the change in the distribution of income from paddy farming among cultivating households. Although the land market is imperfect in that when there is excess demand for leases, access to land for lease depends heavily on whether landlord and tenant are linked by kinship, allocative efficiency is little impaired by this imperfection provided: such access is independent of would-be tenants' farming skills; all farmers can trade in the markets for variable factors at parametrically given prices; and production takes place under certainty, with constant returns to scale. Under these circumstances, unequal access to land when there is excess demand for leases simply involves windfall gains for those who are fortunate to be at the head of the queue: the rents they then pay are below those which would prevail under perfect competition. When the pattern of these (lump-sum) transfers is ignored, the income of each household will be determined by its

Table 4-4. *The Distribution of Nominal Aggregate Income, 1967 and 1974*
(thousands of Malaysian dollars)

Agents	*1967 preproject*	*1974 postproject*	*Percentage increase*
Noncultivating landlords	12,024	63,025	424
Paddy farming households	31,502	124,215	294
Aggregate net income[a]	43,526	187,240	330
Average farm household income from paddy cultivation[b]	592	2,395	305

a. Defined as that income accruing to farm households and landlords. Wage payments to casual laborers are not included.

b. The number of paddy farming households declined slightly between 1967 and 1974.

endowments of labor and land and the set of rentals at which their services are traded. Hence, the much larger proportional increase in land rents than in wage rates implies that the distribution of income would have shifted significantly in favor of the cultivating households that were relatively well endowed with land, unless such a shift were offset by much more work for the family labor that is idle in the 1967 steady state. This offsetting effect is present in the results, but its ultimate scope is inherently limited. Fortunately, actual land rents have not increased as much as predicted by the model, and the actual distribution of income in 1974 was much better as a result. However, the longer-term forces at work in the market may well lead to a continued deterioration of the relative position of paddy farming households that have much labor and relatively little land.

We have already been at pains to emphasize that the main burden of validating TIGER lies on chapter 3. Yet it is still of some interest to compare, where possible, the values of key variables in the steady states provided by TIGER with those actually prevailing in the two years in question. It is clear that some variables will adjust more quickly to their steady-state values than will others, and in these cases the comparisons do imply a test of TIGER's predictive powers. From what we know of the technology and market structure, it is not unreasonable to expect total output, yields, and the use of nitrogen to adjust fairly rapidly to changes in external conditions. However, the use of tractors and especially the level of land rents tend to adjust much more sluggishly.

Table 2-5 provided independent estimates of the actual output of paddy in 1967 and 1974, namely, 4.68 and 11.11 million piculs, respectively. Since these estimates are gross of deductions for chaff, waste, and moisture, however, we must scale them down by 17 percent as we did in TIGER. This leads to estimates of 3.88 and 9.22 million piculs, respectively, which are very close to the comparable estimates of 3.83 and 9.58 million piculs from TIGER. Since Muda is a monoculture, this is not a particularly stringent test, but it is gratifying that TIGER has done well.

The average yield in the main season in the solution for 1967 is 272 gantangs a relong, which is virtually identical to the estimate of 276 gantangs reported in the *Farm Economic Survey* for 1966 (Ministry of Agriculture and Cooperatives, 1967).[5] The average yield in the 1974 solution is 352 gantangs a relong, which is again very close to the average yield of 350 gantangs a relong reported in the 1974–75 agroeconomic survey (Yamashita, Jegatheesan, and Wong, 1976). The use of nitrogen is also predicted fairly accurately. The solution for 1974 gives 30.4 pounds a

5. The actual yield reported is 333 gantangs a relong (see table 2-5), but this is gross of deductions for chaff, waste, and moisture. As before, we have used a 17 percent rate of deduction.

relong in each season, which is not far from the sample average of 35 pounds a relong in the 1974–75 agroeconomic survey. There is no survey evidence on the average dose applied in 1967, though the *Farm Economic Survey* does report that 96 percent of the surveyed area was fertilized. Instead, we shall use the data on fertilizer sales for the region in FAO (1975, table 66). Ignoring possible changes in stocks, we estimate that in 1967 the average dose was about 15 pounds of nitrogen a relong. The model's estimate for that year is 21.5 pounds of nitrogen a relong, but this estimate is for the Muda scheme only, and nitrogen use was probably higher there than in the peripheral paddy-growing areas.

TIGER does not do quite as well in predicting the level of mechanization, but this is hardly surprising. In 1967–74 there were large changes in the underlying conditions, and it takes time for the stocks of capital goods to adjust fully. The *Farm Economic Survey* reported that 5 percent of the total land area was prepared solely by tractors in 1966, with a further 33 percent prepared by tractors in combination with either buffaloes or hand hoeing. By comparison, the area prepared by tractors in the solution for 1967 amounts to 3.3 percent of the total land area. The optimal stock of 39 tractors in that solution is far lower than the actual stock of 304 four-wheel tractors (including the equivalent stock of pedestrian tractors). There were probably more tractors than conditions warranted at the time, because tractor owners were probably anticipating a growth in the demand for mechanized draft power as the project came on stream.

The solution for 1974 calls for all land to be prepared by tractors, with a regional stock of 1,526 tractors. In 1974 the actual stock of tractors (in four-wheel equivalents) by our estimates was 1,071; but the number of tractors increased steadily thereafter until virtually all the land was plowed by tractors in 1977.

The earlier discussion of the land market suggested that the rents actually paid were lower than those emerging from TIGER, which are market-clearing values. In 1966 the average land rent (of all contracts paid in cash) was reported to be $66 a relong in the *Farm Economic Survey*, which is somewhat lower than the average rent of $80 in the solution for 1967. Similarly, MADA's agroeconomic survey of 1974–75 (Yamashita, Jegatheesan, and Wong, 1976) found that land rents were about $100 a relong a year, and by late 1977 rents are said to have reached $220 a relong a year. Nevertheless, even this last figure is only just over half the average rental value of $427 obtained in solution L. Of course, as we have repeatedly emphasized, TIGER does not admit of adjustment problems in any market. It does, however, clearly indicate that, if the high paddy prices of recent years persist, there is likely to be strong and continuing upward pressure on land rents. If social and legal forces are able to dampen this trend, much of the income accruing to landlords in the postproject solution will in fact accrue to farm households.

We have no reliable data on incomes to check the model's results for 1967, but the agroeconomic survey for 1974–75 can be used to provide some check on the solution for 1974, even though the sample for this survey was not regionally representative. The relevant figure is $3,410 a farm (see table 2-16), but the comparable figure from the model is only $2,395 (table 4-4). The latter, however, has been derived on the assumption that all factors are paid the values of their marginal products, whereas the comparison requires the use of actual 1974 factor prices. As noted above, the average rent reported in MADA's agroeconomic survey was $100 a relong a year. Also, the contractors who hire out tractors are known to have charged about $14 a relong for plowing in 1975, or the equivalent of $100 a day. This clearly included an element of profit over and above the full average costs of the service provided. Using these prices, but netting out the fixed and variable costs of four-wheel tractors, we find a distribution of income (in $ thousands) in 1974 rather different from that in table 4-4:

Noncultivating landlords	14,760
Tractor contractors	9,557
Farm households	162,923
Aggregate net income	187,240

The average income of farming households is then $3,141, which is very close to the estimate derived from the 1974–75 survey.

As for real wage rates, the evidence cited in chapter 2 points to a rise of 50 to 100 percent over 1967–74, depending on how much weight is given to transplanting relative to harvesting (table 2-14). The estimate from TIGER is 36 percent, which suggests that the real supply price of labor may have shifted upward at least a little.

The Effect of the Project

Solutions K and O in table 4-3 provide estimates of the key variables for 1974 under the assumption that the project had not been undertaken. By comparing these results with their corresponding values in the 1974 "with project" situation (solutions L and N), we have an estimate of the project's impact on the assumption that the advent of new varieties was anticipated, both at base year (1967) prices and at final year (1974) prices. This is done in the last columns of table 4-3, which set out for each variable the percentage change attributable to the introduction of the project.

In estimating the project's impact, it does not make a great deal of difference whether 1967 or 1974 prices are used as the basis for decomposition, except when taking account of the degree of mechanization, which is rather sensitive to changes in exogenous prices. In general, the project accounts for the lion's share of the change in all variables. For example, paddy production would have been only 4.75 million piculs in 1974 if the project had not been undertaken, a gain of 27 percent over the 3.75 million piculs that would have been achieved in 1967 had 1974 prices prevailed. With the project, paddy production increased to 9.58 million piculs in 1974, thus accounting for 83 percent of the total increase over the level of production in 1967.

As noted above, however, by measuring the project's impact in this way, we have allowed the project to benefit from autonomous changes in the economy. Whether we decompose at 1967 prices or 1974 prices, the increase in aggregate net income attributed to the project is contingent on the simultaneous occurrence of the project, the exogenous changes in the prices of paddy and various inputs, and the introduction of improved varieties. Since these prices and varieties would have changed even if the project had not been undertaken, such interaction effects represent a windfall gain arising from the fortunate timing of the project's maturity. While such gains can be claimed as benefits attributable to the project in an ex post evaluation of the whole scheme, it is instructive to explore what might have happened had the project not been accompanied by such autonomous changes, or if such changes had not been anticipated.

Decomposition of the Sources of Change, 1967–74

To recap, the major external factors affecting the paddy farming economy over 1967–74 were: the irrigation project; the introduction of improved varieties, which had shorter growing periods and better responsiveness to nitrogen; increases in the prices of paddy and most inputs; and a 2.6 percent decline in the number of farms and a 3 percent reduction in the total area under paddy.

Clearly, the changes in the prices of paddy, fertilizer, and tractors would have occurred independently of the project, as would the general inflationary pressures on local wage rates. Improved varieties were already being used by 43 percent of the farmers in 1969, a year before double-cropping had commenced, and 90 percent of the farmers grew them in the main season of 1972 as a rain-fed crop. These facts suggest strongly that the irrigation project was not necessary for the widespread adoption of the new varieties, even though improved extension advice provided by the project may have hastened the process. Hence, the improvements in the varieties available to farmers over that period should be viewed as an autonomous source of change.

The number of farms declined, partly because some 4,000 households operating marginal farms became landless. Presumably this decline was largely in response to the changes in the wage-rental ratio attributable to the project and certain indivisibilities in the new system of cultivation. The farm population might well have remained stationary had the project not been undertaken. Likewise, we have attributed the reduction in the total area under paddy to the project, although there would have been some losses to urban encroachment in any case. In viewing both these changes as entirely attributable to the project, we are persuaded that the small magnitudes involved do not warrant a more detailed analysis.

Next, we must decide on whether to take an ex ante or ex post view in decomposing the overall changes into their constituents. The evidence suggests that the changes in prices were not expected when the project was evaluated, and since it is desirable to separate out the effects of pure good fortune, we shall take an ex ante view. Accordingly, the decomposition will be done at 1967 prices, so that (4.1) is the relevant identity.

The pairs of solutions corresponding to the terms on the right-hand side of (4.1) are as follows. The "pure" project effect[6] is given by the difference between solutions M and J; the variety effect by O and J; the project-variety interaction effect by N, M, O, and J; and the price and price-interaction effects by L and N. All the necessary solutions needed for this analysis are presented in table 4-3, and the results are set out in table 4-5.

Beginning with output and inputs, the introduction of irrigation accounts for about two-thirds of the increase in the output of paddy and inputs of labor, and about half the rise in inputs of nitrogen. The variety effect is much smaller than the "pure" project effect, though still appreciable in the case of inputs of nitrogen, which squares with the fact that the improved varieties are much more responsive to nitrogen than are the traditional strains. The interaction effects of irrigation and improved varieties also accord with intuition. As the improved varieties mature in fewer days, the stringent demands for labor under double-cropping are relaxed and the greater use of labor is encouraged in the face of an upward sloping supply schedule. This leads us to expect quite strong interaction effects on the use of labor—certainly much larger than those stemming from the introduction of improved varieties alone. With irrigation, high-yielding varieties can be grown in two seasons rather than one; their interaction effects on the use of nitrogen and the output of paddy should therefore be about the same magnitude as those of the variety effect alone—and so they are.

It is difficult to have as much a priori confidence in the magnitudes of the price and price-interaction effects, if only because they are a diverse bundle and one element is endogenously determined, namely, the wage rates for casual labor. Among the exogenously determined prices (table

6. In this case, the project effect also includes the impact of induced changes in farm numbers and the total area farmed.

Table 4-5. *Analysis of the Components of Change, 1967–74*

Variable	*1967 value*	+	*"Pure" project effect*	+	*Variety effect*	+	*Project-variety interaction effect*	+	*Price and price-interaction effect*	=	*1974 value*
Aggregate net income ($ thousands)	43,526		32,406		11,709		24,721		74,878		187,240
Percentage change			22.55		8.15		17.20		52.10		
Total paddy production (thousands of piculs)	3,826		3,708		950		1,111		−10		9,584
Percentage change			64.40		16.48		19.30		−0.17		
Total use of labor (thousands of days)	14,440		6,762		1,286		4,041		−2,377		24,152
Percentage change			69.63		13.24		41.61		−24.47		
Total wage payments ($ thousands)	2,285		14,557		561		−4,840		2,052		14,615
Percentage change			118.06		4.55		−39.25		16.64		
Average wage ($/day)	1.46		0.92		0.01		−0.30		1.11		3.20
Percentage change			52.87		0.57		−17.24		63.79		
Land rents ($/relong/year)											
Nonacid soil	85		74		32		67		195		453
Percentage change			20.11		8.70		18.21		52.99		
Acid soil	64		55		25		56		149		349
Percentage change			19.30		8.77		19.65		52.28		
Total use of nitrogen (thousands of pounds)	7,212		6,493		3,358		3,454		−777		19,740
Percentage change			51.83		26.80		27.57		−6.20		
Stock of four-wheel tractors	39		1,697		0		−1,096		886		1,526
Percentage change			114.12				−73.71		59.58		
Stock of buffaloes	15,736		−15,736		188		10,520		−10,708		0
Percentage change			−100.00		1.19		66.85		−68.05		

4-1), the price of paddy in 1974 is almost double that in 1967, the price of nitrogen is exactly double, and the parameters of the labor supply function are some 60 percent higher. It turns out that the average wage rate more than doubles (table 4-3); but this is not known a priori, of course. In any event, the ratio of the price of output to the prices of each of these inputs does not change much between 1967 and 1974. Since the price elasticity of supply is very low,[7] it is hardly surprising that the combined price effects on the output of paddy should be very small. The use of nitrogen is slightly discouraged, as the own-effect of the shift in the price of nitrogen relative to that of paddy would lead one to expect. The use of labor, however, is quite strongly discouraged, largely because tractors displace buffaloes (and their plowmen) from land preparation, the (exogenous) increase in the nominal cost of mechanized draft-power services being only one-third of its absolute level in 1967.

The overall shift in the composition of the stock of draft power is straightforward. Double-cropping makes the cultivation schedule very tight if traditional varieties are grown, and in the present case the associated rise in the wage rate is sufficient for buffaloes to be displaced entirely. Without irrigation, the improved varieties have a negligible effect on the nominal wage, and so the buffaloes keep the tractors at bay. As already noted, however, the improved varieties make for a more relaxed cultivation schedule and hence a lower wage rate. Thus, the interaction effect of irrigation and improved varieties favors buffaloes over tractors, and strongly so in this case.

The fact that wage rates and land rents are endogenously determined makes it hazardous to voice strong expectations about the relative magnitudes of the several effects listed in table 4-5. The tighter schedule imposed by double-cropping raises wage rates, though with the shorter growing period of the improved varieties this effect is significantly attenuated. It so happens that the effects of changes in prices and their interactions are stronger than those of irrigation alone. Annual land rents almost double with irrigation alone, which accords with intuition inasmuch as the farming economy is virtually replicated within the confines of twelve months. The supply of land, however, is perfectly inelastic, with or without irrigation; hence it is not especially surprising that, other things being equal, the much higher levels of exogenous prices in 1974 should lead to much higher rents.

Total wage payments and aggregate net income must also be accounted for. By itself, irrigation generates a strong rise in both the demand for

7. Arc elasticities of paddy supply were calculated from TIGER by varying the paddy price from its 1974 value. The elasticity is 0.003 over the range defined by a 10 percent price increase, and 0.03 over the range defined by a 10 percent price reduction.

Table 4-6. *Decomposition of the Paddy Variety and Project-Variety Interaction Effects, 1967–74*

Variable	Variety effect			Project-variety interaction			Variety plus project-variety interaction		
	Yield effect	*Maturation effect*	*Percentage yield effect*	*Yield effect*	*Maturation effect*	*Percentage yield effect*	*Yield effect*	*Maturation effect*	*Percentage yield effect*
Aggregate net income ($ thousands)	10,756	953	91.86	9,876	14,845	39.95	20,632	15,798	56.6
Total paddy production (thousands of piculs)	950	0	100.00	973	137	87.66	1,923	137	93.3
Total use of labor (thousands of days)	1,267	19	98.52	−310	4,351	—	957	4,370	18.0
Total wage payments ($ thousands)	1,062	−501	—	1,583	−6,423	—	2,645	−6,924	—
Average wage ($/day)	0.03	−0.02	—	0.17	−0.47	—	0.2	−0.49	—
Land rents ($/relong/year)									
Nonacid soil	29	3	90.63	33	34	49.25	62	37	62.6
Acid soil	22	3	88.00	27	29	48.21	49	32	60.5
Total use of nitrogen (thousands of pounds)	3,358	0	100.00	3,182	272	92.13	6,540	272	96.0
Stock of four-wheel tractors	0	0	—	542	−1,638	—	542	−1,638	—
Stock of buffaloes	0	188	0	−6,472	16,992	—	−6,472	17,180	—

— Not relevant.

casual laborers and the wage rate they receive. The pure variety effect is weak: a higher level of output demands more work at harvesttime, and a more relaxed schedule lowers the wage rate, which in turn leads to a strongly negative project-variety interaction. The effects of changes in prices are fairly favorable to casual workers. The effect on aggregate net income of irrigation alone is proportionally much weaker than its effect on output because the associated change in wage payments to casual laborers is large and positive whereas the changes associated with other effects are small or negative, or both. Thus, the effect of project-variety interactions is almost as large as that of irrigation alone, while that of price and price interactions is now dominant. The absolute magnitude of the latter is readily seen. If supply is price-inelastic and relative prices are approximately constant, a doubling of all prices will double the nominal value of outputs and inputs. Moreover, the cost of living index for farming households was about 60 percent higher in 1974 than in 1967 (table 7-1), so that the movement in the terms of trade over that period was favorable to paddy farmers (see chapter 7).

Improved varieties have two distinct advantages over traditional paddy, namely, a shorter growing period and greater responsiveness to nitrogen. To disentangle the effects of these characteristics, we need solutions which are variants of M and O, in which yields are kept at their levels in 1967, but the growing period is reduced from twenty-six to twenty weeks. These results yield the decomposition set out in table 4-6, in which the total variety effect and that of variety-project interactions are taken from table 4-5.

The same considerations that informed the analysis of the results in table 4-5 also apply here. In the absence of irrigation, only one crop is grown, so that varieties with shorter growing periods will not raise output unless they encourage more inputs. In fact, they permit a simple substitution of family for casual labor through a more relaxed cultivation schedule, and both the wage rate and the wage bill for hired workers fall. As expected, rents therefore edge upward. By contrast, varieties offering solely higher yields raise output and the demand for labor and fertilizers. The average wage rate and the bill for casual workers follow in train. Hence, the "pure" variety effect stems overwhelmingly from the improvements in responsiveness to nitrogen.

When two crops are grown in a year, however, varieties that mature in fewer days offer a decided advantage, and this "maturation effect" is likely to play a much more powerful role in the interaction of project and variety effects. Table 4-5 reveals that this is indeed the case. With a less stringent schedule of operations, the wage rate and wage bill are much lower, while much more family labor is used. Thus, the maturation effect is more important than the yield effect in raising aggregate net income, even though the former has but a relatively small effect on output. The maturation effect is now equally important in raising rents, and more than offsets the influence of the yield effect in promoting mechanization.

For the variety and the project-variety interaction effects together, it is no surprise that the maturation effect has a small role in influencing the output of paddy and inputs of nitrogen. It does, however, have a large say in determining the distribution of income, favoring farmers and landowners over casual laborers, and landlords over tenants, while blocking the advance of mechanization by keeping wage rates low.

The Effect of Farm Mechanization

In the above analysis, we allowed the model to determine the optimal stocks of tractors and buffaloes. A comparison of the steady-state solutions for 1967 and 1974 reveals a wholesale switch from buffaloes to four-wheel tractors, which does not unduly exaggerate the actual course of events over that period. A comparison of solutions J and K in table 4-3 suggests that this transition would not have occurred in the absence of the project. In particular, TIGER predicts there would have been only 185 tractors in 1974 if there had been no project, compared with steady-state stocks of 1,526 in that year in the presence of the project, and 39 in 1967.

Treating tractor stocks as endogenously determined seems perfectly reasonable, there being no obvious exogenous factors other than prices affecting their number. If tractors had not been available, however, would their absence have been seriously detrimental to the success of the project? And how much better might casual laborers have fared if the demand for their labor had not been reduced through mechanization?

Table 4-7 reports various steady-state solutions for 1974 in the presence of the project. Solution L is the basic solution for 1974 taken directly from table 4-3 and repeated here for ease of comparison. Solution P is the comparable solution when the region is denied access to four-wheel tractors. Given the choice between buffaloes or pedestrian tractors, farmers choose a stock of 18,035 buffaloes and only 273 pedestrian tractors. A little surprisingly, aggregate net income and total paddy production are not much affected by the absence of four-wheel tractors;[8] they decline by a modest 1 percent or so, which suggests that the unit costs of each technique are similar over the observed range of factor prices. Casual laborers, however, gain substantially as a group. The average wage increases by 12

8. Aggregate net income would be $2.334 million lower, rents would be $5.387 million lower, and there would be no net profits earned on draft-power services. Hence, farmers' incomes would be about $3 million greater in the absence of four-wheel tractors.

Table 4-7. *Impact of Mechanization Options, 1974*

	Model solution			
Variable	*1974 with project (L)*	*No four-wheel tractors (P)*	*Buffaloes only (Q)*	*Pedestrian tractors only (R)*
Aggregate net income ($ thousands)	187,240	184,906	184,488	183,819
Total paddy production (thousands of piculs)	9,584	9,558	9,507	9,569
Total usage				
Nitrogen (thousands of pounds)	19,740	19,697	19,600	19,715
Family labor (thousands of days)	19,586	20,957	20,977	20,268
Day labor (thousands of days)	4,566	7,297	7,252	4,680
Total labor (thousands of days)	24,152	28,254	28,229	24,948
Stock of four-wheel tractors	1,526	—	—	—
Stock of pedestrian tractors	0	273	—	8,020
Stock of buffaloes	0	18,035	18,762	—
Land rents ($/relong/year)				
Nonacid soil	453	415	413	437
Acid soil	349	317	318	335
Monthly wages ($/day)				
Nov. 26–Dec. 23	1.85	1.98	1.93	1.85
Dec. 24–Jan. 20	2.05	1.96	2.04	1.89
Jan. 21–Feb. 17	2.73	2.51	2.56	2.47
Feb. 18–Mar. 17	2.85	3.41	3.30	2.97
Mar. 18–Apr. 14	1.85	2.42	2.44	1.85
Apr. 15–May 12	1.93	2.42	2.35	1.96
May 13–Jun. 9	2.01	2.18	2.18	2.01
Jun. 10–Jul. 7	1.85	1.85	1.85	1.85
Jul. 8–Aug. 4	3.50	3.96	3.96	3.55
Aug. 5–Sep. 1	3.50	4.01	4.18	3.46
Sep. 2–Sep. 29	3.48	4.50	4.36	3.76
Sep. 30–Oct. 27	1.89	2.49	2.47	2.01
Oct. 28–Nov. 25	1.91	2.10	2.08	1.93
Average wage ($/day)	3.20	3.58	3.56	3.29
Annual wage payments ($ thousands)	14,615	26,134	25,846	15,394

— Not applicable.

percent, the total employment of casual labor increases by 60 percent, and total wage earnings increase by almost 80 percent. Clearly, casual laborers were denied significant additional benefits from the project as a result of mechanization involving four-wheel tractors. In the absence of such machines, rents would have been almost 10 percent lower, so that farmers would have been better off too, since aggregate net income would have been only 1 percent lower. Although the absence of tractors would have meant gains to casual laborers and farmers, larger gains accrued instead to rentiers and other sectors of the regional, national, and international economies in the form of demand for tractors, rotivators, and the intermediate inputs necessary for their operation.

Solution Q differs from P in that the region is also denied access to pedestrian tractors. Since only 3 percent of the land area was prepared by pedestrian tractors in solution P, however, a move to buffaloes alone results in only marginal changes.

In solution R the region is denied access to both buffaloes and four-wheel tractors to evaluate the cost to the region of the widespread adoption of pedestrian tractors between 1967 and 1974. Although the actual stocks of these tractors increased from some 250 in 1967 to 2,400 in 1973, and thence to more than 5,000 in 1977, MADA (1970*a*) clearly demonstrated that the cost of plowing by four-wheel tractors is lower than that by pedestrian tractors under trial conditions. Hence, as suggested in chapter 2, the increase in the number of pedestrian tractors must have been due to other considerations.

A comparison of solutions L and R shows that the substitution of pedestrian tractors (R) for four-wheel tractors (L) has a negligible impact on the output of paddy, and, as might be expected, the employment of casual laborers and the wage bill are slightly larger. Although these two forms of mechanization are clearly close substitutes, aggregate net income declines by $3.4 million. Although this drop is relatively small (about 2 percent), it is a real cost to the region. Proceeding as before, the payments to noncultivating landlords at the rents ruling in the steady state amount to $60.6 million in solution R, a decline of $2.47 million (or 3.9 percent) from the comparable figure from solution L (see table 4-4). Hence, the aggregate income accruing to farmers in solution R is about $1 million (less than 1 percent) lower than the comparable figure when four-wheel tractors are permitted. Apparently, the bulk of the cost of using pedestrian rather than four-wheel tractors is borne by landlords, while the farmers lose a little and casual laborers gain a little (there are no longer any contractors).

Conclusions

A comparison of the relevant steady states reveals substantial increases in the incomes of all households engaged in the paddy farming sector during the years from 1967 to 1974. The average income of farm households from the cultivation of paddy increases fourfold when measured in current prices, or two and a half times when adjusted for increases in the local cost of living index. At the rents actually ruling, noncultivating landlords probably doubled their annual rent collections over this period, but the increase would have been much larger had rents adjusted fully to their equilibrium values. In fact, actual land rents were only about 25 percent of their equilibrium values in 1974, although what evidence there is suggests that they increased substantially thereafter.

Casual laborers as a group also benefited substantially as total nominal wage payments increased well over sixfold. Their wage earnings would have been almost twice as large again, however, if tractors were barred by fiat. The model also shows that the combined costs to farmers and landlords of not mechanizing would have been rather small, providing casual workers continued to be available under the same conditions of supply as in 1973. Any significant increase in the real supply price of labor would probably have induced widespread mechanization at an earlier stage.

In reality, the relative incomes of land-abundant and labor-abundant farms probably remained about the same between 1967 and 1974 because average nominal rents and wages increased in the same proportion. If land rents had risen toward the equilibrium values in TIGER, however, land-abundant households would have become relatively more prosperous.

About half the increase in aggregate net income from 1967 to 1974 can be attributed to favorable changes in prices, and less than 10 percent is due to the "pure" effects of improved varieties with shorter growing periods and better response to nitrogen. With these changes in prices, the greatest impact of the project is on paddy production, employment, and the techniques used in land preparation. In particular, there would have been little change in total employment and the stock of tractors over the period had the project not been undertaken, even if improved varieties had been introduced. Nevertheless, although the "pure" effects stemming from the improved varieties appear to be quite small, the interaction effects with the project are appreciable, especially for the distribution of value added.

5

A Social Accounting Matrix for the Regional Economy in 1972

IN THIS CHAPTER, we lay the empirical foundations for the quantitative analysis in chapters 7 and 8. To do so we provide a quantitative description of the structure of the regional economy, defining institutional categories and economic activities so as to serve the subsequent analysis. We begin by estimating a set of regional accounts for 1972 and organizing them into a so-called social accounting matrix (SAM).[1] In choosing this framework, we have been greatly swayed by its capacity to provide a consistent, comprehensive, and detailed picture of the transactions in an economy. Industries, commodities, factors, households, and other institutions can all be accommodated, and the pattern in which incomes are distributed takes its place alongside the sources of income generation. Moreover, although this framework has a healthy, not to say rude, appetite for data, its rigorous enforcement of consistency in estimation is a great help in weighing and reconciling the diverse and patchy evidence commonly available for developing economies, especially when there must be frequent resort to the intelligent "guesstimate." It follows that the questions that can be addressed with a SAM are determined in part by the system of classification and the extent of disaggregation, both of which commonly represent a compromise between the ideal and what the data permit.

Although our main task is to describe the region's economy through its social accounts, we have tried to avoid the tediousness of mere description, which would surely tax the patience of most readers. Fortunately, a careful examination of the accounts provides some illuminating insights into important issues. In the present case, for example, the uses of private savings play a central role in determining the growth of the economy in the period after the public investment project had reached maturity. For if a large part

1. The development of SAMs for use in policy analysis owes much to Pyatt and his collaborators (1973, 1977), who have built on the earlier work of Stone (1977).

of any increase in private savings due to the project is invested outside the region, the impulse to the local economy from the accelerator effects of investment will be the weaker thereby. Although the accounts for a single year cannot establish that such a process was taking place, they allow one to probe into the links between the sources of savings, the uses of investment, and the balance of payments, and at the same time reveal the size of the various flows. Another possibility is that the accounts will reveal some features of the structure of costs or the pattern of final expenditures that can be exploited later when building a quantitative model which draws on the accounts. In such instances we have alerted the reader by digressing from a description of the accounts to a discussion of how we propose to use analytically the insights they afford. More generally, such a set of accounts for a comparatively poor region has wider uses in policy analysis than those considered in this book. Those whose task it is to formulate and assess proposals to reduce regional inequality will generally find that the construction of regional accounts is an invaluable preliminary step.

The Structure of the SAM in Outline

The historical development of economic data systems and the evolution of the social accounting matrix has been well described by Stone (1977), while the structure and potential of a SAM framework has been clearly laid out in Pyatt, Roe, and associates (1977). Nevertheless, for readers not familiar with such frameworks a brief outline of the salient features of a SAM is presented here.

A SAM represents the union of two separate data schemes: national income accounts in matrix form and the Leontief input-output model of production. A SAM is always square, for under every account there are both receipts and expenditures. By convention, the entries in the matrix represent receipts when read across rows and expenditures down columns. Since the principles of double-entry bookkeeping apply, each row total is identical to its corresponding column total. Accordingly, table 5-1 provides a schematic layout, followed by interpretive notes, for a typical, highly aggregated SAM for a small open economy.

Such a scheme is all very well, but it is prudent to ask what can be done with it once the transactions have been estimated. The answer depends, of course, on the questions one wishes to pose. However, there are two general responses. First, a SAM provides a picture of the interdependence of the economy through the circular flows that make up the familiar identities of national income accounting. Second, with suitable assumptions, a SAM can be readily transformed into a linear model of the economy (see chapter 6). Here, however, we concern ourselves with building the

SAM and tackling some issues that require nothing more than an analysis of the regional accounts thus estimated.

The Muda SAM, laid out in schematic form in table 5-2, is relatively "primitive"; unlike that in table 5-1, it has neither factor nor commodity accounts. The SAM for Iran, however, has the same structure as that for

Table 5-1. *An Aggregated Social Accounting Matrix for an Open Economy*

	Factors of production (1)	*Institutions' current accounts* (2)	*Institutions' capital accounts* (3)
1. *Factors of production*	—	—	—
2. *Institutions' current accounts*	Allocation of factor incomes to institutions	Domestic transfers among institutions	—
3. *Institutions' capital accounts*	—	Savings	Capital transfers among institutions
4. *Commodities*	—	Commodity demands by institutions	Gross fixed capital formation and increases in stocks
5. *Production activities*	—	—	—
6. *Rest of world*	—	—	—
7. *Total*	Total factor incomes	Total current outlays by institutions	Aggregate investment

— Not applicable.

Interpretive Notes

Row 1. Production activities provide incomes to factors of production in return for factor services. Factors receive further incomes from abroad. Thus, the sum of columns 5 and 6 in row 1 yields the total in column 7. Total factor incomes may be disaggregated into wage payments to different classes of labor, the returns on capital, and rents for factors such as land.

Col 1. Institutions that own factors receive factorial incomes. For example, households providing wage labor receive wage payments, while governments owning mineral rights receive royalty payments.

Row 2. Institutions also receive other incomes, partly from transfer payments among themselves and partly from abroad. For example, corporate bodies distribute profits to shareholders (households), and governments pay pensions and subsidies.

Col 2. The transfers in row 2 become outlays in their column context. In addition, institutions make outlays on domestic commodities, imported consumer goods, and savings, the latter being a delivery to the capital account.

Row 3. Institutions, apart from their own savings (such as undistributed profits), receive

Muda (Pyatt and others, 1973), whereas the SAM for Sri Lanka has separate accounts for factors and institutions (Pyatt and Roe, 1977), and that for Swaziland also sports accounts for commodities (Pyatt and Round, 1977). Other refinements of classification are possible (Pyatt and Round, 1978), but they are of no interest here. Nevertheless, since the Muda region is a

Commodities (4)	*Production activities* (5)	*Rest of world* (6)	*Total* (7)
—	Payments of value added to factors	Net factor incomes from abroad	Total factor incomes
—	—	Net nonfactor incomes from abroad	Total receipts by institutions net of transfers
—	—	Net receipts of capital from abroad	Aggregate capital receipts
—	Intermediate demand for commodities	Exports	Total commodity demands
Deliveries of domestic commodities by production activities	—	—	Gross outputs
Commodity imports	—	—	Total payments abroad
Total supplies of commodities	Total costs of production	Total receipts from abroad	—

further funds through the operation of the financial markets and the inflow of investment from abroad.

Col 3. Institutions make capital outlays partly through the transfer mechanisms mentioned above, partly by the purchase of fixed assets and changes in the level of stocks, and partly by importing capital goods.

Rows 4, 5. Taken together these rows represent accruals from the sale of consumer goods, capital goods, intermediate goods, and exports broken down by commodity. They represent in sum the value of the gross output bundle of the economy.

Cols 4, 5. The flow of receipts in rows 4 and 5 is expended on the purchase of raw materials (domestically produced or imported) and the distribution of value added (that is, wages, profits, and rent) to the factors of production. By the identity of row and column totals, total costs of production are equal to gross outputs.

Row 6. The flow of payments to the rest of the world takes the form of imports.

Col 6. The outlays on exports by the rest of the world are shown together with the net flows of other receipts from abroad.

Table 5-2. *A Schematic Social Accounts Matrix for the Muda Region*

	Institutions' current accounts							
	Households (1)	*Firms* (2)	*Federal government* (3)	*State government* (4)	*Combined capital account* (5)	*Production activities* (6)	*Rest of world* (7)	*Total* (8)
1. *Households*	Interhousehold transfers and debt repayments	Profits distributed to households	Federal transfer payments to households	State transfer payments to households	—	Value added in production accruing directly to households	Remittance and other income from abroad	Households' total current receipts
2. *Firms*	—	—	—	—	—	Value added in production accruing to firms	Income from abroad	Firms' total current receipts
3. *Federal government*	Taxes paid by households to the federal government	Corporate taxes	—	—	—	Value added in production accruing to federal enterprises	Federal grants to federal agencies in region	Total receipts of regional federal agencies

4. *State government*	Taxes paid by households to the state government	—	—	—	—	Value added in production accruing to state government enterprises	State government receipts from abroad (grants and loans)	Total receipts of state government
5. *Combined capital account*	Household savings	Corporate savings	Federal government savings	State government savings	—	—	Net capital inflow (+) or outflow (−)	Total capital receipts
6. *Production activities*	Household expenditures on domestic goods	—	—	—	Deliveries of domestic output to investment	Interindustry transactions	Exports of goods and services	Gross value of output (uses of production)
7. *Rest of world*	Household expenditures on imports	Firms' payments to agents abroad	—	Repayments of state government debt	Imports of capital goods	Imports of intermediate goods	—	Total payments abroad
8. *Total*	Household total expenditures	Firms' total expenditures	Total outlays of federal agencies in region	Total outlays of state government	Total capital payments	Total cost of production	Total receipts from abroad	—

— Not applicable.

small open economy, with fairly free movement of capital and labor, the absence of separate factor accounts is a drawback.

Our principal reasons for this initial aggregation were the absence of data and the difficulties of making imputations of factor earnings for unincorporated enterprises in both the farm and nonfarm sectors of the economy. In the absence of separate factor accounts, production activities make payments directly to institutions (households, private firms, and government enterprises), so that the problem of mapping the set of factor incomes into the distribution of incomes among households is also avoided. Naturally, this procedure entails a strong simplifying assumption if the tableau is interpreted as a picture of how the economy functions without recourse to an explicit treatment of factor markets. A sufficient condition is that each type of institution is exclusively endowed with a characteristic factor, an assumption which limits to some extent the range of issues the SAM can be used to analyze.

The characteristic factor assumption is not, however, as damaging as it seems. First, the determination of factor prices and household incomes in the paddy sector, which dominates the regional economy, has been analyzed and reported separately in chapter 4 by using a model designed for the purpose. Nevertheless, a weakness remains from our treatment of the earnings of, and local expenditures by, seasonal migrant workers, which are conceptually troublesome transactions. We resolve this problem by treating their wages as imported intermediate goods and ignoring their local expenditures, which are small if not negligible, for they are generally fed and housed by their employers. In addition, we permit only one class of nonfarm households, a severe aggregation discussed below.

The lack of separate commodity accounts should occasion less alarm. In most cases, production activities have been chosen so that each produces a single, relatively homogeneous commodity. The few exceptions produce a bundle of commodities in implicitly fixed proportions.

The fact that taxes on commodities are levied solely by the federal government is of some importance in the context of regional policy analysis, even though it is sensible to take commodity taxes as exogenously given in such cases. However, the reworking of our painfully produced statistics at the fully disaggregated level was beyond our resources at the time. Accordingly, all commodity balances were estimated in prices net of trade and transport margins, but inclusive of commodity taxes (which make up part of the "world" price facing the regional economy). Hence, the federal tax burden on the region cannot be fully inferred solely from the accounts set out below. With the aid of aggregate data for West Malaysia, we provide an estimate of the total indirect tax burden on the region in the following section.

As an illustration of how to read the accounts in table 5-2, consider the current outlays and receipts by the federal government's agencies within the region. Column 3 lists only two classes of expenditure. The first is current transfers to households, not only in cash but also in the form of the imputed value of such public goods and services as education, health, and other administration. The second is the purchase of assets (or "savings"), commonly in the form of capital outlays on roads, dams, buildings, and other structures and equipment. Along the corresponding row, these agencies get revenues from households in the form of income taxes, license fees, and duties; from corporations in the form of corporate taxes; from the profits of federally owned state corporations; and from the federal government in the form of grants. In the nature of accounts such as these, total outlays are equal to total receipts in every category of account.

The Classification of Institutions

The classifications employed are closely linked to the simplification of the accounts structure noted above. For the purposes of this study, which emphasizes project evaluation, a disaggregated treatment of the institutions' accounts is called for. The SAM features five categories of household, one type of firm, two arms of government (federal and state), and a consolidated capital account.

The household classes have been defined to reflect the income distribution aspects of regional activities, while avoiding the worst of the difficulties posed by the absence of separate factor accounts. Thus, the definitions rest on socioeconomic criteria: endowments of labor and land, access to irrigation, and sector of employment. Three of the four classes of agricultural households reside in the project's command area and specialize in the production of paddy. They are: landless households, which derive most of their income from employment on other paddy farms;[2] labor-abundant farm households, which possess a high ratio of family labor to area operated; and land-abundant households, which hire in substantial amounts of nonfamily labor.[3] Plainly the household's endowments of land and labor are intimately connected with its labor market transactions, which underlie the definitions adopted here.

The fourth class of farm households resides outside the project's command area and is heavily engaged in "other agriculture" as well as in producing rubber and raising a single annual crop of unirrigated paddy. They also supply labor to households in the project area at times of peak activity in paddy cultivation. As their economic activities differ from those

2. Very few of them are without a small parcel of land which they cultivate, although such parcels are invariably rented.

3. A two-way tabulation of paddy land per adult family worker against hired labor costs per unit of paddy land shows a strong block-diagonal pattern, the critical land endowment being 1.2 relongs per adult family worker.

of households within the project's command area and there is an intrinsic interest in what happens to households on the periphery of large development projects, they merit the status of a distinct household category.

The fifth and final household group comprises all nonfarm households, which account for 35 percent of the region's population and display wide variations in income and wealth. Since the project's effects on income distribution are central to this study, these households should be subdivided into further categories. But there were no data to place them in more delicately defined classes, and the absence of separate factor accounts would have rendered such disaggregation largely spurious unless we also undertook a formal analysis of the determination of factor prices in the nonfarm economy, a task we carefully eschewed. Nevertheless, two points can be made. First, the majority of nonagricultural households are ethnic Chinese, who generally enjoy higher per capita incomes than the predominantly rural Malays. Second, the SAM provides a solitary indirect pointer to inequality within the class of nonfarm households in the form of the profits of incorporated enterprises.

The existence of only one kind of firm arises out of the way in which households enter into the production and income accounts. Enterprises employing mainly family capital, expertise, and labor distribute the whole of their net revenues back to the household(s) concerned. As the enterprise grows, however, it may acquire a different legal status—and an independent institutional existence—by retaining some operating profits as an enterprise. We have employed the distinction between incorporated and unincorporated enterprises to separate private "firms" from nonfarm households at the evident cost of splitting in one place the graduated range from backyard (or front room) production to public company. State corporations are not separately treated, partly because there were no data and partly because, at the time, the relevant institutions (for example, the LPN[4]) were so constrained by federal development policies that their autonomy as separate economic entities was seriously in question. In short, we treat such agencies as arms of the federal or state governments, to which their operations are a fiscal resource or liability as the case may be.

For the purposes of regional analysis, there is much to be said in favor of distinguishing between the state and federal levels of the government accounts. On the revenue side, the state government is heavily dependent on grants from the federal, and local taxes take the form mainly of licenses and fees, property taxes, and water rates. All income and virtually all commodity taxation are the sole preserve of the federal government, though the size of the corresponding leakage from the regional economy

4. Lembaga Padi dan Beras Negara (State Paddy and Rice Board).

depends both on the level of regional economic activity and on the (exogenously determined) tax rates.[5] On the expenditure side, there is a fairly well-defined pattern. The federal government has exclusive responsibility for education, health, electrification, telecommunications, and law enforcement, while the state government plays the primary role in public works, roads, and low-cost housing.

To round off the institutions' accounts, there is only a consolidated capital account, because it was not possible to estimate the relevant flows of funds among households, firms, and government.

The Classification of Production Activities

As noted above, the production and commodity accounts have been combined into one. The cost of this simplification has been slight inasmuch as only three sectors (food processing, other agriculture, and manufacturing not elsewhere classified) are ragbags of different goods, and they make up only about 9 percent of regional value added. In the remaining cases, the nature of the characteristic commodity is clear-cut. We undertook our disaggregation in accordance with the following rationale, which led to the thirty-five sectors in table 5-4.

When a particular natural commodity is produced under different institutional arrangements, we have assigned a separate characteristic commodity to each arrangement if the institutional differences appeared important in terms of production technology or the distribution of sectoral income among households. A priori, therefore, distinctions have been made between: commercial and small-scale rice mills, which have different technologies and serve different markets; smallholder and estate rubber production; private and public health services; organized and informal distribution services, the latter being supplied by farm households in the slack part of the agricultural cycle; and modern and traditional financial services. Such distinctions are obviously vital if the future balance among the various forms of organization in production is important in making public policy.

A natural commodity may take different forms according to its use or the type of demand it can satisfy, often with differences in technology and the organization of production. This criterion led us to distinguish between: road and rail transport, construction for residential and nonresidential purposes, and rural and urban housing services.

Given their wide scope and importance, we have chosen to treat the

5. The traditional Islamic tax, *zakat*, redistributes income among households directly and accordingly does not figure in this discussion.

government sectors endogenously within the production accounts, in recognition of the fact that they meet intermediate as well as final demand.

Some disaggregations rest on specific features of the economy, such as the workings of markets and the ownership of productive assets. To give an example, there is a thriving and competitive market for agricultural machinery services. The lion's share is supplied by non-Malays, most of whom own a single four-wheeled tractor but, being non-Malays, they are not legally permitted to own land—a fact that underpins our distinction between paddy production and agricultural machinery services.

Clearly, the foregoing disaggregation of production activities has led to pairs of characteristic commodities that seem to be close, or even perfect, substitutes in demand. This is certainly so in the case of rubber produced by smallholders and estates, and arguably so in the case of road and rail transport. In the remaining cases, however, spatial and institutional factors do much to ensure that the sources of demand for such characteristic commodities are substantially separable. The most important example of this is the two sectors producing rice (small and commercial mills). Apart from basic differences in technology, they are required by law to serve different markets. Small mills produce only for domestic consumption, while the commercial mills serve the wider market. Thus, while the set of production activities in the SAM entails some awkward features in the expenditure side of the accounts, there are no difficulties that cannot be surmounted by careful formulation of the questions addressed.

The Estimation of the SAM

The estimation of the SAM is documented in detail in appendix C; in consequence, the discussion below is brief. We had access to a variety of data sources of varying quality, which formed a somewhat patchy basis for the estimation of a set of initial values for the cells of the matrix. The most important primary data (in the sense that they were specific to the region and collected with considerable care) were the results of the surveys of several important sectors undertaken by FAO in collaboration with the World Bank (FAO, 1975). In addition, the diverse publications of the state and federal governments were a rich source of secondary material.

Nevertheless, unlike import-export flows for national economies, which are usually among the more reliable statistical series available, movements of goods across regional boundaries are rarely measured, even by spot surveys along roads, and must therefore be inferred from other data.[6] But

6. The FAO/World Bank studies did spot surveys of the flow of goods by both road and rail, but they were not conducted frequently enough to capture the heavy seasonal variations. They have therefore been used only as checks on the indirect calculations of annual flows.

this lacuna is not very troublesome in practice because regions usually have heavily specialized patterns of trade in commodities, which the commodity classification scheme must take into account. In the present case, imports were determined, in the main, from final and intermediate expenditures in the light of whether a particular commodity was produced in the region. Since noncompetitive imports accounted for about 85 percent of all imports, the assumptions concerning competitive imports cannot be too damaging. Similarly, in most cases, exports do not compete with imports and were determined as residuals, although the FAO/World Bank material and a priori considerations provided useful qualitative checks on their values.

There were, of course, other gaps where the logic was less compelling and where we necessarily had recourse to less reliable material, which had often been produced for purposes quite different from ours. In extreme circumstances, we were obliged to rely on the informed "guesstimate," albeit based on detailed general knowledge of the region.

The estimation process, which was both time-consuming and exhausting, resulted, at length, in a full but (not surprisingly) inconsistent set of prior estimates. Using a technique developed by Byron (1978) from ideas formulated earlier by Stone, we transformed this set of prior estimates into a balanced (that is, consistent) SAM. This process involved the largely subjective estimation of a matrix of the variances of the initial estimates. If, then, we have some confidence (as manifested in the variance matrix) in our prior estimates, the resulting final estimates are consistent not only in the accounting sense, but also with our prior judgments.

The Consolidated Regional Accounts for 1972

We begin in table 5-3 with the consolidated accounts, which correspond exactly to the labeled cells of the tableau in table 5-2.

Output and Its Uses

Regional gross output was \$711.40 million (row 8, column 6) of which \$209.97 million represented outlays on domestically produced intermediate goods (6, 6) and \$85.09 million, purchases of imported intermediate goods (7, 6). Hence, regional value added (gross of depreciation) in market prices was \$416.34 million, which was distributed among households, firms, and federal and state government enterprises as follows: \$373.21 (1, 6), \$40.36 (2, 6), \$1.61 (3, 6), and \$1.16 (4, 6) million, respectively. As noted above, payments to seasonal migrant workers, which were \$3.55 million, have been classified as imported intermediate goods. Hence, the

Table 5-3. *Consolidated Regional Accounts for 1972*
(millions of Malaysian dollars)

	Institutions' current accounts							
	Households (1)	*Firms* (2)	*Federal government* (3)	*State government* (4)	*Combined capital account* (5)	*Production activities* (6)	*Rest of world* (7)	*Total* (8)
1. *Households*	5.88	5.43	70.04	24.52	—	373.21	2.04	481.11
2. *Firms*	—	—	—	—	—	40.36	0.99	41.35
3. *Federal government*	29.76	19.02	—	—	—	1.61	36.70	87.09
4. *State government*	15.25	—	—	—	—	1.16	14.22	30.63
5. *Combined capital account*	62.65	10.25	17.05	5.61	—	—	-35.18	60.38
6. *Production activities*	261.95	—	—	—	43.23	209.97	196.24	711.40
7. *Rest of world*	105.62	6.65	—	0.50	17.15	85.09	—	215.01
8. *Total*	481.11	41.35	87.09	30.63	60.38	711.40	215.01	—

— Not applicable.

Note: Some entries may differ slightly from those in table 5-4, and some rows and columns do not sum exactly to their corresponding totals, because of rounding errors.

region's gross domestic product (GDP) was $419.89 million, implying that per capita GDP was $611.

For West Malaysia, per capita GDP in 1972 was $1,100. Thus, the region was (and remains) a comparatively poor one by the standards of the Malaysian economy as a whole. The only earlier estimate available to us (Hoerr, 1970) put Kedah's per capita GDP in 1965 at $518, compared with $850 for West Malaysia. These figures suggest that, notwithstanding the introduction of irrigation on a large scale, the ratio of the regional to national levels of per capita GDP slipped slightly between 1965 and 1972. In view of the possible differences between northern and southern Kedah—the latter is heavily specialized in rubber production—and the inevitable (if unknown) differences between Hoerr's methods and our own, perhaps not too much should be made of this shift. It appears unlikely, however, that the living standards of the region caught up very much with those of the national economy.

Consumption by households (6, 1), interindustry transactions (6, 6), and exports (6, 7) were roughly equally important uses of domestic gross output. Deliveries to investment activities (6, 5) were relatively small at $43.23 million, or 6 percent of the total. The economy was highly open to trade, as one would expect, and commodity exports were $196.24 million, or about 47 percent of the region's GDP.

Household Incomes and Outlays

In addition to their direct earnings (actual and imputed) from production activities (1, 6), households received $5.43 million from firms in the form of distributed profits (1, 2), $2.04 million in net remittances from relatives living outside the region (1, 7), and $70.04 million and $24.51 million in the form of "transfers" from the federal (1, 3) and state governments (1, 4). As pensions and other direct subsidy payments were rather small, most of these transfers from government were largely notional ones needed to balance the "purchases" of public goods and services by households. In total, these purchases were $88.16 million, of which only $1.69 million were actual cash outlays.[7] Hence the value of cash transfers from both state and federal governments to households was $70.04 + 24.51 - 88.16 + 1.69 = \$8.08$ million. Accordingly, the total money income accruing to households was $5.43 + 8.08 + 373.21 + 2.04 = \$388.76$ million. Interhousehold transfers have, of course, been ignored.

The expenditures against this income were as follows: $29.76 and $15.25 million were paid out in direct taxes to the federal (3, 1) and state (4, 1)

7. The figures mentioned here are derived by simple arithmetic from table 5-4.

governments, respectively, yielding a disposable income of $343.75 million. Of the latter, $62.65 million were saved (5, 1), $175.48 million (261.95 − 88.16 + 1.69) were spent on domestically produced goods (6, 1), and the remaining $105.62 million were spent on imported consumer goods (7, 1). The savings rate out of disposable household income was, therefore, about 18 percent, which is quite high but not grossly out of line with that for West Malaysia (Treasury of Malaysia, 1979).

Firms

The profits of incorporated private enterprises (2, 6) were $40.36 million, which also received $0.99 million in income from parent companies and investors resident "abroad" (2, 7), that is, outside the region. Their outgoings took the form of $5.43 million in distributed profits to households (1, 2), $19.02 million in corporate taxes paid to the federal government (3, 2), and $6.65 million in remittances to parent companies "abroad" (7, 2). The balance, $10.25 million, was retained to augment the enterprises' assets (5, 2) in the local economy.

Government

The incomings of the agencies of the federal government were made up of $29.76 million from direct taxes paid by households (3, 1), $19.02 million in corporate taxes (3, 2), $1.61 million from the gross profits of state enterprises (3, 6), and $36.70 million in grants from the federal government (3, 7). Against this income, federal agencies made outlays of $70.04 million in the form of "transfers" to households for the purchase of public goods and services (1, 3) and $17.05 million in the form of capital expenditures (5, 3). As for the state government, it took in $15.25 million from households in the form of direct taxes (4, 1), $1.16 million in gross profits from its own enterprises (4, 6), and $14.22 million in grants and loans from the federal government (4, 7). It made outlays of $24.52 million in "transfers" to households and $0.50 million in repayments of principal and interest on past loans (7, 4). This left $5.61 million in savings (5, 4), which were committed to outlays on the capital account.

It is now natural to consider the size of the federal subsidy to the region. This subsidy was not the $36.70 million (3, 7) in grants mentioned above, plus the $10.20 million grant to the state government that made up part of (4, 7), because no allowances have been made for commodity taxes levied on regional production and consumption activities by the federal government. These, it will be recalled, do not appear explicitly in the tableau, because we were unable to accomplish the task of estimating them for the disaggregated matrix. But as we are now dealing with aggregates, there is a basis for arriving at an estimate of the net indirect taxes levied by the federal government on the Muda region. The estimate of such taxes made by Pyatt and Round (1978) for West Malaysia in 1970 is $1,549 million. The Muda region had about 7.1 percent of West Malaysia's population in 1972, although, as we have seen, the region's average income per capita was only about 55 percent of that for West Malaysia. Hence, if the incidence of indirect taxes was neutral with respect to outlays, the total revenue yielded by indirect taxation in the Muda region would have been about $60 million, which in turn implies a net tax *burden* of $13 million or so. This may be an overestimate, since the domestic price of paddy was well above its world price in that year (see table 7-1) and the paddy sector accounted for nearly 28 percent of regional value added (table 5-5). However, the incidence of indirect taxation on consumers is not usually highly progressive. Thus, on balance, it would be difficult to argue that the federal government was unduly generous toward this comparatively poor province.

The Combined Capital Account

Reading across row 5, we see that savings by households (5, 1), firms (5, 2), and the federal (5, 3) and state governments (5, 4) summed to $95.56 million. The outlays on this account were $43.23 million on domestically produced investment goods (6, 5), virtually all in the form of construction, and $17.15 million on imported investment goods (7, 5). To balance the account, $35.18 million flowed out of the region to uses elsewhere in the national economy (5, 7).

The Balance of Payments and the Flow of Funds

Commodity imports by households for final consumption (7, 1), by investment activities (7, 5), and by production activities (7, 6) were $207.86 million. Against this, commodity exports (6, 7) were $196.24 million, so that there was a deficit in commodity trade of $11.62 million.

The remaining outflows were remittances of profits by regional subsidiaries to their parent companies (7, 2) and debt service payments by the state government to "foreign" (out-of-region) bondholders (7, 4), which sum to $7.15 million. As for other incomings, households received $2.04 million in net remittances from abroad (1, 7) and subsidiaries of "foreign" firms received $0.99 million from their parents (2, 7). The really large "foreign" receipts, however, were grants and loans from the federal government to its local agencies (3, 7) and the state government (4, 7), which together amounted to $50.92 million. Thus, excluding commodity exports, total incomings were 50.92 + 2.04 + 0.99 = $53.95 million. Subtracting the commodity trade deficit and other net payments outside the region

(53.95 − 11.62 − 7.15), we obtain, as before, a net capital outflow from the region of $35.18 million (5, 7).

This outflow from the region was fully 8 percent of regional GDP, an outflow of investible funds which could have created much extra income and employment locally (see chapter 9). Although matters could have been worse, this was still an unwelcome state of affairs for a poor region, and it calls for a deeper probe into the acquisition of assets by the various institutions in the regional economy.

The state government made outlays of $5.62 million on its capital account (5, 4); its net borrowings from the federal government were $4.02 million, on the assumption that (7, 4) represents only interest payments. Hence, its net assets increased by $1.6 million. The local agencies of the federal government made capital outlays (5, 3) of $17.05 million. Thus the value of local investments made by households and firms was 60.38 − 5.62 − 17.05 = $37.71 million. However, some of the region's private firms are subsidiaries of parent companies based elsewhere, and although we have been able to estimate their net remittances of profits—(7, 2) minus (2, 7)—it was not possible to arrive at an estimate of their (gross) investments. Such evidence as we have suggests that the latter were very small, and we ignore them without blushing. Hence, the whole of retained corporate savings (5, 2) are attributable to local private firms, so that the aggregate private savings of the region's households and firms was $72.90 million. Setting this sum against their local investments indicates they placed $35.18 million of their savings in uses outside the region. To put matters slightly differently, only 48 percent of the region's private savings found its way into investments in the region. Moreover, this outflow was much larger than the inflow on capital account from the federal government in the form of grants ($17.05 million) and loans ($4.02 million).

This analysis of the flow of funds has been quite revealing. There is some circumstantial evidence that a large part of the outflow occurred through the banking system, with banks channeling deposits out of the region to satisfy demands for loans elsewhere in Malaysia, subject to an overall liquidity ratio. In addition, some of the affluent households in the region were almost certainly investing outside directly, probably in real estate. Thus an extremely large flow of investible resources, which could have contributed much to raising output and employment in the comparatively poor regional economy, was being allocated to various forms of investment (and consumption) in other, no doubt richer, regions of Malaysia. It is no small irony that the annual outflow of regional private savings in 1972 was about 15 percent of the total investment in the Muda project.

One must be wary, however, of trying to read too much into the figures for a single year. The outflow of private capital from the region may have been large before the advent of irrigation. Moreover, since 1972 was a year of transition, with the project rapidly approaching maturity, households and firms were unlikely to have fully adjusted their investment plans to their newly established, higher levels of income. In such circumstances the ultimate placement of any savings in excess of the regional demand for investment resources would then be in the hands of financial institutions whose horizons would extend beyond the boundaries of our region. We return to this issue in chapter 7, where we construct SAMs for 1967 and 1974.

A Social Accounts Matrix for 1972

The disaggregated set of accounts featuring five classes of household and thirty-five production sectors permits us to describe the sources and distribution of income among households together with the composition of their final demands. We can also outline the structures of production and interindustry transactions. The complete matrix is set out in table 5-4; but to simplify presentation, some of the key elements of the production activities expenditure accounts for all thirty-five sectors are set out in tables 5-5 and 5-7.

The Structure of Value Added and Its Distribution among Institutions

As one would expect of an economy such as Muda, the agricultural complex (cols. 10–20 of table 5-4) accounted for almost half the regional value added, and just over two-thirds of this subtotal resulted from paddy production and rice milling. Other significant groupings were distributive trades and petty trading (9.3 percent), residential and other construction (5.2 percent), the four government service sectors (18.9 percent), and two sectors providing real estate services (10.0 percent).

Some or all classes of farm households derived incomes directly from nonfarm activities: manufacturing not elsewhere classified, informal distribution (petty trading), residential construction, and rural housing services. The distribution of value added from paddy production is, however, the key element in the distribution of income among farm households. Note also (table 5-5) that 6.7 percent of value added in paddy production accrued to nonfarm households; that is, paddy cultivators paid \$7.75 million to absentee landlords and seasonal workers from other nonfarm households. Also noteworthy is the pattern and balance of incorporated and unincorporated enterprises across sectors, which is likely to change a good deal in the 1980s. The production of state corporations accounted for less than 1 percent of regional value added, and the gross profits of public

(*Text continues page 124.*)

Table 5-4. *Disaggregated Social Accounts Matrix for the Muda Region, 1972*
(millions of Malaysian dollars)

		Institutions				
		Households				
	Institutions and production activities	*Paddy farm workers* (1)	*Labor-abundant paddy farms* (2)	*Land-abundant paddy farms* (3)	*Non-project farms* (4)	*Nonfarm* (5)
	Institutions					
1.	Paddy farm workers			0.103		
2.	Labor-abundant paddy farms					
3.	Land-abundant paddy farms					
4.	Nonproject farms					
5.	Nonfarm		1.249	4.441	0.092	
6.	Firms					
7.	Federal government	0.025	1.070	4.591	1.315	22.765
8.	State government	0.002	0.127	0.555	0.100	14.466
9.	*Combined capital account*	0.010	3.132	22.736	7.681	29.090
	Production activities					
10.	Commercial rice mills					10.060
11.	Small rice mills	0.693	4.988	9.418	6.150	
12.	Food processing	0.010	0.052	0.128	0.188	0.115
13.	Fish processing	0.041	0.178	0.364	0.273	0.352
14.	Paddy production					
15.	Fishing	0.446	1.718	3.669	2.707	6.502
16.	Estates rubber					
17.	Smallholder rubber					
18.	Rubber processing					
19.	Other agriculture	0.492	2.170	4.575	3.244	5.687
20.	Sawmilling	0.009	0.057	0.179	0.147	
21.	Agricultural machinery services					
22.	Machinery repair					0.278
23.	Manufacturing n.e.c.	0.020	0.093	0.242	0.159	7.981
24.	Road transport	0.094	0.496	1.417	0.937	8.175
25.	Rail transport	0.002	0.014	0.044	0.028	0.314
26.	Hotels and restaurants	0.034	0.229	0.699	0.448	10.484
27.	Entertainment	0.006	0.043	0.131	0.084	4.415
28.	Services n.e.c.	0.019	0.128	0.398	0.258	0.376
29.	Private health	0.017	0.114	0.356	0.224	2.725
30.	Distributive trades	0.256	1.011	2.217	1.572	13.277
31.	Petty trading	0.099	0.398	0.988	0.793	0.985
32.	Electricity	0.010	0.045	0.088	0.067	1.462
33.	Water	0.009	0.057	0.174	0.111	1.136
34.	Posts, telecommunications	0.002	0.014	0.044	0.028	0.365
35.	Residential construction					
36.	Other construction					
37.	Government, irrigation					
38.	Government, education	1.125	4.755	8.895	6.212	20.268
39.	Government, health	0.361	1.543	2.936	2.199	3.520
40.	Government, other	1.221	5.241	9.770	7.370	12.742
41.	Traditional financial services	0.026	0.476	1.939	0.287	0.100
42.	Modern financial services					0.126
43.	Urban housing					13.786
44.	Rural housing	0.556	3.397	9.597	5.290	6.913
45.	*Rest of the world*	1.421	7.389	19.724	13.190	63.397
46.	*Total*	7.006	40.684	110.419	61.154	261.862

Institutions						
	Government			Production activities		
Firms (6)	Federal (7)	State (8)	Combined capital account (9)	Commercial rice mills (10)	Small rice mills (11)	Food processing (12)
	1.997	0.843				
	9.201	3.524				
	19.987	6.563				
	10.461	4.945				
5.429	28.398	8.647		2.954	5.619	0.926
				6.054		1.130
19.021						
10.250	17.050	5.610				
				68.054	51.395	
			−3.533			0.195
			0.616	0.059	0.021	0.048
				1.041	0.298	0.042
				5.007	2.801	0.194
					0.021	0.251
				0.014		0.031
				0.078	0.010	0.023
			16.228			
			29.907	0.131	0.016	0.005
					0.132	
				0.297	0.040	0.031
						0.040
6.650		0.501	17.161	4.009	1.221	4.115
41.351	87.093	30.633	60.379	87.698	61.606	7.255

(Table continues on the following pages.)

Table 5-4 *(continued)*

	Institutions and production activities	Production activities: Fish processing (13)	Paddy production (14)	Fishing (15)	Estates rubber (16)	Smallholder rubber (17)
	Institutions					
1.	Paddy farm workers		2.730	0.181	0.027	
2.	Labor-abundant paddy farms		20.705	0.233	0.141	
3.	Land-abundant paddy farms		67.128	0.507	0.078	
4.	Nonproject farms		17.059	0.382	2.212	12.891
5.	Nonfarm	0.464	7.749	16.472		
6.	Firms	0.054			0.118	
7.	Federal government					
8.	State government					
9.	*Combined capital account*					
	Production activities					
10.	Commercial rice mills					
11.	Small rice mills					
12.	Food processing			1.044		
13.	Fish processing					
14.	Paddy production		1.202			
15.	Fishing	4.432				
16.	Estates rubber				0.025	
17.	Smallholder rubber					0.112
18.	Rubber processing					
19.	Other agriculture		1.249			
20.	Sawmilling					
21.	Agricultural machinery services		2.571			
22.	Machinery repair		0.938	0.414	0.100	0.359
23.	Manufacturing n.e.c.		0.269	0.106		
24.	Road transport	0.023	0.227	0.044	0.146	0.075
25.	Rail transport		0.002			
26.	Hotels and restaurants					
27.	Entertainment					
28.	Services n.e.c.					
29.	Private health					
30.	Distributive trades	0.004	1.230	0.349	0.071	0.312
31.	Petty trading					
32.	Electricity				0.075	
33.	Water	0.045	0.001			
34.	Posts, telecommunications				0.010	
35.	Residential construction					
36.	Other construction		0.127		0.025	
37.	Government, irrigation		8.469			
38.	Government, education					
39.	Government, health					
40.	Government, other					
41.	Traditional financial services		0.458	0.018		0.049
42.	Modern financial services	0.001	0.212	0.017	0.057	
43.	Urban housing					
44.	Rural housing					
45.	*Rest of the world*	0.058	11.258	8.794	0.479	2.006
46.	*Total*	5.037	143.584	28.561	3.564	15.804

Production activities							
Rubber processing (18)	*Other agriculture (19)*	*Sawmilling (20)*	*Agricultural machinery services (21)*	*Machinery repair (22)*	*Manufacturing n.e.c. (23)*	*Road transport (24)*	*Rail transport (25)*
	0.193				0.160		
	1.726				0.315		
	4.789						
	6.088				0.563		
0.129	2.316	1.573	1.433	2.060	4.772	8.664	0.261
0.347		2.189		0.282	5.111	2.010	
							0.293
	0.390						
	0.025						
1.567							
	0.130				0.352		
		0.181		0.020	0.199		
	0.003	0.030	0.579		0.057	1.605	
0.010	0.124			0.134	0.648	0.899	
0.018	0.038	0.718		0.041	0.053	0.005	0.006
	0.009						
							0.006
							0.003
						0.016	
0.117	0.205	0.018	0.094	0.518	0.397	0.674	
0.014		0.068		0.075	0.221	0.021	0.004
0.001		0.009		0.038	0.016	0.021	0.009
0.003		0.029		0.033	0.002	0.110	0.007
0.011	0.174			0.003		0.016	0.173
	0.148						0.001
	0.150						
0.003	0.067	0.047	0.192	0.010	0.013	0.104	
		0.045			0.053		0.005
0.097	1.243	4.990	0.273	2.737	8.475	2.196	0.271
2.317	17.819	9.897	2.571	5.950	21.405	16.340	1.039

(Table continues on the following pages.)

Table 5-4 *(continued)*

	Production activities			
Institutions and production activities	*Motels and res-taurants (26)*	*Enter-tainment (27)*	*Services n.e.c. (28)*	*Private health (29)*
Institutions				
1. Paddy farm workers				
2. Labor-abundant paddy farms				
3. Land-abundant paddy farms				
4. Nonproject farms				
5. Nonfarm	4.296	0.270	0.604	2.227
6. Firms	0.249	3.934		
7. Federal government				
8. State government				
9. *Combined capital account*				
Production activities				
10. Commercial rice mills	0.723			
11. Small rice mills				
12. Food processing	2.007			0.014
13. Fish processing				
14. Paddy production				
15. Fishing	0.963			
16. Estates rubber				
17. Smallholder rubber				
18. Rubber processing				
19. Other agriculture	0.509			
20. Sawmilling				
21. Agricultural machinery services				
22. Machinery repair		0.025	0.001	
23. Manufacturing n.e.c.	0.213	0.030		0.009
24. Road transport	0.265	0.001	0.006	0.021
25. Rail transport				
26. Hotels and restaurants		0.009		0.054
27. Entertainment				
28. Services n.e.c.	0.002			0.001
29. Private health				0.032
30. Distributive trades	0.946	0.030	0.050	0.123
31. Petty trading				
32. Electricity	0.326	0.075	0.045	0.035
33. Water	0.050	0.025	0.023	0.015
34. Posts, telecommunications	0.031	0.050	0.019	0.034
35. Residential construction				
36. Other construction	0.130	0.049		0.025
37. Government, irrigation				
38. Government, education				
39. Government, health				
40. Government, other				
41. Traditional financial services	0.030			
42. Modern financial services	0.037	0.013	0.009	0.076
43. Urban housing	1.439		0.176	
44. Rural housing				
45. *Rest of the world*	2.130	0.168	0.295	0.820
46. *Total*	14.347	4.679	1.228	3.486

Production activities						
Dis-tributive trades (30)	*Petty trading (31)*	*Elec-tricity (32)*	*Water (33)*	*Posts and telecom-munica-tions (34)*	*Resi-dential con-struction (35)*	*Other con-struction (36)*
	0.061				0.104	
	0.758				0.486	
	0.759				0.636	
	0.495				0.465	
27.544	0.954	2.700	0.483	1.039	2.487	7.640
8.236					2.612	7.504
		0.354		0.862		
			1.165			
					2.263	1.567
	0.003		0.001		0.099	1.211
0.010			0.010	0.002	1.679	1.281
0.353		0.093	0.016	0.009	0.270	0.656
			0.006	0.007	0.001	0.003
				0.005		
0.006					1.487	0.835
0.698			0.014	0.038	0.011	0.058
0.012		0.009	0.057	0.006	0.009	0.042
0.059		0.016	0.007	0.172	0.004	0.031
0.050		0.005	0.098	0.064		
			0.001			
0.007			0.149		0.038	0.328
0.565	0.233					
0.233		0.049	0.023	0.009	0.094	0.150
1.871		0.001		0.011		
0.617						
0.117		1.625	0.426	0.062	5.595	15.083
40.378	3.263	4.853	2.455	2.285	18.340	36.389

(Table continues on the following pages.)

Table 5-4 *(continued)*

Institutions and production activities		Production activities: Government, Irrigation (37)	Education (38)	Health (39)	Other (40)
	Institutions				
1.	Paddy farm workers				
2.	Labor-abundant paddy farms				
3.	Land-abundant paddy farms				
4.	Nonproject farms				
5.	Nonfarm	3.656	38.064	8.767	27.922
6.	Firms				
7.	Federal government				
8.	State government				
9.	*Combined capital account*				
	Production activities				
10.	Commercial rice mills			0.096	
11.	Small rice mills				
12.	Food processing		0.137	0.140	0.006
13.	Fish processing				
14.	Paddy production				
15.	Fishing			0.042	
16.	Estates rubber				
17.	Smallholder rubber				
18.	Rubber processing				
19.	Other agriculture			0.002	
20.	Sawmilling		0.053	0.002	
21.	Agricultural machinery services				
22.	Machinery repair	0.062		0.036	0.138
23.	Manufacturing n.e.c.	0.118		0.029	0.075
24.	Road transport	0.010	0.096	0.120	0.189
25.	Rail transport	0.002	0.008	0.025	0.088
26.	Hotels and restaurants	0.005	0.891	0.092	1.358
27.	Entertainment				
28.	Services n.e.c.		0.012	0.030	
29.	Private health				
30.	Distributive trades	0.074	0.107	0.103	0.095
31.	Petty trading				
32.	Electricity	0.120	0.299	0.208	0.460
33.	Water	0.020	0.118	0.190	0.191
34.	Posts, telecommunications	0.125	0.159	0.061	0.631
35.	Residential construction	0.119			
36.	Other construction	3.300	0.581	0.198	1.245
37.	Government, irrigation				
38.	Government, education				
39.	Government, health	0.127	0.090		0.018
40.	Government, other	0.122	0.025	0.015	0.063
41.	Traditional financial services				
42.	Modern financial services		0.037	0.020	
43.	Urban housing		0.033	0.011	0.762
44.	Rural housing		0.033		
45.	*Rest of the world*	0.609	0.513	0.606	4.005
46.	*Total*	8.469	41.255	10.794	37.246

Production activities					
Traditional financial services (41)	*Modern financial services (42)*	*Urban housing (43)*	*Rural housing (44)*	*Rest of the world (45)*	*Total (46)*
			0.525	0.082	7.006
			3.253	0.343	40.684
			9.335	0.636	110.419
			5.113	0.480	61.154
4.338	1.231	16.659	6.834	0.500	261.862
	0.524			0.995	41.351
	0.100			36.697	87.093
				14.219	30.633
				−35.180	60.379
				76.629	87.698
				39.948	61.606
				3.413	7.255
				3.804	5.037
				22.933	143.584
				8.082	28.561
				3.539	3.564
				14.124	15.804
				2.317	2.317
				2.747	17.819
				5.220	9.897
					2.571
0.010					5.950
	0.011			6.509	21.405
0.010				0.331	16.340
				0.486	1.039
	0.034				14.347
					4.679
					1.228
	0.002				3.486
0.026	0.008			6.165	40.378
					3.263
	0.033	0.012			4.853
	0.017				2.455
0.010	0.116				2.285
		0.693	1.300		18.340
0.005	0.039	0.012			36.389
					8.469
					41.255
					10.794
		0.006			37.246
			0.044		4.507
0.033	0.001	0.324			2.327
	0.022				18.254
					26.403
0.074	0.190	0.548			215.020
4.507	2.327	18.254	26.403	215.020	

Table 5-5. *Structure and Distribution of Value Added*

			Percentage distribution of value added						
			Households						
Sector[a]	*Value added (millions of Malaysian dollars)*	*Percent of regional value added*	*Paddy farm workers*	*Labor-abundant paddy farms*	*Land-abundant paddy farms*	*Non-project farms*	*Nonfarm*	*Firms*	*State enter-prises*
10. Commercial rice mills	9.008	2.2	0	0	0	0	32.8	67.2	0
11. Small rice mills	5.619	1.4	0	0	0	0	100.0	0	0
12. Food processing	2.056	0.5	0	0	0	0	45.1	54.9	0
13. Fish processing	518	0.1	0	0	0	0	89.5	10.5	0
14. Paddy production	115.371	27.7	2.4	17.9	58.2	14.8	6.7	0	0
15. Fishing	17.775	4.3	1.0	1.3	2.9	2.2	92.6	0	0
16. Estates rubber	2.576	0.6	1.1	5.5	3.0	85.9	0	4.5	0
17. Smallholder rubber	12.891	3.1	0	0	0	100.0	0	0	0
18. Rubber processing	476	0.1	0	0	0	0	27.1	72.9	0
19. Other agriculture	15.112	3.6	1.3	11.4	31.7	40.3	15.3	0	0
20. Sawmilling	3.762	0.9	0	0	0	0	41.8	58.2	0
21. Agricultural machinery services	1.433	0.3	0	0	0	0	100.0	0	0
22. Machinery repair	2.342	0.6	0	0	0	0	87.9	12.1	0
23. Manufacturing n.e.c.	10.921	2.6	1.5	2.9	0	5.2	43.7	46.7	0
24. Road transport	10.673	2.6	0	0	0	0	81.2	18.8	0

25. Rail transport	554	0.1	0	0	0	0	47.2	0	52.8
26. Hotels and restaurants	4.546	1.1	0	0	0	0	94.5	5.5	0
27. Entertainment	4.204	1.0	0	0	0	0	6.4	93.6	0
28. Services n.e.c.	604	0.2	0	0	0	0	100.0	0	0
29. Private health	2.227	0.5	0	0	0	0	100.0	0	0
30. Distributive trades	35.780	8.6	0	0	0	0	77.0	23.0	0
31. Petty trading	3.027	0.7	2.0	25.0	25.1	16.4	31.5	0	0
32. Electricity	3.054	0.7	0	0	0	0	88.4	0	11.6
33. Water	1.648	0.4	0	0	0	0	29.3	0	70.7
34. Posts and telecommunications	1.901	0.5	0	0	0	0	54.7	0	45.3
35 Residential construction	6.789	1.6	1.5	7.2	9.4	6.8	36.6	38.5	0
36 Other construction	15.141	3.6	0	0	0	0	50.5	49.5	0
37 Government, irrigation	3.656	0.9	0	0	0	0	100.0	0	0
38 Government, education	38.064	9.2	0	0	0	0	100.0	0	0
39 Government, health	8.767	2.1	0	0	0	0	100.0	0	0
40. Government, other	27.922	6.7	0	0	0	0	100.0	0	0
41. Traditional financial services	4.338	1.0	0	0	0	0	100.0	0	0
42. Modern financial services	1.855	0.5	0	0	0	0	66.3	28.4	5.3
43. Urban housing	16.659	4.0	0	0	0	0	100.0	0	0
44. Rural housing	25.060	6.0	2.1	13.0	37.2	20.4	27.3	0	0
Total	416.329	100.0	1.0	6.6	20.0	10.8	51.2	9.7	0.7

a. The numbering of the sectors corresponds to the column numbers in table 5-4.

and private enterprises totaled 10.4 percent. Slightly more than half of regional value added accrued to nonfarm households in the form of wages, salaries, and earnings from unincorporated enterprises.

Household Incomes and Outlays

There is no need to go through the accounts in table 5-4 in detail, for this would merely repeat the discussion of the aggregate accounts above. But since the disaggregated accounts contain a few elements not found in table 5-3, it is worth illustrating them by taking the case of the (precarious) fortunes of landless paddy farm worker households, as revealed by the tableau in table 5-4.

Reading across row 1, we see that they received transfers of $0.1 million (1, 3) from land-abundant households, and $2.0 million (1, 7) and $0.84 million (1, 8) from the federal and state governments, respectively. The first of these is the zakat payments (religious taxes), which are channeled directly from relatively affluent to needy farm households without going through the official Zakat Board. In effect, these are informal, albeit minor, redistributive transfers within the village community. As already noted, the transfers from the federal and state governments are largely notional. The total value of such transfers was $2.84 million, while the "purchases" of associated services provided by government (37 to 40, 1) were $2.71 million, $0.02 million of which were actual cash outlays. Hence, the value of government transfers in cash was $0.15 million. Continuing across row 1 into the production accounts, we see that these households derived incomes from paddy production (1, 14), fishing (1, 15), rubber estates (1, 16), other agriculture (1, 19), manufacturing (1, 23), petty trading (1, 31), residential construction (1, 35), and housing services (1, 44). The amounts were $2.73, $0.18, $0.03, $0.19, $0.16, $0.06, $0.1, and $0.52 million, respectively. In addition, they received $0.08 million in net remittances from outside the region (1, 45), that is, from relatives in other parts of Malaysia.

Their outgoings are found by tracing down column 1. Although they made no repayments of debt to other households, they did pay taxes to the federal (7, 1) and state governments (8, 1). Of the remaining income, a tiny bit was saved (9, 1) and the rest spent on domestically produced goods (10–44, 1) and imports (45, 1).

It follows that pretax household incomes may be obtained by adding income from production (value added) to *net* receipts from all other households and firms (in the form of distributed profits). In the case of landless paddy worker households, aggregate income thus defined was $3.97 million from production activities, plus $0.18 million in transfers from land-abundant households and outside the region, or $4.15 million all told.

Disposable incomes may be obtained by deducting net direct taxes, which are given by the excess of "free" government services "purchased" over government transfers plus the tax payments in rows 7 and 8. Since not all "purchases" of government services are "free," the tableau cannot be made to yield disposable incomes defined as we have done here. Subsidiary calculations give the following estimates of net direct taxes for the five household classes, respectively: \$0.13, \$0.01, \$0.20, \$1.78, and \$36.72 million, from which the relevant calculations can be performed.

A summary picture of population and incomes by household class is set out in table 5-6. In all cases, the distribution of incomes within each class had a high variance and positive skewness. Almost all "landless" and a very high proportion of both labor-abundant paddy farm households and non-project farm households were probably in poverty on the criterion adopted by the Malaysian government at that time, namely, an annual income of less than \$315 per capita (Anand, forthcoming). Hence, although the Muda project had already done much to raise incomes by 1972, it had not solved the region's poverty problem. Moreover, even when the project had reached maturity in 1974, many of the region's households remained below this poverty line (see chapter 7).

As one would expect in the case of a relatively poor economy specializing in agricultural products, the direct import content of consumption increases with income; but the rise is modest, increasing from 0.33 for landless households to 0.40 for nonfarm households. The goods involved (principally durables) had high income elasticities of demand and were not produced locally on any scale. Indeed, the ratio of noncompetitive to competitive imports entering into household consumption exceeded seven to one.

In regard to the expenditure patterns of households, the average share of outlays devoted to nontradables (sectors 24–44, but excluding the four government services) for the five household classes were 26.5, 27.2, 32.1, 28.0, and 40.7 percent, respectively. These shares were sizable, they increased with income, and they were the main determinants of the project's downstream effects (see chapter 7).

The Uses of Output

We have already seen that households' consumption expenditures were the most important end use of regional gross output, with deliveries to satisfy intermediate and "foreign" demand not far behind. The aggregate figure for intermediate demand, however, gives a rather misleading indication of the strength of the region's interindustry connections, for deliveries of paddy to rice milling accounted for no less than 60 percent of all interindustry transactions. Three specific points are of interest: there was a

Table 5-6. *Socioeconomic Characteristics of Household Groups, 1972*

	Household class					
Item	*Landless paddy workers*	*Labor-abundant paddy farms*	*Land-abundant paddy farms*	*Nonproject farms*	*Nonfarm*	*All*
Number of households	4,367	18,045	33,426	25,333	44,047	125,239
Number of persons (thousands)	21.3	103.3	184.7	131.8	246.0	687.1
Average family size	4.88	5.72	5.53	5.20	5.58	5.49
Income per family ($)						
Pretax	950	1,548	2,506	1,806	4,973	3,040
Posttax	980	1,548	2,500	1,735	4,139	2,732
Income per head ($)						
Pretax	195	270	454	347	890	554
Posttax	201	270	452	334	741	498

sharp fall in livestock numbers in 1972, which is reflected in (19, 9); 16 percent of paddy output was being exported to mills outside the region (14, 45); and, as one would expect, the agricultural complex accounted for virtually all the region's commodity exports.

Gross Output and the Structure of Costs

By definition, the value of gross output equals payments to factors (value added) plus outlays on domestically produced and imported intermediate goods. From table 5-7 we deduce that interindustry linkages were generally rather feeble, in the sense that an expansion in the output of most sectors would not result in large increases in their outlays on the outputs of other sectors if the level of income, and hence domestic final demand, were kept constant (see chapter 7, the subsection "Production and Consumption Linkages"). The notable exceptions are rice milling, rubber processing, hotels and restaurants, irrigation, and, to a lesser extent, agricultural machinery services and residential construction. In some cases, outlays on imported intermediates, which were overwhelmingly of the noncompetitive variety, outweighed those on domestically produced intermediates.

Payments to factors and outlays on noncompetitive imported intermediate inputs dominated the cost structures of the region's industries. This carries an important message: the downstream effects of projects in rural regions will make themselves felt, if at all, through the links between production and household incomes (and thence final demand) rather than through the structure of interindustry deliveries. Hence, the analysis of the effects of such projects must treat household final demand (and perhaps private investment outlays) as endogenous. This issue will be taken up at length in subsequent chapters.

Conclusions

The existence of a large outflow of private capital casts some doubt on the proposition that the Muda region will be able to sustain rapid growth once a few, well-chosen public projects have come to fruition. There are several reasons for this skepticism. In chapter 7 we estimate that about one-third of the total increase in household incomes arising directly and indirectly from the Muda project at maturity accrued to nonfarm households, and such households are much more likely to make direct long-term investments outside the region than are farm households. More important, the scope for private on-farm investments is limited in this paddy monoculture. Hence, farm households may simply save less over the longer run, spending more lavishly on (imported) consumer durables, such as motorcycles

Table 5-7. *The Structure of Output and Costs*

Sector[a]	Gross output (millions of Malaysian dollars)	Proportion of gross output		
		Factor payments	Domestically produced intermediates	Imported intermediates
10. Commercial rice mills	87.698	0.103	0.851	0.046
11. Small rice mills	61.606	0.091	0.889	0.020
12. Food processing	7.255	0.283	0.149	0.567
13. Fish processing	5.037	0.103	0.885	0.012
14. Paddy production	143.584	0.804	0.118	0.078
15. Fishing	28.561	0.620	0.070	0.310
16. Estates rubber	3.564	0.723	0.143	0.134
17. Smallholder rubber	15.804	0.816	0.057	0.127
18. Rubber processing	2.317	0.206	0.752	0.042
19. Other agriculture	17.819	0.848	0.082	0.070
20. Sawmilling	9.897	0.380	0.116	0.504
21. Agricultural machinery services	2.571	0.558	0.336	0.106
22. Machinery repair	5.950	0.393	0.147	0.460
23. Manufacturing n.e.c.	21.405	0.510	0.094	0.396
24. Road transport	16.340	0.654	0.212	0.134

25. Rail transport	1.039	0.533	0.206	0.261
26. Hotels and restaurants	14.347	0.317	0.534	0.149
27. Entertainment	4.679	0.898	0.066	0.036
28. Services n.e.c.	1.228	0.492	0.268	0.240
29. Private health	3.486	0.639	0.126	0.235
30. Distributive trades	40.378	0.886	0.111	0.003
31. Petty trading	3.263	0.928	0.072	0.000
32. Electricity	4.853	0.629	0.036	0.335
33. Water	2.455	0.671	0.155	0.174
34. Posts and telecommunications	2.285	0.832	0.141	0.027
35. Residential construction	18.339	0.370	0.325	0.305
36. Other construction	36.388	0.416	0.169	0.415
37. Government, irrigation	8.468	0.432	0.496	0.072
38. Government, education	41.254	0.923	0.065	0.012
39. Government, health	10.794	0.812	0.132	0.056
40. Government, other	37.245	0.750	0.142	0.108
41. Traditional financial services	4.507	0.963	0.021	0.016
42. Modern financial services	2.327	0.797	0.121	0.082
43. Urban housing	18.254	0.913	0.057	0.030
44. Rural housing	26.403	0.949	0.051	0.000
Total	711.400	0.585	0.295	0.120

a. The numbering of sectors corresponds to the column numbers in table 5-4.

and televisions. Over and above on-farm investment, there are two alternatives to leaving investment decisions entirely to financial institutions, which pay very low or even negative real rates of interest. Either farmers must branch out on their own as entrepreneurs in nonfarm activities, or special institutions must be created to mobilize farm savings and invest them locally, thereby providing new jobs and, one hopes, a stream of dividends. The workings of the capital market probably hinder the former, and there is little sign so far of special institutions. Farmers' Associations are, however, beginning to diversify their activities, and the Majlis Amanah Ra'ayat (MARA), a specifically Malay institution designed to mobilize savings and promote Malay entrepreneurs, has steadily expanded its operations in the region since the early 1970s (FAO, 1975). These considerations suggest that the outflow of private capital from the region will not be readily staunched. As long as rates of return outside the region remain higher than those obtainable from regional investment activities—for whatever reason—and the distribution of incomes generated by public projects favors those with a high propensity to save, then most of the people of the Muda region will benefit comparatively little from the induced accumulation effects of public projects. Of course, the notion that there are large capital (or "surplus") outflows from agriculture and from poor regions will not startle anyone acquainted with economic history or the ideas of stucturalists, such as Kaldor and Myrdal. The magnitude of the outflow in the Muda case (more than 8 percent of GDP) is sobering, however, to say nothing of the difficulties of stemming it.

The quantitative structure of the SAM suggests that the region's income-expenditure linkages may play a more important role in determining the magnitude of a project's downstream effects than the commonly stressed interindustry linkages. This theme recurs constantly in the formulation of, and the results from, the models in chapters 6 and 8. A project's secondary, or downstream, effects must be estimated within a closed-loop system in which household incomes and expenditures are determined endogenously along with production. If such effects are significant and if they spring largely from households' expenditures,[8] it is obviously more important to investigate how people spend their incomes than the niceties of interindustry relations. To put it within the setting of cost-benefit analysis, short-cut methods for estimating downstream effects are likely to involve household consumption parameters rather than interindustry coefficients.

This latter conclusion provides considerable justification for the procedures used to define the region (outlined in chapter 2) and provides strong

8. In a closed model, production, incomes, and demand are inseparable; but by leaving the system "open" in certain ways, we are able to get some insight into which classes of loops are the more powerful (see chapter 7).

support for the decisions of the FAO and the World Bank to include the detailed measurement of household incomes and expenditures in the early surveys.

It must be stressed, however, that building a SAM is not a substitute for grasping how the economy in question works, simply because a set of accounts does not constitute a model. Certainly, devising the structure of the accounts and its categories, as well as estimating the accounts in a consistent way, did advance our understanding of the workings of the Muda regional economy. But although the analytical frameworks laid out in the succeeding chapters are based on the SAM, they stand or fall by the plausibility of the assumptions made about the technology available to the economy, the nature of the markets for goods, and the behavior of its institutions in earning incomes and making outlays.

6

A Model of the Regional Economy

THIS CHAPTER and the one that follows analyze the sources of growth in the regional economy between 1967 and 1974. As in chapters 3 and 4, our first need is a model tailored closely to the task at hand. Then, building on the results from TIGER, we use the model to decompose the set of changes in that period into its constituents: the effects of the project at maturity, the effects of changes in the prices which are set exogenously to the region, the effects of other autonomous forces, and all interactions among them. In so doing, this analysis completes our attempt to quantify the effects of the various sources of change on the whole of the regional economy. As such, these two chapters bring us to the central questions addressed in this book.

Our regional model is a variant of Tinbergen's (1966) semi-input-output method,[1] with the SAM for 1972 as its empirical base. This approach enables us to combine the structural detail of the SAM with an analytical formulation which is both simple and computationally tractable. Like TIGER, this model permits nothing more than comparative statics—though we offer no particular apologies for that. Once more it must be emphasized that when we use the model to construct "pictures" of the economy in 1967 and 1974 from the evidence available to us, these are "pictures" not of the economy as it was, but of the steady-state equilibria corresponding to the exogenous conditions characterizing those years. The same interpretation applies, of course, to the model's simulation of various hypothetical circumstances.

The model is developed in this chapter, and its application to an analysis of the changes that occurred in the regional economy between 1967 and 1974 follows in chapter 7.

To simplify the presentation, the discussion is based on a small, illustrative SAM (table 6-1) rather than the full version in chapter 5. This smaller

1. Kuyvenhoven (1978) provides a useful and comprehensive account of the semi-input-output method, which Tinbergen developed in the early 1960s. As the nomenclature suggests, this method is a particular case of Leontief's input-output scheme, its salient feature being that changes in exogenous demand affect only the output of nontraded goods, the output of traded goods being given exogenously. The relation between input-output and social accounts was, of course, evident from the outset in Leontief's work.

SAM has a structure sufficiently close to that of the Muda SAM that it generalizes to the latter in an obvious way. Since the model and the SAM are intimately related, it is useful to begin with a more formal statement of the accounting identities than was offered in chapter 5. For pedagogic reasons, we begin by formulating a fixed-price variant of the model. This enables us to establish the general structure and method of the semi-input-output approach without being unduly burdened by the complexities of endogenous prices. In addition to simplifying the presentation, the fixed-price model is also of methodological interest in its own right, for the decomposition analysis requires simulations of the economy on the premise that prices stayed at their levels in the base year. Subsequently, we develop the full model, in which some prices are endogenously determined.

The SAM and Its Accounting Identities

The SAM in table 6-1 lays out a tableau for a simple regional economy. There are two types of households, the government, three production activities, a combined capital account, and the rest of the world. Each of the three production activities produces a single characteristic commodity: a tradable good (T), a nontradable good (N), or a distributive service (D), which is of course a nontraded good, too. These three categories are a natural and useful way of classifying the detailed sectoral structure of the SAM for Muda when we turn to empirical analysis.

Two important features of this structure call for some remarks. First, since there are no firms, production activities pay value added directly to households, which make all decisions concerning private savings. Although the region's firms do retain some of their posttax profits (chapter 5), as long as their investment outlays are independent of retained profits, the analysis is not affected in a qualitative way.

Second, the government is treated as a federal institution in that its outlays are independent of its revenues. As state government revenues, and hence outlays, are largely independent of the level of activity in the region, this aggregation is permissible; any excess of revenues over expenditures simply winds up in the coffers of the treasury in the national capital. Expenditures take the form of investments and the costs of producing goods and services, which are sold like any other. Profits or losses can be made in these activities, of course, but the considerations are no different from those arising when private firms retain part of their profits. We will return to these matters as they arise in the sections that follow. It must also be stressed that the simplifications discussed here are not made in the full models on which the analysis in the next chapter rests. Meanwhile, the exposition is appreciably simplified.

Table 6-1. *The Schematic SAM*

	Expenditures								
	Households			*Production activities*					
Receipts	*1*	2	*Government*	*T*	*N*	*D*	*Capital account*	*Rest of world*	*Row sums*
Households									
1				V_{1T}	V_{1N}	V_{1D}		R_1	Y_1
2				V_{2T}	V_{2N}	V_{2D}		R_2	Y_2
Government	T_1	T_2							G
Production activities									
T	C_{T1}	C_{T2}		Z_{TT}	Z_{TN}	Z_{TD}	J_T	E_T	X_T
N	C_{N1}	C_{N2}		Z_{NT}	Z_{NN}	Z_{ND}	J_N	E_N	X_N
D	C_{D1}	C_{D2}		Z_{DT}	Z_{DN}	Z_{DD}	J_D	E_D	X_D
Capital account	S_1	S_2	G					$-F$	$\Sigma_i J_i$
Rest of world	M_1	M_2		M_T	M_N	M_D			M
Column sums	Y_1	Y_2	G	X_T	X_N	X_D	$\Sigma_i J_i$	M	

Note: T = tradable good, N = nontradable good; D = distributive service.

The notation is as follows, where the subscript $k = 1,2$ denotes the class of household and $i,j = T,N,D$ denote the sectors:

Y_k = total income of households of type k
X_j = value of gross output of sector j
V_{kj} = payments to households of type k from value added of sector j
R_k = income transfers from "abroad" to households of type k
C_{ik} = expenditure by households of type k on output of sector i
Z_{ij} = value of purchases of intermediates by sector j from sector i
J_i = value of investment demand for output of sector i
E_i = value of exports of output of sector i
S_k = savings by households of type k
T_k = tax payments by households of type k
G = total tax revenues
M_k = expenditure on imported goods by households of type k
M_j = value of imported intermediates purchased by sector j
F = capital outflow from region
M = total value of regional imports

Since a SAM is a double-entry accounting system, we can utilize either the row or column accounts. The rows provide the statement of receipts for each account, whereas the columns provide the statement of expenditures. For example, the row for household type 1 in table 6-1 reads:

$$\underset{[\text{Income}]}{Y_1} = \underset{[\text{Value added received}]}{V_{1T} + V_{1N} + V_{1D}} + \underset{[\text{Income from abroad}]}{R_1}$$

whereas the corresponding column is:

$$\underset{\begin{bmatrix}\text{Total}\\\text{expend-}\\\text{iture}\end{bmatrix}}{Y_1} = \underset{\begin{bmatrix}\text{Expenditure}\\\text{on domestic}\\\text{goods}\end{bmatrix}}{C_{T1} + C_{N1} + C_{D1}} + \underset{[\text{Savings}]}{S_1} + \underset{[\text{Taxes}]}{T_1} + \underset{\begin{bmatrix}\text{Expenditure}\\\text{on imported}\\\text{goods}\end{bmatrix}}{M_1}.$$

By definition, total outlays must equal total receipts for each and every account, so that the row and column sums are identical in each and every case. Since we shall be building models which are defined in the primal variables X and Y, it will be more useful to work with the row accounts. These are stated below.

Institutional accounts

Households

$$\underset{[\text{Income}]}{Y_k} = \underset{[\text{Value added receipts}]}{\sum_j V_{kj}} + \underset{[\text{Income from abroad}]}{R_k} \qquad (6.1)$$

Government

$$(6.2) \qquad G = \sum_k T_k$$

[Revenue] = [Total tax collections]

Production accounts

$$(6.3) \qquad X_i = \sum_k C_{ik} + \sum_j Z_{ij} + J_i + E_i$$

$$\begin{bmatrix}\text{Gross}\\ \text{output}\end{bmatrix} = \begin{bmatrix}\text{House-}\\ \text{hold}\\ \text{demand}\end{bmatrix} + \begin{bmatrix}\text{Inter-}\\ \text{mediate}\\ \text{demands}\end{bmatrix} + \begin{bmatrix}\text{Invest-}\\ \text{ment}\\ \text{demands}\end{bmatrix} + \begin{bmatrix}\text{Export}\\ \text{demands}\end{bmatrix}$$

Capital account

$$(6.4) \qquad F = \sum_k S_k + G - \sum_i J_i$$

$$[\text{Capital outflow}] = \begin{bmatrix}\text{Household}\\ \text{savings}\end{bmatrix} + \begin{bmatrix}\text{Tax}\\ \text{collections}\end{bmatrix} - \begin{bmatrix}\text{Total}\\ \text{investment}\end{bmatrix}$$

Rest of world account

$$(6.5) \qquad M = \sum_k M_k + \sum_j M_j$$

$$[\text{Total imports}] = \begin{bmatrix}\text{Imports by}\\ \text{households}\end{bmatrix} + \begin{bmatrix}\text{Imports by}\\ \text{sectors}\end{bmatrix}$$

In all, there are 8 identities and 46 variables in this schematic system. Hence, if 38 variables are given preassigned values, the remaining 8 can be uniquely determined from the system (6.1) to (6.5). This property can be very useful in the empirical estimation and balancing of a SAM, since the data on some variables, such as household savings, may be unreliable or very difficult to obtain. Provided a sufficient number of variables can be estimated with some confidence, these data can be used with the SAM equations to provide a consistency check or, if necessary, indirect estimates of any missing or dubiously valued variables. However, because the number of variables that need to be predetermined is large in relation to the number of equations, the SAM identities have limited value as a device for estimating variables for years other than the one on which the SAM is based.

Since we need a more versatile and powerful estimating framework than this, we must have recourse to assumptions about how certain variables are determined in the regional economy. In the parlance of econometrics, these assumptions might be regarded as providing identifying restrictions on the system. To an economist, these assumptions convert the SAM from a set of accounting identities into an economic model that purports to characterize the workings of the economy.

An Extended Input-Output Model

As in a Leontief system, we assume that all the structural relations in the economy (both behavioral and technological) are linear, or at least that they can be usefully approximated by linear functions. Our specific assumptions are stated below.

The first two assumptions impose proportional relationships between inputs and outputs in all sectors.

$$Z_{ij} = a_{ij} X_j, \text{ all } i,j \tag{A.1}$$

$$M_j = a_{mj} X_j, \text{ all } j \tag{A.2}$$

where a_{ij} and a_{mj} are, respectively, the values of domestic and imported intermediates required for the production of one unit of value of gross output in sector j.

Households are assumed to receive constant shares of gross output from each sector:

$$V_{kj} = \omega_{kj} X_j, \text{ all } k,j \tag{A.3}$$

where ω_{kj} is the share of gross output from sector j received by households of type k. If constant returns to scale are assumed, value added is proportional to gross output:

$$V_j = X_j - \sum_i a_{ij}X_j - a_{mj}X_j = (1 - \sum_i a_{ij} - a_{mj})X_j.$$

Hence, $\Sigma_k \omega_{kj} = (1 - \Sigma_i a_{ij} - a_{mj})$,[2] so that households receive constant shares of value added, as well as constant shares of gross output.

At this juncture, it is natural to ask what conditions are sufficient to ensure the constancy of the technological parameters from one equilibrium to another. Since all the variables have been defined in nominal terms, we know that if prices should change, the constancy of cost-shares can hold if and only if the technology for producing each and every good is characterized by both constant returns to scale and unitary elasticities of substitution between all intermediate inputs and primary factors. Under these conditions, the parameters a_{ij}, a_{mj}, and ω_{kj} are independent of prices, and any model based on them will solve for all variables in nominal terms. Thus, nothing can be said about changes in quantities unless the supply functions for factors are specified.

2. If firms or the government retain a share ρ_j of the gross output produced in sector j as profits, we have

$$\sum_k \omega_{kj} + \rho_j = 1 - \sum_i a_{ij} - a_{mj}.$$

Alternatively, and this is the case that concerns us for the moment, if relative prices do not change from one equilibrium to another, any set of linearly homogeneous production functions will do. The assumption of constant returns to scale with no joint production implies that the prices of goods are independent of the scale of output, provided primary factors (including noncompetitive imports) are available to the economy in perfectly elastic supplies. Since we are interested in comparing various steady-state equilibriums in a regional economy, in which all adjustments to changes in exogenous variables have taken place, the assumption of perfectly elastic supplies of primary factors does not seem especially strong. Capital (in the form of finance) and workers are mobile, and the rates of interest and wages will not vary with changes in the level of the region's economic activity if the region is sufficiently small. This being so, the interpretation of the model's results is clear. By definition, all changes in variables, whatever their source, are real ones, so it is convenient to choose physical units of measure such that all prices are unity. We are now in the familiar territory of fixed-price, linear economic models.

Tax collections from households are assumed to be linear functions of income:

$$T_k = \overline{T}_k + t_k Y_k, \text{ all } k \tag{A.4}$$

where $\overline{T}_k$ is a lump sum tax, and t_k is the marginal tax rate levied on income for households of type k.

Households are assumed to save a fixed proportion of after-tax income:

$$S_k = s_k(Y_k - T_k), \text{ all } k \tag{A.5}$$

where s_k is the marginal propensity to save for households of type k.

Further, we assume that households' preferences for goods may be expressed by the linear expenditure system, so that:

$$\begin{aligned} C_{ik} &= \alpha_{ik} + \beta_{ik}(Y_k - T_k - S_k - \sum_i \alpha_{ik} - \alpha_{mk}) \\ &= \gamma_{ik} + \beta_{ik}(Y_k - T_k - S_k) \end{aligned} \tag{A.6}$$

where $\gamma_{ik} = \alpha_{ik} - \beta_{ik} \sum_i \alpha_{ik} - \beta_{ik}\, \alpha_{mk}$, and

$$M_k = \gamma_{mk} + \beta_{mk}(Y_k - T_k - S_k) \tag{A.7}$$

where α_{ik} and α_{mk} are the values of "subsistence" consumption of commodity i and imports by households of type k, and β_{ik} and β_{mk} are the corresponding marginal expenditure coefficients for domestic and imported goods, respectively. Since all household income must be either saved or consumed, we have the restriction that $\Sigma_i\, \beta_{ik} + \beta_{mk} = 1$, for all k. Hence, (A.7) can be written as:

$$M_k = \gamma_{mk} + (1 - \sum_i \beta_{ik})(Y_k - T_k - S_k), \text{ all } k.$$

Income received by households from abroad is also assumed to be exogenously given, so that

$$R_k = \overline{R}_k, \text{ all } k. \quad (A.8)$$

Using these assumptions, we obtain:

$$Y_k - T_k = (1 - t_k)Y_k - \overline{T}_k, \text{ all } k,$$

and
$$Y_k - T_k - S_k = Y_k(1 - t_k)(1 - s_k) - (1 - s_k)\overline{T}_k, \text{ all } k.$$

By substituting these results and assumptions (A.1) to (A.8) into the SAM identities (6.1) to (6.5), we get the following linear economic model:

$$Y_k = \sum_j \omega_{kj} X_j + \overline{R}_k, \text{ all } k \quad (6.6)$$

$$G = \sum_k (\overline{T}_k + t_k Y_k) \quad (6.7)$$

$$X_i = \sum_k [\gamma_{ik} + \beta_{ik}(1 - t_k)(1 - s_k)Y_k - \beta_{ik}(1 - s_k)\overline{T}_k] + \sum_j a_{ij} X_j + J_i + E_i, \text{ all } i \quad (6.8)$$

$$F = \sum_k s_k[(1 - t_k)Y_k - \overline{T}_k] + G - \sum_i J_i \quad (6.9)$$

$$M = \sum_k \gamma_{mk} + \sum_k (1 - \sum_i \beta_{ik})[(1 - t_k)(1 - s_k)Y_k - (1 - s_k)\overline{T}_k] + \sum_j a_{mj} X_j. \quad (6.10)$$

In the illustrative example, this system has eight equations; but unlike the accounting identities (6.1) to (6.5), there are now only fourteen variables: X_i, Y_k, F, G, M, J_i, and E_i. There are, of course, forty parameters (ω_{kj}, γ_{ik}, γ_{mk}, β_{ik}, t_k, s_k, a_{ij}, a_{mj}, $\overline{R}_k$ and $\overline{T}_k$), and these have to be estimated; but once obtained, their values are invariant with respect to the values of the variables in the system. The strength of the model is therefore apparent. By fixing six variables (say, J_i and E_i, all i), the model can be solved to provide the values of the remaining eight variables in the system, and thence, through assumptions (A.1) to (A.8), to provide estimates of all the original forty-six variables in the illustrative SAM. Put another way, once the parameters have been estimated, the model can be used to estimate a full SAM corresponding to the state of the economy for any given values of six of the variables appearing in equations (6.6) to (6.10). We shall make frequent use of this property.

For subsequent purposes, it will be more convenient to transform the model into matrix form. We define the following matrices and vectors of model parameters:

A = the 3×3 matrix $\{a_{ij}\}$

B = the 3×2 matrix $\{\beta_{ik}\}$

Ω = the 2×3 matrix $\{\omega_{kj}\}$

Γ = the 3×2 matrix $\{\gamma_{ik}\}$

Γ_m = the 2×1 vector $\{\gamma_{mk}\}$

s = the 2×2 diagonal matrix with diagonal elements $\{s_k\}$

t = the 2×2 diagonal matrix with diagonal elements $\{t_k\}$

b = the 2×2 diagonal matrix with diagonal elements $\{\sum_i \beta_{ik}\}$

$\bar{T}$ = the 2×1 vector $\{\bar{T}_k\}$

$\bar{R}$ = the 2×1 vector $\{\bar{R}_k\}$

A_m = the 3×1 vector $\{a_{mj}\}$

With X, Y, T, J, and E denoting the vectors of corresponding variables, the equations of the model can be expressed as:

$$(6.6)' \qquad Y = \Omega X + \bar{R}$$

$$(6.7)' \qquad T = \bar{T} + tY \text{ and } G = u'T$$

$$(6.8)' \qquad X = AX + \Gamma u + BWY - B(I - s)\bar{T} + J + E$$

$$(6.9)' \qquad F = u's[(I - t)Y - \bar{T}] + G - u'J$$

$$(6.10)' \qquad M = u'\Gamma_m + u'(I - b)[WY - (I - s)\bar{T}] + A_m'X$$

where $W = (I - s)(I - t)$ is a 2×2 matrix, and u denotes a vector of 1's of dimension appropriate to its context.

Now G, F, and M are dependent on X, Y, J, and E, but not conversely. Consequently, the model can be partitioned into two. The first part, equations (6.6)′ and (6.8)′, can be collected and rearranged as:

$$(6.11) \qquad \begin{bmatrix} I - A & -BW \\ -\Omega & I \end{bmatrix} \begin{bmatrix} X \\ Y \end{bmatrix} = \begin{bmatrix} \Gamma u - B(I - s)\bar{T} + J + E \\ \bar{R} \end{bmatrix}.$$

In our illustrative example, (6.11) comprises five equations, the solution to which provides the values of any five elements of X, Y, J, or E, given predetermined values of the remaining six elements. In a standard Leontief system, of which (6.11) is a closed-loop variant, J and E would be given exogenously, and (6.11) would then be solved for the corresponding values of X and Y. Given all the predetermined and derived values of the elements of X, Y, J, and E from (6.11), it remains to substitute these into equations (6.7)′, (6.9)′, and (6.10)′ to obtain the values of G, F, and M. Further, using the relations (A.1) to (A.8), a full SAM can in fact be constructed as a final step in the analysis.

To illustrate these ideas, we will now work through a numerical example based on the SAM in table 6-2. Suppose the matrices of the relevant parameters are as follows:[3]

$$A = \begin{bmatrix} 0.38303 & 0.05713 & 0.36926 \\ 0.19478 & 0.07923 & 0.25000 \\ 0.06819 & 0.38845 & 0.08335 \end{bmatrix}$$

$$B = \begin{bmatrix} 0.19565 & 0.12402 \\ 0.56900 & 0.39665 \\ 0.12449 & 0.22310 \end{bmatrix}, \quad \Omega = \begin{bmatrix} 0.25 & 0.1429 & 0.0417 \\ 0 & 0.2857 & 0.1667 \end{bmatrix}$$

$$\Gamma = \begin{bmatrix} -1 & 0 \\ 3 & -1 \\ -2 & 1 \end{bmatrix}, \quad \Gamma_m = \begin{bmatrix} 0 \\ 0 \end{bmatrix}, \quad s = \begin{bmatrix} 0.08202 & 0 \\ 0 & 0.33143 \end{bmatrix}$$

$$t = \begin{bmatrix} 0.10496 & 0 \\ 0 & 0.15836 \end{bmatrix}, \quad \overline{T} = \begin{bmatrix} 20 \\ 19 \end{bmatrix}, \quad \overline{R} = \begin{bmatrix} 10 \\ 20 \end{bmatrix}$$

$$A'_m = [0.104 \quad 0.04658 \quad 0.08901].$$

Note that A, A_m, Ω, and s are obtained directly from the SAM. In particular, $a_{ij} = Z_{ij}/X_j$, $a_{mj} = M_j/X_j$, $\omega_{kj} = V_{kj}/X_j$, and $s_k = S_k/(Y_k - T_k)$. All the other parameter values must be obtained from additional data on expenditure patterns and the structure of taxation. (Here, a valuable source would be income and expenditure data for a sample covering different types of households.)

Using these parameter values, (6.11) specializes to:

$$\text{(6.12)} \quad \begin{bmatrix} 0.61697 & -0.05713 & -0.36926 & -0.16075 & -0.06979 \\ -0.19478 & 0.92077 & -0.25000 & -0.46750 & -0.22319 \\ -0.06819 & -0.38845 & 0.91665 & -0.10228 & -0.12554 \\ -0.25000 & -0.14290 & -0.04170 & 1 & 0 \\ 0 & -0.28570 & -0.16670 & 0 & 1 \end{bmatrix} \begin{bmatrix} X_T \\ X_N \\ X_D \\ Y_1 \\ Y_2 \end{bmatrix}$$

$$= \begin{bmatrix} -6.16746 \\ -13.48519 \\ -3.83404 \\ 10 \\ 20 \end{bmatrix} + \begin{bmatrix} J_T + E_T \\ J_N + E_N \\ J_D + E_D \\ 0 \\ 0 \end{bmatrix}$$

3. Our apparently extravagant use of five decimal places is designed to banish rounding errors in the numerous calculations that follow.

Table 6-2. *An Illustrative SAM*

	Expenditures								
	Households			*Production activities*					
Receipts	*1*	2	*Government*	*T*	*N*	*D*	*Capital account*	*Rest of world*	*Row sums*
Households									
1				76.92	37.12	8.10		10.00	132.14
2					74.21	32.40		20.00	126.61
Government	33.87	39.05							72.92
Production activities									
T	16.65	7.26		117.85	14.84	71.77		79.31	307.68
N	54.33	22.22		59.93	20.58	48.59	54.10		259.75
D	9.23	14.06		20.98	100.90	16.20	25.06	7.93	194.36
Capital account	8.06	29.02	72.92					−30.84	79.16
Rest of world	10.00	15.00		32.00	12.10	17.30			86.40
Column sums	132.14	126.61	72.92	307.68	259.75	194.36	79.16	86.40	

If the values of J and E are fixed at their values in table 6-2, the right-hand side of (6.12) becomes:

$$\begin{bmatrix} -6.16746 \\ -13.48519 \\ -6.11964 \\ 10 \\ 20 \end{bmatrix} + \begin{bmatrix} 79.31 \\ 54.10 \\ 32.99 \\ 0 \\ 0 \end{bmatrix} = \begin{bmatrix} 73.14254 \\ 40.61481 \\ 26.87036 \\ 10 \\ 20 \end{bmatrix},$$

and the solution to the system is:

$$\begin{bmatrix} X_T \\ X_N \\ X_D \\ Y_1 \\ Y_2 \end{bmatrix} = \begin{bmatrix} 2.634973 & 1.199126 & 1.558211 & 1.143517 & 0.647139 \\ 1.417730 & 2.209670 & 1.376639 & 1.401698 & 0.764946 \\ 0.975193 & 1.214108 & 1.963469 & 0.925168 & 0.585528 \\ 0.902002 & 0.666172 & 0.668151 & 1.524761 & 0.295512 \\ 0.567610 & 0.833695 & 0.720616 & 0.554691 & 1.316153 \end{bmatrix} \begin{bmatrix} 73.14254 \\ 40.61481 \\ 26.87036 \\ 10 \\ 20 \end{bmatrix}$$

$$= \begin{bmatrix} 307.68 \\ 259.75 \\ 194.36 \\ 132.14 \\ 126.61 \end{bmatrix},$$

which are exactly the values of X and Y to be found in table 6-2.

Then, using equations (6.7)′, (6.9)′, and (6.10)′, the corresponding values of T, F, and M are as follows:

$$T = \begin{bmatrix} 20 \\ 19 \end{bmatrix} + \begin{bmatrix} 0.10496 & 0 \\ 0 & 0.15836 \end{bmatrix} \begin{bmatrix} 132.14 \\ 126.61 \end{bmatrix} = \begin{bmatrix} 33.87 \\ 39.05 \end{bmatrix},$$

from which

$$G = [1 \quad 1] \begin{bmatrix} 33.87 \\ 39.05 \end{bmatrix} = 72.92$$

$$F = [1 \quad 1] \begin{bmatrix} 0.08202 & 0 \\ 0 & 0.33143 \end{bmatrix} \left\{ \begin{bmatrix} 0.89504 & 0 \\ 0 & 0.84164 \end{bmatrix} \begin{bmatrix} 132.14 \\ 126.61 \end{bmatrix} - \begin{bmatrix} 20 \\ 19 \end{bmatrix} \right\}$$

$$+\ 72.92 - [1 \quad 1 \quad 1] \begin{bmatrix} 0 \\ 54.1 \\ 25.06 \end{bmatrix} = 30.84$$

$$M = [1 \quad 1]\begin{bmatrix} 0.11086 & 0 \\ 0 & 0.25623 \end{bmatrix}$$

$$\times \left\{ \begin{bmatrix} 0.82162 & 0 \\ 0 & 0.5627 \end{bmatrix} \begin{bmatrix} 132.14 \\ 126.61 \end{bmatrix} - \begin{bmatrix} 0.91798 & 0 \\ 0 & 0.66857 \end{bmatrix} \begin{bmatrix} 20 \\ 19 \end{bmatrix} \right\}$$

$$+ [0.104 \ 0.04658 \ 0.08901] \begin{bmatrix} 307.68 \\ 259.75 \\ 194.36 \end{bmatrix} = 86.40.$$

The remaining elements of the SAM in table 6-2 can then be calculated using assumptions (A.1) to (A.8). For example, by (A.1),

$$Z_{TT} = a_{TT}X_T = (0.38303)(307.68) = 117.85.$$

In this example, the values of J and E are set equal to those in table 6-2. Consequently, all the calculations lead to values for the remaining variables which are also the same as those in table 6-2. It should be clear, however, that any values of J and E could have been chosen, and the same kinds of calculations would have led to a SAM depicting the new equilibrium of the economy consistent with those predetermined values.

A Semi-Input-Output Model

Equation (6.11) is written in the standard closed Leontief form, which solves for all gross outputs (X) and household incomes (Y) given the levels of exogenous demands—in this case, deliveries to investment activities (J) and exports (E). However, any six elements of the vectors X, Y, J, or E could be chosen as the predetermined variables, for this would leave five endogenous variables, exactly one for each equation. More generally, in the case of n sectors and h household types, (6.11) would contain $n + h$ equations, and $3n + h$ variables. Thus, $2n$ variables must be predetermined. Making J and E exogenous, as we did in the above example, is but one of many possible arrangements.[4]

The choice of elements in X, Y, J, and E that are to be predetermined will depend on the purpose for which the model is to be used. If the aim is to create a comprehensive set of social accounts for a year in the past for which only fragmentary information exists, then the elements of X, Y, J, and E chosen should obviously be those for which the best independent estimates can be made.

4. With Y endogenous, there are $3n$ configurations of exogenous variables.

If, however, a set of social accounts is to be constructed for some hypothetical situation (for example, the Muda economy in 1974 as it would have been in the absence of the project), the choice of predetermined variables is not a matter of convenience or reliability in estimation, but rather of one's view as to how the economy works. For example, choosing an element of X as a predetermined variable is equivalent to assuming that supply from the sector in question is perfectly inelastic. This would come about if the sector in question faced perfectly elastic supplies of variable inputs and perfectly elastic demand for its output, for it would then produce at the full capacity permitted by its fixed factors. Since we are dealing with a small, open economy, this would be an appropriate way of treating the sectors producing traded goods, provided the production of nontraded goods expands (or contracts) at constant average costs, as would be the case under the conditions discussed above. Conversely, fixing an element of E is equivalent to assuming that "foreign" demand for that commodity is perfectly inelastic, which is the appropriate formulation in the case of nontraded goods. Thus, following the established tradition of semi-input-output, the output of traded goods is parametrically given and exports are endogenously determined, whereas the output of nontraded goods is endogenously determined and exports are parametrically given (at zero).[5]

We depart from the standard formulation of semi-input-output models in that, in our analysis, the demand for investment goods (J) is always assumed to be exogenously given. In this case, investment activity is determined by entrepreneurs' "animal spirits" and the government's development programs, for if funds are mobile across the economy's boundaries, investment is independent of domestic savings. In the standard formulation, investment is endogenously determined, being the result of the expansion in the capacities of the sectors producing nontraded goods that is required to match a parametrically given expansion in the output of traded goods, an expansion usually referred to as the "complementary bunch" of investments.[6]

Since we are concerned with the indirect effects of the Muda project, we shall want household incomes to be endogenous. This leaves a final choice of n predetermined variables from among the $2n$ elements of X and E; for each sector, either X_i or E_i must be fixed exogenously.

5. It may happen that some sectors face completely inelastic, albeit positive, "foreign" demand. In that case, they are producing nontraded goods at the margin and are treated as such in the analysis, the vector of these foreign demands being E_N.

6. Let Ξ be the matrix of partial incremental capital-output ratios and ΔX be the vector of changes in sectoral outputs during the period of time in question. Then $J = \Xi \Delta X$. For an extensive discussion of complementary bunches, see Kuyvenhoven (1978).

Our choice of predetermined variables for each solution to the model will be discussed at length in chapter 7 when we return to the analysis of the region's growth. At this point, we merely wish to lay out the implications of choosing some predetermined variables from X, as opposed to assuming that all X_i are endogenous as in equation (6.11). Let $K = \Gamma u - B(I - s)\bar{T} + J$, which, given that J is now predetermined, is a constant. Then (6.11) can be more conveniently arranged as:

$$\begin{bmatrix} I - A & -BW \\ & \\ -\Omega & I \end{bmatrix} \begin{bmatrix} X \\ \\ Y \end{bmatrix} = \begin{bmatrix} K \\ \\ \bar{R} \end{bmatrix} + \begin{bmatrix} I \\ \\ O \end{bmatrix} E , \tag{6.13}$$

or in the case of our illustrative example as:

$$\begin{bmatrix} 1 - a_{TT} & -a_{TN} & -a_{TD} & -\beta_{T1}W_{11} & -\beta_{T2}W_{22} \\ -a_{NT} & 1 - a_{NN} & -a_{ND} & -\beta_{N1}W_{11} & -\beta_{N2}W_{22} \\ -a_{DT} & -a_{DN} & 1 - a_{DD} & -\beta_{D1}W_{11} & -\beta_{D2}W_{22} \\ -\omega_{1T} & -\omega_{1N} & -\omega_{1D} & 1 & 0 \\ -\omega_{2T} & -\omega_{2N} & -\omega_{2D} & 0 & 1 \end{bmatrix} \begin{bmatrix} X_T \\ X_N \\ X_D \\ Y_1 \\ Y_2 \end{bmatrix}$$

$$= \begin{bmatrix} K_T \\ K_N \\ K_D \\ \bar{R}_1 \\ \bar{R}_2 \end{bmatrix} + \begin{bmatrix} 1 & 0 & 0 \\ 0 & 1 & 0 \\ 0 & 0 & 1 \\ 0 & 0 & 0 \\ 0 & 0 & 0 \end{bmatrix} \begin{bmatrix} E_T \\ E_N \\ E_D \end{bmatrix} .$$

With X and Y endogenously determined, the solution to this system is obtained, of course, by inverting the matrix on the left-hand side.

In the light of the discussion above, X_T and E_N and E_D are the predetermined variables in the semi-input-output model. To solve the model for E_T, X_N, X_D, Y_1, and Y_2, we must now rearrange the variables so that E_T appears on the left-hand side of (6.13), and X_T appears on the right. As the reader can easily verify, the resulting rearrangement is:

$$\begin{bmatrix} -1 & -a_{TN} & -a_{TD} & -\beta_{T1}W_{11} & -\beta_{T2}W_{22} \\ 0 & 1 - a_{NN} & -a_{ND} & -\beta_{N1}W_{11} & -\beta_{N2}W_{22} \\ 0 & -a_{DN} & 1 - a_{DD} & -\beta_{D1}W_{11} & -\beta_{D2}W_{22} \\ 0 & -\omega_{1N} & -\omega_{1D} & 1 & 0 \\ 0 & -\omega_{2N} & -\omega_{2D} & 0 & 1 \end{bmatrix} \begin{bmatrix} E_T \\ X_N \\ X_D \\ Y_1 \\ Y_2 \end{bmatrix} \tag{6.14}$$

$$= \begin{bmatrix} K_T \\ K_N \\ K_D \\ \bar{R}_1 \\ \bar{R}_2 \end{bmatrix} + \begin{bmatrix} -(1 - a_{TT}) & 0 & 0 \\ a_{NT} & 1 & 0 \\ a_{DT} & 0 & 1 \\ \omega_{1T} & 0 & 0 \\ \omega_{2T} & 0 & 0 \end{bmatrix} \begin{bmatrix} X_T \\ E_N \\ E_D \end{bmatrix}$$

Because of the way in which (6.13) has been arranged, (6.14) is obtained simply by interchanging the two column vectors marked by arrows in (6.13), and then changing the signs of all the entries in those two columns. This process generalizes to the interchange of any pair of X_i and E_i; one simply interchanges the ith columns of the two matrices in (6.13), at the same time reversing the signs of all the entries in those columns.

Equation (6.14) is a variant of the semi-input-output approach developed by Tinbergen. It also generalizes very easily to the full-scale version used in the empirical analysis, since the production sectors can be classified into tradables, nontradables and distributive services, and the elements of (6.14), for example, a_{TN} and X_T, become matrix partitions rather than single scalars.

In applying this model we shall use two specializations. First, exports of nontradables (E_N) are, by definition, always zero, so that E_N will always be selected as a predetermined variable. Second, because the Muda SAM is measured in producer prices, exports of distributive and transport services (E_D) take the form of the trade and distribution margins on the exports of tradables. Let μ denote these trade and distribution margins per unit of tradable exports, so that $E_D = \mu E_T$.[7]

With these specializations, (6.13) and (6.14) become:

$$(6.13)' \quad \begin{bmatrix} 1-a_{TT} & -a_{TN} & -a_{TD} & -\beta_{T1}W_{11} & -\beta_{T2}W_{22} \\ -a_{NT} & 1-a_{NN} & -a_{ND} & -\beta_{N1}W_{11} & -\beta_{N2}W_{22} \\ -a_{DT} & -a_{DN} & 1-a_{DD} & -\beta_{D1}W_{11} & -\beta_{D2}W_{22} \\ -\omega_{1T} & -\omega_{1N} & -\omega_{1D} & 1 & 0 \\ -\omega_{2T} & -\omega_{2N} & -\omega_{2D} & 0 & 1 \end{bmatrix} \begin{bmatrix} X_T \\ X_N \\ X_D \\ Y_1 \\ Y_2 \end{bmatrix}$$

$$= \begin{bmatrix} K_T \\ K_N \\ K_D \\ \bar{R}_1 \\ \bar{R}_2 \end{bmatrix} + \begin{bmatrix} 1 & 0 & 0 \\ 0 & 0 & 0 \\ \mu & 0 & 0 \\ 0 & 0 & 0 \\ 0 & 0 & 0 \end{bmatrix} \begin{bmatrix} E_T \\ 0 \\ 0 \end{bmatrix}$$

and

$$(6.14)' \quad \begin{bmatrix} -1 & -a_{TN} & -a_{TD} & -\beta_{T1}W_{11} & -\beta_{T2}W_{22} \\ 0 & 1-a_{NN} & -a_{ND} & -\beta_{N1}W_{11} & -\beta_{N2}W_{22} \\ -\mu & -a_{DN} & 1-a_{DD} & -\beta_{D1}W_{11} & -\beta_{D2}W_{22} \\ 0 & -\omega_{1N} & -\omega_{1D} & 1 & 0 \\ 0 & -\omega_{2N} & -\omega_{2D} & 0 & 1 \end{bmatrix} \begin{bmatrix} E_T \\ X_N \\ X_D \\ Y_1 \\ Y_2 \end{bmatrix}$$

7. Distributive and transport margins are earned on gross rather than net flows, so that this formulation is strictly correct only when there are no competitive imports of goods that are also exported. In the case of the Muda economy, that is a defensible approximation to the observed patterns of trade.

$$= \begin{bmatrix} K_T \\ K_N \\ K_D \\ \bar{R}_1 \\ \bar{R}_2 \end{bmatrix} + \begin{bmatrix} -(1-a_{TT}) & 0 & 0 \\ a_{NT} & 0 & 0 \\ a_{DT} & 0 & 0 \\ \omega_{1T} & 0 & 0 \\ \omega_{2T} & 0 & 0 \end{bmatrix} \begin{bmatrix} X_T \\ 0 \\ 0 \end{bmatrix}.$$

These two systems characterize the fixed price models used in the analysis in chapter 7. Equation (6.13)′ is the extended input-output model, and (6.14)′ is our variant of the semi-input-output approach.

A Price-endogenous Extension of the Model

The model developed above assumes that all relative prices remain constant at base year values. This means that in any analysis in which the model's parameters are drawn directly from the 1972 SAM, any experiments conducted with that model are valid only if all prices remained at their 1972 levels. Since we shall need to obtain solutions for 1967 and 1974, when relative prices were different from those ruling in 1972, the model must be extended to deal with these changes. Failure to do so would clearly lead to incorrect estimates of gross outputs and household incomes in the years in question.

There are two problems to be dealt with when incorporating changes in prices. First, if prices change, then, in general, the parameters of the model (a_{ij}, a_{mj}, γ_{ik}, and so forth) will also change. Second, unless all prices are given exogenously, the model must include rules to determine the values of all endogenously determined prices. For the purposes of exposition, we shall assume that the prices of tradable goods are exogenously given, whereas the prices of nontradables and distributive services are determined endogenously. In reality, of course, the prices of certain nontradables also may be fixed exogenously; for example, they may be regulated by government. Such cases appear in chapter 7, but they need not detain us here.

Our exposition concentrates first on the problem of expressing the coefficients of the model as functions of a set of known prices. We then turn to the question of how prices are determined, and formulate a simple price endogenous version of the model. To fulfill our first task we need a variant of the model which is expressed in terms of prices and quantities of all inputs and outputs. We obtain this by first reformulating some of the assumptions (A.1) to (A.8) in price and quantity terms.

All notations used so far have been defined in value (money) units. Let a "hat" (^) on variables and coefficients now denote their physical counterparts. For example, $\hat{X}_j$ will denote the gross output of sector j measured in quantity or physical units, whereas X_j measures the money value of gross output. Let p_j denote the price of a unit of $\hat{X}_j$; then $X_j = p_j\hat{X}_j$.

In money terms, (A.1) and (A.2) are written as

(A.1)′ $$Z_{ij} = (\hat{a}_{ij}p_i)\hat{X}_j$$

(A.2)′ $$M_j = (\hat{a}_{mj}p_{mj})\hat{X}_j$$

where p_{mj} denotes the price of a unit of intermediate imports used by sector j. In physical terms, $\hat{Z}_{ij} = \hat{a}_{ij}\,\hat{X}_j$ and $\hat{M}_j = \hat{a}_{mj}\,\hat{X}_j$. We again assume that households receive fixed shares of value added in the various sectors, that is, $V_{kj} = \phi_{kj}V_j$, where ϕ_{kj} is the share of value added in sector j received by households in class k. The value added in sector j is:

$$V_j = p_j\hat{X}_j - \sum_i p_i\hat{a}_{ij}\hat{X}_j - p_{mj}\hat{a}_{mj}\hat{X}_j.$$

Let λ_j be the nominal value added per unit of physical gross output in sector j, so that

$$\lambda_j = p_j - \sum_i p_i\hat{a}_{ij} - p_{mj}\hat{a}_{mj}.$$

Then,

(A.3)′ $$V_{kj} = \phi_{kj}\lambda_j\hat{X}_j = \omega_{kj}\hat{X}_j$$

where $\omega_{kj} = \phi_{kj}\,\lambda_k$. Thus, the incomes received by households from production activities now depend on all prices.

Assumptions (A.4) and (A.5) remain unchanged:

(A.4)′ $$T_k = \overline{T}_k + t_kY_k,\ \text{all}\ k$$

(A.5)′ $$S_k = s_k(Y_k - T_k),\ \text{all}\ k.$$

Assumption (A.6) can be expanded as:

(A.6)′ $$\begin{aligned} C_{ik} &= p_i\hat{\alpha}_{ik} + \beta_{ik}\,(Y_k - T_k - S_k - \sum_i p_i\hat{\alpha}_{ik} - p_{mk}\hat{\alpha}_{mk}) \\ &= \gamma_{ik} + \beta_{ik}\,(Y_k - T_k - S_k) \end{aligned}$$

where $\gamma_{ik} = p_i\hat{\alpha}_{ik} - \beta_{ik}\sum_i p_i\hat{\alpha}_{ik} - \beta_{ik}\hat{\alpha}_{mk}$.

Further, in physical terms,

$$\begin{aligned} \hat{C}_{ik} &= \gamma_{ik}/p_i + (\beta_{ik}/p_i)(Y_k - T_k - S_k) \\ &= \hat{\gamma}_{ik} + \hat{\beta}_{ik}\,(Y_k - T_k - S_k). \end{aligned}$$

Assumptions (A.7) and (A.8) also remain unchanged.

(A.7)′ $$M_k = \gamma_{mk} + (1 - \sum_i \beta_{ik})(Y_k - T_k - S_k),\ \text{all}\ k$$

(A.8)′ $$R_k = \overline{R}_k,\ \text{all}\ k.$$

Reworking the identities (6.1) to (6.5), we now obtain the following variant of the model:

(6.15) $$Y_k = \sum_j \phi_{kj}\lambda_j\hat{X}_j + \bar{R}_k, \text{ all } k$$

(6.16) $$G = \sum_k (\bar{T}_k + t_k Y_k)$$

(6.17) $$\hat{X}_i = \sum_k \hat{C}_{ik} + \sum_j \hat{a}_{ij}\hat{X}_j + \hat{J}_i + \hat{E}_i$$
$$= \sum_k [\hat{\gamma}_{ik} + \hat{\beta}_{ik}(1 - t_k)(1 - s_k)Y_k - \hat{\beta}_{ik}(1 - s_k)\bar{T}_k]$$
$$+ \sum_j \hat{a}_{ij}\hat{X}_j + \hat{J}_i + \hat{E}_i, \text{ all } i$$

(6.18) $$F = \sum_k s_k[(1 - t_k)Y_k - \bar{T}_k] + G - \sum_i J_i$$

(6.19) $$M = \sum_k \gamma_{mk} + \sum_k (1 - \sum_i \beta_{ik})[(1 - t_k)(1 - s_k)Y_k - (1 - s_k)\bar{T}_k]$$
$$+ \sum_j \hat{a}_{mj} p_{mj} \hat{X}_j$$

where the production accounts (6.3) are now defined as material balances.

As before, only equations (6.15) and (6.17) are required to solve for $\hat{X}_i$ and Y_k (all i,k), and this subsystem can be arranged in matrix notation to obtain the following counterpart to (6.11):

(6.20) $$\begin{bmatrix} I - \hat{A} & -\hat{B}W \\ -\Phi\Lambda & I \end{bmatrix} \begin{bmatrix} \hat{X} \\ Y \end{bmatrix} = \begin{bmatrix} \hat{\Gamma}u - \hat{B}(I - s)\bar{T} + \hat{J} + \hat{E} \\ \bar{R} \end{bmatrix}$$

where a hat now denotes the physical counterparts of the matrices previously defined in value terms, and Λ is the diagonal matrix with jth diagonal element λ_j. Given a solution for $\hat{X}_j$ from (6.20) and knowing p_j, then X_j can be computed as $p_j\hat{X}_j$. The values of G, F, and M can also be computed once (6.20) has been solved.

Note that, in general, Λ varies with the prices of sectoral outputs and imported intermediates, since

$$\lambda_j = p_j - \sum_i p_i \hat{a}_{ij} - p_{mj}\hat{a}_{mj}, \text{ all } j.$$

Further, $\hat{\beta}_{ik} = \beta_{ik}/p_i$ and $\hat{\gamma}_{ik} = \hat{\alpha}_{ik} - \hat{\beta}_{ik}\Sigma_i p_i \hat{\alpha}_{ik} - p_{mk}\hat{\alpha}_{mk}$ also vary with prices. Thus, before (6.20) is solved for any set of prices, the elements of Λ, $\hat{\Gamma}$, and $\hat{B}$ must first be calibrated to those prices. These calculations reflect the changes in the nominal value added paid to households per unit of real gross output as a result of changes in the costs of intermediates and the prices of output, and the changes in marginal budget shares in household expenditure as relative prices change.

To return to our specialized models in (6.13)′ and (6.14)′, the extended input-output model becomes:

$$(6.13)'' \quad \begin{bmatrix} 1-\hat{a}_{TT} & -\hat{a}_{TN} & -\hat{a}_{TD} & -\hat{\beta}_{T1}W_{11} & -\hat{\beta}_{T2}W_{22} \\ -\hat{a}_{NT} & 1-\hat{a}_{NN} & -\hat{a}_{ND} & -\hat{\beta}_{N1}W_{11} & -\hat{\beta}_{N2}W_{22} \\ -\hat{a}_{DT} & 1-\hat{a}_{DN} & 1-\hat{a}_{DD} & -\hat{\beta}_{D1}W_{11} & -\hat{\beta}_{D2}W_{22} \\ -\phi_{1T}\lambda_T & -\phi_{1N}\lambda_N & -\phi_{1D}\lambda_D & 1 & 0 \\ -\phi_{2T}\lambda_T & -\phi_{2N}\lambda_N & -\phi_{2D}\lambda_D & 0 & 1 \end{bmatrix} \begin{bmatrix} \hat{X}_T \\ \hat{X}_N \\ \hat{X}_D \\ Y_1 \\ Y_2 \end{bmatrix}$$

$$= \begin{bmatrix} \hat{K}_T \\ \hat{K}_N \\ \hat{K}_D \\ \overline{R}_1 \\ \overline{R}_2 \end{bmatrix} + \begin{bmatrix} 1 & 0 & 0 \\ 0 & 0 & 0 \\ \mu & 0 & 0 \\ 0 & 0 & 0 \\ 0 & 0 & 0 \end{bmatrix} \begin{bmatrix} \hat{E}_T \\ 0 \\ 0 \end{bmatrix}$$

where $\hat{K}_i = \Sigma_i \hat{\gamma}_{ik} - \Sigma_k \hat{\beta}_{ik}(1-s_k)\overline{T}_k + \hat{J}_i$. When the appropriate elements of X and $\hat{E}$ are interchanged, the semi-input-output model becomes:

$$(6.14)'' \quad \begin{bmatrix} -1 & -\hat{a}_{TN} & -\hat{a}_{TD} & -\hat{\beta}_{T1}W_{11} & -\hat{\beta}_{T2}W_{22} \\ 0 & 1-\hat{a}_{NN} & -\hat{a}_{ND} & -\hat{\beta}_{N1}W_{11} & -\hat{\beta}_{N2}W_{22} \\ -\mu & -\hat{a}_{DN} & 1-\hat{a}_{DD} & -\hat{\beta}_{D1}W_{11} & -\hat{\beta}_{D2}W_{22} \\ 0 & -\phi_{1N}\lambda_N & -\phi_{1D}\lambda_D & 1 & 0 \\ 0 & -\phi_{2N}\lambda_N & -\phi_{2D}\lambda_D & 0 & 1 \end{bmatrix} \begin{bmatrix} \hat{E}_T \\ \hat{X}_N \\ \hat{X}_D \\ Y_1 \\ Y_2 \end{bmatrix}$$

$$= \begin{bmatrix} \hat{K}_T \\ \hat{K}_N \\ \hat{K}_D \\ \overline{R}_1 \\ \overline{R}_2 \end{bmatrix} + \begin{bmatrix} -(1-\hat{a}_{TT}) & 0 & 0 \\ \hat{a}_{NT} & 0 & 0 \\ \hat{a}_{DT} & 0 & 0 \\ \phi_{1T}\lambda_T & 0 & 0 \\ \phi_{2T}\lambda_T & 0 & 0 \end{bmatrix} \begin{bmatrix} \hat{X}_T \\ 0 \\ 0 \end{bmatrix}.$$

If all the prices p_j and p_{mj} are known for a given year, (6.13)″ or (6.14)″ can be used to generate all the data necessary to construct a SAM for that year. In some cases, though, not all the prices are, or could be, known. If it is desired to construct a SAM for some past year in which there is no information on certain prices, they must then be estimated. Also, when we want to look at the effects of hypothetical changes in the prices exogenous to the region, we need a schema that will yield the corresponding changes in the prices endogenously determined in the regional economy. We assume that the prices of tradables and noncompetitive imports are exogenously given, but that the prices of nontradables and distributive services are determined endogenously. In the notation of our illustrative example, p_N and p_D are to be treated endogenously, while p_T, p_{mT}, p_{mN}, and p_{mD} are given. It is now necessary to formulate the way in which the endogenous prices are to be determined.

From the definition of value added for sector j,

$$V_j = p_j\hat{X}_j - \sum_i p_i\hat{a}_{ij}\hat{X}_j - p_{mj}\hat{a}_{mj}\hat{X}_j,$$

the price of output is:

$$p_j = (V_j/\hat{X}_j) + \sum_i p_i \hat{a}_{ij} + p_{mj}\hat{a}_{mj}.$$

Now $V_j/\hat{X}$ measures the value added per unit of physical output from sector j, and it will be useful to write this as

$$V_j/\hat{X}_j = p_{vj}\hat{v}_j$$

where p_{vj} denotes the "price" of value added and $\hat{v}_j$ denotes the "quantity" of value added required to produce one unit of $\hat{X}_j$. When value added is interpreted as the payment to primary factors, such as labor and capital, then $\hat{v}_j$ represents the amount of the "characteristic" bundle of factors used to produce one unit of $\hat{X}_j$, and p_{vj} is the price of that "characteristic" bundle. Collecting terms,

$$p_j = p_{vj}\hat{v}_j + \sum_i p_i \hat{a}_{ij} + p_{mj}\hat{a}_{mj}. \tag{6.21}$$

We must now be more precise about the units in which such coefficients as $\hat{a}_{ij}$ and $\hat{a}_{mj}$ are defined. In empirical work, sectors must often be defined in such a way that they produce a composite bundle of outputs, so that the concept of a "physical" unit is not well defined. Fortunately, we can now avoid these problems by choosing to measure these units in base year prices. This is done by arbitrarily defining physical units of measure such that all base year prices are unity. Hence, (6.21) becomes

$$1 = \hat{v}_j^o + \sum_i \hat{a}_{ij}^o + \hat{a}_{mj}^o \tag{6.22}$$

where $\hat{a}_{ij}^o$, $\hat{a}_{mj}^o$ and $\hat{v}_j^o$ are now equal to their nominal values in the base year. The base year would, of course, be the year for which the SAM was estimated (which in Muda is 1972). Further, the "physical" units for $\hat{X}$, $\hat{a}_{ij}$, $\hat{v}_j$, and so forth are then measured in constant base year prices.

With these interpretations, we now introduce two key assumptions, which determine all endogenous prices within the model. The first is that in sectors with endogenous prices, nominal value added is a constant share of nominal gross output, irrespective of the prevailing price ratios. That is,

$$V_j = \theta_j X_j, \text{ all } j \in D,N \tag{A.9}$$

where θ_j is a constant. The second assumption is that the matrices $\hat{A}$ and $\hat{A}_m$ are independent of relative prices, which is equivalent to assuming that in each sector intermediate inputs are combined in bundles whose physical composition is fixed.

These strong assumptions require a few remarks. We have not found any evidence to test the constancy of θ for sectors producing nontradable goods in Malaysia, much less the Muda region. Time series data for Canada and Egypt, however, suggest that the ratio of nominal value added to nominal gross output was fairly constant in the sectors producing nontradables,[8] so there are grounds for accepting the first assumption, indirect though the supporting evidence may be. In regard to the constancy of $\hat{A}$ and $\hat{A}_m$, the relevant papers in Carter and Bródy (1972) and Bródy and Carter (1972)[9] arrive at the finding that these technical coefficients do change somewhat with the passage of time. However, since the sources of such changes include technical progress, changes in the composition of output in individual sectors, and economies of scale, as well as changes in prices, the assumption that there is no substitutability among intermediate goods may be considered a tolerable approximation, especially if there is sufficient aggregation of inputs. The sector about which one would have the strongest reservation is agriculture, for which techniques of production are fairly flexible. Yet here we have a tailor-made sectoral model in TIGER, which yields an aggregate technology vector for each vector of exogenous prices.

Expanding (A.9),

$$p_{vj}\, \hat{v}_j\, \hat{X}_j = \theta_j\, p_j\, \hat{X}_j, \text{ or}$$

$$p_{vj}\, \hat{v}_j = \theta_j\, p_j, \text{ all } j \,\varepsilon\, D,N. \tag{6.23}$$

But this relationship must also hold in the base year, for which we have the normalization $p_{vj} = p_j = 1$. Hence, $\hat{v}_j^o = \theta_j$, and substituting into (6.23), we have $p_{vj}\, \hat{v}_j = p_j\, \hat{v}_j^o$. Substituting this into (6.21), we obtain

$$p_j(1 - \hat{v}_j^o) = \sum_i p_i \hat{a}_{ij}^o + p_{mj}\hat{a}_{mj}^o, \text{ all } j \,\varepsilon\, D,N \tag{6.24}$$

where we have used the fact that $\hat{a}_{ij}$ and $\hat{a}_{mj}$ are constant at their base year values. It is clear from (6.24) that in those sectors whose output prices are endogenously determined, the price of output is given by a proportional markup over the total cost of intermediate inputs.

To return to our illustrative example, only p_N and p_D are endogenously determined:

$$p_N(1 - \hat{v}_N^o) = p_T\hat{a}_{TN}^o + p_N\hat{a}_{NN}^o + p_D\hat{a}_{DN}^o + p_{mN}\hat{a}_{mN}^o$$

$$p_D(1 - \hat{v}_D^o) = p_T\hat{a}_{TD}^o + p_N\hat{a}_{ND}^o + p_D\hat{a}_{DD}^o + p_{mD}\hat{a}_{mD}^o.$$

These two equations can be arranged as:

8. The series for Canada covers the years 1961–71 and the following sectors: construction, transport, communications, electricity and gas, wholesale trades, retail trades, finance and insurance, and personal and cultural services (Canada, 1977). The series for Egypt covers the years 1967–72 and the following sectors: construction, transport and communications, finance and trade, housing, public utilities, and other services (Arab Republic of Egypt, 1975).

9. Forssell (1972), Middelhoek (1972), and Sevaldson (1972).

$$(6.25)\quad \begin{bmatrix} (1-\hat{v}_N^o - \hat{a}_{NN}^o) & -\hat{a}_{DN}^o \\ -\hat{a}_{ND}^o & (1-\hat{v}_D^o - \hat{a}_{DD}^o) \end{bmatrix} \begin{bmatrix} p_N \\ p_D \end{bmatrix} = \begin{bmatrix} \hat{a}_{TN}^o p_T + \hat{a}_{mN}^o p_{mN} \\ \hat{a}_{TD}^o p_T + \hat{a}_{mD}^o p_{mD} \end{bmatrix}.$$

Then, for given values of p_T, p_{mN}, and p_{mD}, (6.25) can be solved to obtain the corresponding values of p_N and p_D.

A key feature of (6.25) is that the endogenous prices depend solely on a set of invariant technical coefficients and the prices which are exogenously determined. In particular, they are independent of the levels of sectoral outputs. This is precisely what our assumption (A.9) was designed to contrive, for it has enabled us to avoid tackling a general equilibrium problem of a much more daunting kind.

To summarize, (6.13)″ and (6.14)″ are, respectively, statements of our extended input-output and semi-input-output models, which permit key variables to be calculated under different sets of prices. They provide the tools for constructing SAMs for both past years and hypothetical situations, given a set of prices for all inputs and outputs in the economy. Recall that the prices of tradables and noncompetitive imports are exogenously given, but the prices of nontradables and distributive services are endogenous. Equation (6.25) provides the rules by which these prices are determined, given the values of all exogenously fixed prices. Since the prices in (6.25) are determined independently of sectoral output levels, the endogenous prices for any given situation are first determined, and then used in (6.13)″ or (6.14)″ to obtain the values of sectoral outputs, exports, and household incomes.

To illustrate this method, we return to our numerical example. Letting $\mu = 0.1$, the numerical version of (6.14)″ for the base year is:

$$\begin{bmatrix} -1 & -0.05713 & -0.36926 & -0.16075 & -0.06979 \\ 0 & 0.92077 & -0.25000 & -0.46750 & -0.22319 \\ -0.1 & -0.38845 & 0.91665 & -0.10228 & -0.12554 \\ 0 & -0.14290 & -0.04170 & 1 & 0 \\ 0 & -0.28570 & -0.16670 & 0 & 1 \end{bmatrix} \begin{bmatrix} \hat{E}_T \\ \hat{X}_N \\ \hat{X}_D \\ Y_1 \\ Y_2 \end{bmatrix}$$

$$= \begin{bmatrix} -6.16746 \\ 40.61481 \\ 18.94036 \\ 10 \\ 20 \end{bmatrix} + \begin{bmatrix} -0.61697 & 0 & 0 \\ 0.19478 & 0 & 0 \\ 0.06819 & 0 & 0 \\ 0.25000 & 0 & 0 \\ 0 & 0 & 0 \end{bmatrix} \begin{bmatrix} \hat{X}_T \\ 0 \\ 0 \end{bmatrix}.$$

Setting $\hat{X}_T = 307.68$, its base year value, the solution values of $\hat{E}_T$, $\hat{X}_N$, $\hat{X}_D$, Y_1, and Y_2 are the same as in table 6-2.

We shall need separate estimates of the coefficients ϕ_{kj} and λ_j, since the latter are not invariant with respect to price changes. Using the definition of λ_j, that is, $\lambda_j = p_j - \Sigma_i p_i \hat{a}_{ij} - p_{mj} \hat{a}_{mj}$, all j, and the fact that we have defined all base year prices to be unity, then:

$$\lambda_T = 1 - 0.38303 - 0.19478 - 0.06819 - 0.10400 = 0.25000$$
$$\lambda_N = 1 - 0.05713 - 0.07923 - 0.38845 - 0.04658 = 0.42860$$
$$\lambda_D = 1 - 0.36926 - 0.25000 - 0.08335 - 0.08901 = 0.20838.$$

From the relation $\Phi = \Omega\Lambda^{-1}$, it follows that

$$\Phi = \begin{bmatrix} 1 & 0.33341 & 0.20012 \\ 0 & 0.66659 & 0.79988 \end{bmatrix}.$$

We also need estimates of the $\hat{\alpha}_{jk}$'s in (A.6)′ in order to be able to adjust the elements of $\hat{\Gamma}$ for price changes. From the definition of γ_{jk},

$$\gamma_{jk} = p_j \hat{\alpha}_{jk} - \beta_{jk} \sum_i p_i \hat{\alpha}_{ik}, \text{ all } j,k,$$

and in the base year, when all prices are unity,

$$\gamma_{jk}^o = \hat{\alpha}_{jk} - \beta_{jk} \sum_i \hat{\alpha}_{ik}.$$

But from the restrictions on the estimates of the parameters in the linear expenditure system (LES), $\Sigma_i p_i \hat{\alpha}_{ik} = 0$. In base year prices, this is $\Sigma_i \hat{\alpha}_{ik} = 0$, so that $\gamma_{jk}^o = \hat{\alpha}_{jk}$. Our Γ matrix for the base year is therefore also the relevant estimate of the matrix of $\hat{\alpha}_{jk}$'s.

Suppose now that the price of the tradable increases by 25 percent, but that the prices of imported intermediates remain constant. We continue to assume that the total output of tradables remains constant when measured in physical units (that is, $\hat{X}_T = 307.68$), though, of course, the nominal value of this output will also increase by 25 percent. Our problem is to assess the impact of the change in p_T on the prices of nontradables (p_N) and distributive services (p_D), and thence its impact on the levels of $\hat{E}_T$, $\hat{X}_N$, $\hat{X}_D$, Y_1, and Y_2.

The first task is to compute the new values of p_N and p_D. Turning to (6.25), and observing that $\hat{v}_N^o = 0.4286$ and $\hat{v}_D^o = 0.20838$, we have

$$\begin{bmatrix} 0.49217 & -0.38845 \\ -0.25 & 0.70827 \end{bmatrix} \begin{bmatrix} p_N \\ p_D \end{bmatrix} = \begin{bmatrix} 0.05713\, p_T + 0.04658 \\ 0.36926\, p_T + 0.08901 \end{bmatrix}.$$

As a check on these calculations, we note that if $p_T = 1$, then the solution to the equation is $p_N = p_D = 1$. Now, setting $p_T = 1.25$, and solving for p_N and p_D, we obtain

$$\begin{bmatrix} p_N \\ p_D \end{bmatrix} = \begin{bmatrix} 2.816440 & 1.544676 \\ 0.994128 & 0.957119 \end{bmatrix} \begin{bmatrix} 0.11799 \\ 0.55059 \end{bmatrix} = \begin{bmatrix} 1.1828 \\ 1.1949 \end{bmatrix}.$$

A 25 percent increase in p_T is therefore seen to induce an increase of approximately 20 percent in the prices of nontradables and distributive services.

The next step is to calculate the effect of this change in prices on the λ_j's, and thence on the shares (ω_{kj}'s) of value added received by different households. Again, using the definition of λ_j, we have

$$\lambda_T = 1.25 - (1.25)(0.38303) - (1.1828)(0.19478) - (1.1949)(0.06819) - 0.104 = 0.3553$$

$$\lambda_N = 1.1828 - (1.25)(0.05713) - (1.1828)(0.07923) - (1.1949)(0.38845) - 0.04658 = 0.5069$$

$$\lambda_D = 1.1949 - (1.25)(0.36926) - (1.1828)(0.25) - (1.1949)(0.08335) - 0.08901 = 0.249.$$

Hence,

$$\begin{bmatrix} \omega_{1T} & \omega_{1N} & \omega_{1D} \\ \omega_{2T} & \omega_{2N} & \omega_{2D} \end{bmatrix} = \begin{bmatrix} 1 & 0.33341 & 0.20012 \\ 0 & 0.66659 & 0.79988 \end{bmatrix} \begin{bmatrix} 0.3553 & 0 & 0 \\ 0 & 0.5069 & 0 \\ 0 & 0 & 0.249 \end{bmatrix}$$

$$= \begin{bmatrix} 0.3553 & 0.16900 & 0.04983 \\ 0 & 0.33789 & 0.19917 \end{bmatrix}.$$

We also need to revise $\hat{B}$ and $\hat{\Gamma}$ to match the new prices. The new $\hat{B}$ is obtained simply by dividing the base year β_{jk}'s by the new prices p_j, for example

$$\hat{\beta}_{11} = \beta^o_{11}/p_1 = 0.19565/1.25 = 0.15652.$$

This process yields:

$$\hat{B} = \begin{bmatrix} 0.15652 & 0.09922 \\ 0.48106 & 0.33535 \\ 0.10418 & 0.18671 \end{bmatrix}$$

Now $\hat{\gamma}_{jk} = \hat{\alpha}_{jk} - \hat{\beta}_{jk} \Sigma_i p_i \hat{\alpha}_{ik}$, and with the base year estimates of $\gamma^o_{jk} = \hat{\alpha}_{jk}$, this becomes $\hat{\gamma}_{jk} = \gamma^o_{jk} - \hat{\beta}_{jk} \Sigma_i p_i \gamma^o_{jk}$. Using the available values of γ^o_{jk}, $\hat{\beta}_{jk}$, and p_i, we find $\hat{\gamma}_{11} = (-1) - (0.15652)(-0.0914) = -0.98569$, and by similar calculations,

$$\hat{\Gamma} = \begin{bmatrix} -0.98569 & -0.00120 \\ 3.04397 & -1.00406 \\ -1.99048 & 0.99774 \end{bmatrix}.$$

The next step in the analysis is to substitute the new values of ω_{kj}, $\hat{\gamma}_{jk}$, and $\hat{\beta}_{jk}$ into (6.14)″, and to solve for $\hat{E}_T$, $\hat{X}_N$, $\hat{X}_D$, Y_1, and Y_2. We note that

$$\begin{bmatrix} \hat{K}_T \\ \hat{K}_N \\ \hat{K}_D \end{bmatrix} = \begin{bmatrix} -0.98569 & -0.0012 \\ 3.04397 & -1.00406 \\ -1.99048 & 0.99774 \end{bmatrix} \begin{bmatrix} 1 \\ 1 \\ 1 \end{bmatrix}$$

$$- \begin{bmatrix} 0.15652 & 0.09922 \\ 0.48106 & 0.33535 \\ 0.10418 & 0.18671 \end{bmatrix} \begin{bmatrix} 0.91799 & 0 \\ 0 & 0.66857 \end{bmatrix} \begin{bmatrix} 20 \\ 19 \end{bmatrix}$$

$$+ \begin{bmatrix} 0 \\ 54.1 \\ 25.06 \end{bmatrix} = \begin{bmatrix} -5.1209 \\ 43.0479 \\ 19.7828 \end{bmatrix}.$$

Then

$$\begin{bmatrix} -1 & -0.05713 & -0.36926 & -0.1286 & -0.05583 \\ 0 & 0.92077 & -0.25 & -0.39525 & -0.1887 \\ -0.1 & -0.38845 & 0.91665 & -0.0856 & -0.10506 \\ 0 & -0.169 & -0.04983 & 1 & 0 \\ 0 & -0.33789 & -0.19917 & 0 & 1 \end{bmatrix} \begin{bmatrix} \hat{E}_T \\ \hat{X}_N \\ \hat{X}_D \\ Y_1 \\ Y_2 \end{bmatrix}$$

$$= \begin{bmatrix} -5.1209 \\ 43.0479 \\ 19.7828 \\ 10 \\ 20 \end{bmatrix} + \begin{bmatrix} -0.61697 & 0 & 0 \\ 0.19478 & 0 & 0 \\ 0.06819 & 0 & 0 \\ 0.3553 & 0 & 0 \\ 0 & 0 & 0 \end{bmatrix} \begin{bmatrix} 307.68 \\ 0 \\ 0 \end{bmatrix} = \begin{bmatrix} -194.9502 \\ 102.9778 \\ 40.7635 \\ 119.319 \\ 20 \end{bmatrix}.$$

The solution to this problem is:

$$\begin{bmatrix} \hat{E}_T \\ \hat{X}_N \\ \hat{X}_D \\ Y_1 \\ Y_2 \end{bmatrix} = \begin{bmatrix} 73.537 \\ 273.423 \\ 202.258 \\ 175.606 \\ 152.671 \end{bmatrix}.$$

Multiplying $\hat{E}_T$, $\hat{X}_N$, and $\hat{X}_D$ by their respective prices, we have the solution values of all variables in nominal terms:

$$\begin{bmatrix} E_T \\ X_N \\ X_D \\ Y_1 \\ Y_2 \end{bmatrix} = \begin{bmatrix} 91.921 \\ 323.405 \\ 241.678 \\ 175.606 \\ 152.671 \end{bmatrix}.$$

Comparing these results with their base year values in table 6-2, E_T has increased by 16 percent, X_N by 25 percent, and X_D by 24 percent. House-

Table 6-3. *The SAM for the Case $p_T = 1.25$*
(nominal Malaysian dollars)

Receipts	Expenditures: Households 1	Households 2	Government	Production activities T	Production activities N	Production activities D	Capital account	Rest of world	Row sums
Households									
1				109.32	46.21	10.08		10.00	175.61
2					92.39	40.28		20.00	152.67
Government	38.43	43.17							81.60
Production activities									
T	23.41	9.08		147.31	19.53	93.36		91.92	384.60
N	75.25	27.85		70.89	25.62	59.81	63.99		323.41
D	13.30	17.53		25.07	126.91	20.14	29.94	8.79	241.68
Capital account	11.25	36.29	81.60					−35.25	93.89
Rest of world	13.96	18.76		32.00	12.74	18.00			95.46
Column sums	175.60	152.68	81.60	384.59	323.40	241.67	93.93	95.46	

hold incomes have also increased: by 33 percent for H_1 and by 21 percent for H_2, both in nominal terms, of course.

As a final step in the analysis, it is now possible to use assumptions (A.1)′ to (A.8)′ to construct a revised SAM for the economy. Since the nature of these calculations has already been indicated, we leave it to the interested reader to verify that our calculations lead to table 6-3.

7

The Sources of Change in the Regional Economy

IN THIS CHAPTER, we analyze the changes that occurred in the regional economy from 1967 to 1974. In particular, we provide quantitative estimates of both the direct and indirect effects of the irrigation project. To do so, we utilize the semi-input-output models developed in the preceding chapter, and the regional SAM for 1972 described in chapter 5.

As in the analysis of the paddy farming sector in chapter 4, we use the regional model to compute the values of relevant variables in the steady state characterized as "project maturity" with their counterparts in the hypothetical case that the project had not been undertaken. This set of variables comprises all the elements of a complete SAM for the region in each situation, as well as the endogenously determined prices for nontradables and distributive services. (For the sake of brevity, however, only a part of the complete set of results will be presented here.) Since many variables are taken as exogenous to the model—for example, sectoral gross output or export levels, investment deliveries, and government expenditure—it is necessary to select a particular year to represent the underlying conditions of project maturity. For this purpose we have again chosen 1974, the year immediately following the completion of the project.

In addition, we seek to isolate the effects of various autonomous changes in the regional economy from those of the project. Changes in population, government spending, and private investment affected the outputs of many sectors (and hence incomes and demand) but had nothing to do with the irrigation project. There were also important changes in relative prices, particularly those entering into the terms of trade for the paddy farming sector (the price of paddy almost doubled between 1972 and 1974). To disentangle the effects of these changes from each other and from those of the project, we use the regional model to conduct the same kind of comparative static analysis as that in chapter 4. Once again, it must be emphasized that the model provides steady-state configurations of the economy under the prevailing conditions defined by the exogenous variables. In particular, our preproject situation is taken as the state corresponding to the values of the exogenous variables for 1967.

The Growth of the Regional Economy, 1967–74

Beginning with the 1972 SAM, it is necessary to make three kinds of changes in order to construct SAMs for 1967 and 1974 with the aid of our semi-input-output model. First, allowance must be made for price changes. Taking the prices of tradables as exogenously given at their historic values, and using the price determination equation (6.25) in chapter 6, we can compute the values of all endogenous prices. Then, with the use of the semi-input-output model defined in (6.14)″, the entries in the SAM can be updated to allow for all changes in prices. Second, since for each sector either its gross output, X_i, or its level of exports, E_i, must be exogenously given (chapter 6), it is necessary to estimate these values for the years in question, together with the levels of investment and other exogenous variables. Third, changes in some of the structural parameters are required to reflect changes in technology and population.

Price Changes

The set of all prices for 1967 and 1974, together with the associated cost of living indices for the different classes of households, are reported in table 7-1. In keeping with the year for which the regional SAM was estimated, the base year is 1972. With the exception of other agriculture, electricity (in 1967), and government services, all sectors producing nontradables have endogenously determined prices. The sectoral prices treated exogenously are marked by asterisks in table 7-1.

With the exception of electricity and government services, the prices of both exportables and nontradable goods rose sharply over the period, with fairly wide variations within each group. Within the agricultural complex (sectors 1–11), which produces exportables, the rise in the price of rubber was very modest, whereas that for fish was the steepest of all. As a group, the prices of imports destined for intermediate use in the agricultural complex increased proportionally more than those of other intermediate imports and imported consumer goods; they also increased more rapidly than the prices of outputs in the agricultural complex. In the aggregate, the ratio of the price of tradables to that of nontradables increased somewhat over the period, as the weight of government services in the latter is large. In 1967 the region's barter terms of trade stood at 120 (1972 = 100). This relatively favorable position stemmed equally from higher prices for exportables and lower prices for imports. The index was virtually unchanged

Table 7-1. *Price and Cost of Living Indexes, 1967 and 1974*
(1972 = 100)

	1967		
Sector[a]	*Output price* (p_i)	*Price of imported intermediates* (p_{mi})	*Nominal value added per unit of real gross output* (λ_1)
1. Commercial rice mills	112.3*	80.0	0.125
2. Small rice mills	112.3*	82.0	0.097
3. Food processing	96.2*	96.2	0.274
4. Fish processing	57.7*	82.0	0.055
5. Paddy production	115.2*	91.0	1.025
6. Fishing	57.7*	88.0	0.241
7. Estates rubber	139.6*	91.0	1.146
8. Smallholder rubber	139.6*	91.0	1.228
9. Rubber processing	127.7*	86.0	0.229
10. Other agriculture	91.6*	91.0	0.775
11. Sawmilling	86.8*	92.0	0.303
12. Agricultural machinery services	84.3	90.0	0.470
13. Machinery repair	78.6	75.0	0.310
14. Manufacturing n.e.c.	94.3	95.0	0.481
15. Road transport	87.3	90.0	0.570
16. Rail transport	84.2	85.0	0.449
17. Hotels and restaurants	91.0	96.0	0.288
18. Entertainment	88.8	90.0	0.798
19. Services n.e.c.	85.1	80.0	0.419
20. Private health	94.6	97.0	0.605
21. Distributive trades	89.4	95.0	0.793
22. Petty trading	88.7	100.0	0.822
23. Electricity	94.0*	94.0	0.594
24. Water	84.7	85.0	0.568
25. Posts and telecommunications	86.5	85.0	0.719
26. Residential construction	84.9	80.0	0.314
27. Other construction	82.1	80.0	0.342
28. Government, irrigation	85.0*	93.0	0.371
29. Government, education	82.6*	90.0	0.758
30. Government, health	82.6*	90.0	0.658
31. Government, other	82.6*	90.0	0.604
32. Traditional financial services	88.8	90.0	0.855
33. Modern financial services	90.7	95.0	0.723
34. Urban housing	89.6	95.0	0.818
35. Rural housing	85.0	100.0	0.807
		1967 cost-of-living index	
Household		*Imports*	*Overall*
Landless paddy workers		94.2	90.8
Labor-abundant paddy farms		93.5	92.6
Land-abundant paddy farms		92.8	91.5
Nonproject farms		93.0	91.5
Nonfarm		93.7	91.4

Note: An asterisk denotes an exogenously fixed price. The prices of all imports are exogenous.

1972 Value added per unit of real gross output ($\hat{v}_i$)	1974 Output price (p_i)	1974 Price of imported intermediates (p_{mi})	1974 Nominal value added per unit of real gross output (λ_i)
0.103	179.1*	163.1	0.223
0.091	179.1*	163.1	0.188
0.283	146.1*	146.1	0.426
0.103	146.4*	163.1	0.150
0.804	179.7*	193.5	1.528
0.621	146.4*	160.0	0.872
0.723	167.6*	193.5	1.234
0.816	167.6*	193.5	1.354
0.206	191.8*	178.3	0.610
0.848	149.5*	193.5	1.240
0.490	149.4*	166.3	0.507
0.558	123.7	110.0	0.690
0.394	127.0	125.4	0.500
0.510	154.3	157.6	0.787
0.653	125.3	110.0	0.819
0.535	130.7	111.5	0.706
0.317	144.1	146.1	0.457
0.899	122.9	110.0	1.104
0.492	145.8	163.1	0.717
0.639	124.0	120.0	0.793
0.886	130.7	120.0	1.158
0.928	121.8	100.0	1.130
0.629	102.7	100.0	0.647
0.671	124.7	125.4	0.837
0.832	132.1	125.4	1.099
0.370	153.2	163.1	0.567
0.416	155.6	163.1	0.648
0.432	132.0*	102.7	0.507
0.923	100.9*	110.0	0.906
0.812	100.9*	110.0	0.769
0.750	100.9*	110.0	0.694
0.963	121.8	110.0	1.172
0.797	130.0	123.9	1.036
0.913	137.9	123.9	1.258
0.949	152.2	100.0	1.444

1974 cost-of-living index

Imports	Overall
134.6	146.5
132.6	147.5
131.1	145.0
132.1	145.5
129.7	137.4

a. Sectors are numbered consecutively, and not to correspond to table 5-4.

at 121 in 1974, the relative contributions of movements in export and import prices away from their levels in 1972 being roughly equal once more.

The consumer prices indices for the various socioeconomic classes of households rose at slightly different rates. Among farm households, the index for the landless class increased a little more than those for the rest, while the index for nonfarm households rose perceptibly less than for farm households as a group.

To complete the picture, table 7-1 also displays the values of the nominal value added per unit of real gross output in each sector. These parameters—$\lambda_i = V_i/\hat{X}_i$, to use the notation of chapter 6—reflect the changes in the own-price of outputs relative to the prices of the intermediate inputs needed to produce them. Their departures from their base year (1972) values are summary measures of the changes in nominal incomes (arising from unit production activity levels) which are brought about by changes in the prices of tradables. For those sectors whose output prices are endogenously determined, λ_i is proportional to the price of output. Among the remainder, the proportional increase in λ was about the same as that in the price of output in other agriculture and food processing; it was significantly greater in the cases of rice milling, fishing, and rubber processing; and it was rather less in paddy production, smallholder rubber, sawmilling, and the government sectors. In the light of what has already been said about price movements, it seems that, in the aggregate, the direct effect of the changes in exogenous prices was to shift nominal purchasing power away from nonfarm households, which derive much of their income from government employment, and nonproject farm households, which depend heavily on income from rubber production, toward households engaged mainly in paddy production.

Changes in Exogenous Variables

As discussed in chapter 6, any attempt to construct a detailed picture of the economy as it was in the past need take account only of data availability when it comes to choosing between fixing export or output levels for each sector. The sectors comprising the agricultural complex (1–11) all produce tradable goods. With the exception of sector 10, it was possible to derive estimates of gross output from the same sources used to estimate the priors of the 1972 SAM. Fortunately, in the case of sector 10 there were relevant data for estimating exports in 1967 and 1974. The only other sector producing tradables is 14, and once again data were available on its export components. In the remaining sectors, of course, net exports have been set at zero, although the output levels of the four government sectors are known and given independently of demand.[1] Deliveries to investment activities are made only by sectors 10, 14, 26, and 27. Again, the sources for the 1972 estimates provided the data for their counterparts in 1967 and 1974. It should also be noted, however, that the treatment of sawmilling for 1974 differs from that for 1967 in that net exports have been set at their level in 1972. This choice reflects the fall in the volume of Malaysia's timber exports during the world recession in 1974 and accords well both with the views expressed by local businessmen and with the enclave nature of the forestry sector (see chapter 2). In effect, then, the output of this sector in 1974 was demand-driven. The values of all exogenously fixed gross output levels are marked by asterisks in table 7-6.

Of the remaining exogenous variables, net private transfers to households have been set at their 1972 levels, these being very small and there being no other basis for a set of 1967 or 1974 estimates. Finally, "exogenous" taxes ($\bar{T}_k$) on households have been estimated in the same way as for the 1972 SAM. These taxes take the form of licenses, fees, and duties and are levied independently of income levels.

Wherever possible, we have derived the values for the exogenous variables in 1967 and 1974 in the same way, and using the same sources, as the prior estimates of their 1972 counterparts. This is desirable in itself, for what matters here is the *difference* in the values of variables between one set of circumstances and another. But if there is also a keen interest in the absolute levels of a particular set of variables, there is a second reason for insisting, whenever possible, on a comparable basis for estimation. As shown in chapter 5, the 1972 SAM is not a tableau of priors, but a transformation thereof to produce a consistent set of accounts using a technique developed by Byron (1978). In a broad sense, these consistent ("posterior") estimates of the values of all flows are our best. Hence, before fixing the exogenous variables for the model, each of the prior estimates for 1967 has been scaled by the ratio of the posterior to the prior value of its 1972 counterpart.[2]

1. To be exact, government expenditures are completely independent of household incomes, but slightly dependent on activity levels in other sectors through weak interindustry linkages. The allocations to households are given exogenously and the (small) supplies to satisfy interindustry demands are endogenously determined.

2. Devarajan has provided the following statistical rationale for this procedure: Let $a_1, \ldots, a_n$ be prior estimates of parameters $\alpha_1, \ldots, \alpha_n$ which are constrained by $\Sigma\alpha_i = 0$. Then the constrained likelihood function is of the form

$$L(\alpha_1, \ldots, \alpha_{n-1}, -\alpha_1 - \alpha_2, \ldots, -\alpha_{n-1} | a_1, \ldots, a_n, s_1, \ldots, s_n)$$

where $s_1, \ldots, s_n$ are the estimates of the standard errors. The maximum likelihood estimators (MLE) of $\alpha_1, \ldots, \alpha_n$ are found by solving the equations

(Note continues on the following page.)

Changes in Structural Parameters

In general, we have assumed that the structural parameters of the model did not change over the 1967–74 period. There are, however, two important exceptions: first, the changes in paddy production technology and the distribution of the sector's value added following the introduction of irrigation and new paddy varieties; second, changes in the region's population and in the pattern of seasonal migration associated with paddy cultivation.

For the paddy sector, we have derived the estimates of the technology (A_5) and distribution (Ω_5) vectors largely from the programming model we set out in chapter 3. (Hereafter, number subscripts refer to the sector numbers in table 7-6.) The important changes in technology between 1967 and 1974 were the substitution of mechanical for animal draft power and the more intensive use of fertilizer on new, high-yielding varieties. There was also a marked increase in the (small) proportion of value added paid as wages to migrant laborers from the nonproject area and other regions, although the real wage rate rose only modestly. The distribution of value added among households underwent more radical changes, the most noteworthy being the fall in the share of nonproject farm households, which continued to cultivate a single crop, although they did benefit from additional wage earnings within the project area. The shares of output paid as wages to workers from landless and nonfarm households increased somewhat, as did the share of output paid as rents (some of which accrued to rentiers among nonfarm households). These shifts are summarized in table 7-2.

Population growth affects the demand system by altering the intercept terms $\{\gamma_{ik}\}$, though leaving the marginal expenditure proportions $\{\beta_{ik}\}$ unchanged.[3] The only population data available are contained in the two

---

$$\frac{\partial L}{\partial \alpha_i} = 0 \qquad i = 1, \ldots, n-1$$

which yields MLE's $\hat{\alpha}_i = \hat{\alpha}_i\,(a_1, \ldots, a_n, s_1, \ldots, s_n). \qquad i = 1, \ldots, n-1$

If and only if these functions $\hat{\alpha}_i$ are linear, so that $\hat{\alpha}_i = c_i a_i$, then for another prior estimate a_i' of another parameter α_i' with the same likelihood, the MLE of α_i' can be obtained by

$$\hat{\alpha}_i' = c_i a_i' = \hat{\alpha}_i\,(a_i'/a_i).$$

3. In period t, the expenditure of the kth class of household on the ith commodity is

$$C_{ik}(t) = \gamma_{ik} + \beta_{ik} C_k(t).$$

If the population grows by g_k percent and expenditure per family stays constant, then $C_{ik}(t)$ grows by g_k percent also. If, however, the population is stationary and expenditure per family grows by ε_k percent, then

(Note continues on the following page.)

Table 7-2. *Household Shares in Gross Output from the Paddy Sector* (Ω_5)

Household	*1967*	*1972*	*1974*	*1974*[a]	*1974*[b]
Landless paddy workers	0.0143	0.0190	0.0190	0.0145	0.0190
Labor-abundant paddy farms	0.1456	0.1442	0.1534	0.1485	0.1534
Land-abundant paddy farms	0.4724	0.4675	0.4975	0.4813	0.4975
Nonproject farms	0.1814	0.1188	0.1000	0.1590	0.1000
Nonfarm	0.0437	0.0540	0.0586	0.0442	0.0586
Total	0.8574	0.8035	0.8285	0.8474	0.8285

a. In the absence of the project.
b. In the absence of autonomous changes elsewhere in the economy.

censuses of 1957 and 1970. The quinquennial growth rates for the farm and nonfarm subpopulations were 4.6 percent and 3.7 percent, respectively. It has been assumed that these rates held for 1967–72, because only after 1971 did farm incomes begin to rise strongly in response to the arrival of irrigation, and it is unlikely that expectations about alternative income levels, which figure heavily in the decision to migrate, would have been revised sharply in the immediate aftermath. It seems plausible that the surge in real incomes between 1971 and 1974 should have done something to stem the outflow of permanent migrants from the region. But in the absence of any evidence, we have been reduced to guessing that the farm and nonfarm populations were rising at 1 percent and 2 percent a year, respectively, between 1972 and 1974.

Model Results for 1967 and 1974

Complete SAMs have been generated by the model for 1967 and 1974 corresponding to the steady states of the economy, given the exogenous variables for those years. Tables 7-3 and 7-4 contain aggregated versions of these SAMs, the production activities having been aggregated into a single account. The details of sectoral outputs at the thirty-five-sector level are to be found in columns D and H in table 7-6.

In nominal terms, regional gross output rose by 146 percent over the seven-year period, the largest absolute contribution to this increase being

---

$$C_{ik}(t+1)|_{g=0} = \gamma_{ik} + \beta_{ik}(1+\varepsilon_k)C_k(t).$$

Combining the two, we get

$$C_{ik}(t+1) = (1+g_k)\gamma_{ik} + \beta_{ik}[(1+g_k)(1+\varepsilon_k)]C_k(t)$$
$$= (1+g_k)\gamma_{ik} + \beta_{ik}C_k(t+1),$$

and the marginal expenditure remains the same.

Table 7-3. *Estimated SAM for 1967*
(millions of Malaysian dollars in current prices)

	Expenditures				
	Institutions' current accounts				
	Households				
Receipts	*Landless paddy workers (1)*	*Labor-abundant paddy farms (2)*	*Land-abundant paddy farms (3)*	*Nonproject farms (4)*	*Nonfarm (5)*
1. Landless paddy workers			0.100		
2. Labor-abundant paddy farms					
3. Land-abundant paddy farms					
4. Nonproject farms					
5. Nonfarm households					
6. Firms					
7. Federal government	0.025	0.846	3.578	0.888	10.529
8. State government	0.002	0.113	0.493	0.100	8.215
9. Combined capital account	0.006	2.208	10.266	6.911	19.090
10. Production activities	3.371	18.874	44.697	30.446	85.981
11. Rest of world	0.757	5.703	16.615	12.026	35.906
12. Total	4.161	27.744	75.749	50.371	159.721

Table 7-4. *Estimated SAM for 1974*
(millions of Malaysian dollars in current prices)

	Expenditures				
	Institutions' current accounts				
	Households				
Receipts	*Landless paddy workers (1)*	*Labor-abundant paddy farms (2)*	*Land-abundant paddy farms (3)*	*Nonproject farms (4)*	*Nonfarm (5)*
1. Landless paddy workers			0.100		
2. Labor-abundant paddy farms					
3. Land-abundant paddy farms					
4. Nonproject farms					
5. Nonfarm households					
6. Firms					
7. Federal government	0.087	2.656	11.492	2.085	27.434
8. State government	0.002	0.113	0.493	0.100	16.190
9. Combined capital account	0.020	6.799	32.411	15.504	47.378
10. Production activities	9.167	50.936	124.613	67.876	202.465
11. Rest of world	2.942	19.092	57.882	27.228	100.237
12. Total	12.218	79.596	226.991	112.793	393.704

Expenditures						
Institutions' current accounts						
Firms (6)	*Federal government* (7)	*State government* (8)	*Combined capital account* (9)	*Production activities* (10)	*Rest of world* (11)	*Total* (12)
	1.090	0.565		2.321	0.085	4.161
	5.099	2.413		19.889	0.343	27.744
	10.897	4.510		59.709	0.633	75.749
	6.054	3.381		40.456	0.480	50.371
3.569	15.180	5.887		134.585	0.500	159.721
				33.937		33.937
15.340				0.880	50.470	82.556
				0.825	11.731	21.479
7.561	44.236	4.523			−17.854	77.037
			55.564	164.654	117.802	521.389
7.377		0.200	21.473	64.133		164.190
33.937	82.556	21.479	77.037	521.389	164.190	

Expenditures						
Institutions' current accounts						
Firms (6)	*Federal government* (7)	*State government* (8)	*Combined capital account* (9)	*Production activities* (10)	*Rest of world* (11)	*Total* (12)
	2.639	1.117		8.277	0.085	12.218
	12.948	4.770		61.535	0.343	79.596
	29.627	8.915		187.816	0.633	226.991
	14.307	6.685		91.321	0.480	112.793
8.856	40.844	11.638		331.866	0.500	393.704
				66.488		66.488
30.053				2.740	34.639	111.186
				1.991	20.500	39.389
18.899	10.821	5.564			−73.941	63.455
			45.959	395.319	387.606	1283.941
8.680		0.700	17.496	136.588		370.845
66.488	111.186	39.389	63.455	1283.941	370.845	

Table 7-5. *Per Capita Household Incomes, 1967 and 1974*
(current Malaysian dollars)

Household	*1967*	*1974*	*Percentage increase*
Landless paddy workers	123	390	217.0
Labor-abundant paddy farms	205	587	186.3
Land-abundant paddy farms	341	1,001	193.5
Nonproject farms	325	682	109.8
Nonfarm	585	1,333	127.9
All households	399	980	145.6

made by the paddy sector, which accounted for about 24 percent of the total in 1974. Regional value added grew slightly faster than gross output, because output increased at a greater than average rate in certain large sectors which have a high ratio of value added to gross output, such as paddy, smallholder rubber, and government. At a more disaggregated level, gross output from the agricultural complex (sectors 1–11) expanded by 178 percent, noteworthy details being a tripling in the value of paddy production, big jumps in smallholder rubber output (8) and rubber processing (9), and the more rapid expansion of output by small-scale rice mills than by commercial ones. Elsewhere in the economy, there were broad gains in transport (15, 16), services and distributive trades (17–22), and utilities (23–25). Residential construction (26) boomed, but there was a steep decline in nonresidential construction (27) as work on the Muda irrigation project tailed off from its 1968 peak. Current outlays by government (28–31) increased substantially.

Nominal household incomes grew by 163 percent in the aggregate over this period, which was more than twice the increase in the local cost of living.[4] Table 7–5 shows the per capita income levels in 1967 and 1974 for each household class. These figures suggest a definite improvement in the distribution of income between agricultural and nonagricultural households within the irrigation scheme area, with landless paddy farm workers doing particularly well. Despite a 109 percent increase, however, it is clear that the incomes of agricultural households outside the project's command area slipped substantially in relative terms.

Chapter 5 indicates a capital outflow from the region of \$35.2 million in 1972, and table 7-3 shows that even in 1967 there was a capital outflow of \$17.9 million. This figure increased to \$73.9 million in 1974 (table 7-4),

4. The household income levels in tables 7-3 and 7-4 differ from the values given in table 7-6 because the former include the value of subsidized services (education, health, and irrigation) provided by the government. See chapter 5 for a full discussion of the treatment of government services.

buoyed by the substantial rise in the price of paddy between 1972 and 1974. It is clear, however, that the region did not offer investment opportunities for local private capital that were competitive with the rest of the world. Incidentally, the capital outflow was equivalent to 46 percent of total household savings in 1967, and no less than 72 percent of total household savings in 1974.

The nominal value of regional imports increased by \$206.7 million (126 percent) between 1967 and 1974, while the value of regional exports of goods and services increased by \$269.8 million (229 percent). As a result, an import surplus of \$46.4 million in 1967 was transformed into an export surplus of \$16.7 million in 1974.

The Effect of the Project

In measuring the project's impact, we encounter the same "path-dependence" problem discussed in chapter 4. Once again, we measure the effect at constant prices and in the presence of autonomous sources of growth (the latter parallel the effect of improved varieties in chapter 4). That is to say, in estimating the impact of the project, we assume that it was reasonable to foresee the autonomous changes. As the rise in government expenditure was far and away the most important of them, this assumption does not seem unduly strong. "Project" here means irrigation plus the new varieties, whereas in chapter 4 they are kept separate. Since the SAM for 1972 has been estimated directly, the analysis will be done at 1972 prices rather than 1967 prices, the basis provided by the SAM being the more solid.

As before, let Z denote the vector of variables of interest and p the vector of prices. Then the project's impact at maturity is:

$$Z_{74}(p_{72}) - Z^{o}_{74}(p_{72}) \tag{7.1}$$

where the superscript ^{o} again denotes the situation in which there is no project. Now, the total change in Z from 1967 to 1974, in constant 1972 prices, is:

$$Z_{74}(p_{72}) - Z^{o}_{67}(p_{72}). \tag{7.2}$$

Hence, the project's contribution to the total change in Z can be expressed as:

$$\frac{Z_{74}(p_{72}) - Z^{o}_{74}(p_{72})}{Z_{74}(p_{72}) - Z^{o}_{67}(p_{72})} \times 100\%. \tag{7.3}$$

The analysis also requires a solution for the hypothetical case of 1967 in the absence of the project. Project construction actually began that year, and though the early work on the irrigation infrastructure had no immediate effects on the paddy sector, the construction expenditures would have provided a stimulus to some sectors of the regional economy. We have made certain assumptions about the exogenous variables of the model for the 1967 and 1974 situations without the project. Since the analysis is to be conducted in 1972 prices, the first step is to transform the models for 1967 and 1974 with the project, discussed in the preceding section, into variants at 1972 prices. The changes discussed below represent departures from those transformed values.

1967 without the Project

It may seem that we have complicated the task unnecessarily by choosing a starting year in which project construction was already under way, but the defense is a simple one. The data set on which the estimation of the exogenous variables would be based is very sketchy for earlier years, and we think it better to rest our (minor) hypothetical modifications to 1967 on the relatively secure foundations of the estimated "actuals" than to estimate "actuals" for, say, 1965 on a far shakier data base. We arrived at our hypothetical picture of 1967 by making a change in one exogenous variable of the set generating Z_{67} (p_{67}): a cut in investment deliveries for sector 27 of $40 million, the direct construction demands of the project in that year.

Several assumptions implicitly accompany this change. First, the outputs of sector 1–9 in the agricultural complex are held fixed at their actual levels in 1967, which implies that they were unaffected by the demand for factors generated by the project construction. In essence it is assumed that labor was in perfectly elastic supply, an assumption made plausible by the heavy out-migration from the region between 1957 and 1970. Second, it is unlikely that current government outlays (28–31) would have changed from their 1967 levels. Third, in the remaining sectors, since output is demand-driven, there is no shortage of capacity because a cut in exogenous demand will reduce output levels. Fourth, private transfers and lump sum taxes have been left alone. Estimates of the "actual" values of the former in 1967 are already tenuous. License fees for vehicles and businesses might have been a little lower in the absence of project construction work, but it is very difficult to gauge by how much. Last, a more subtle point: if there had been no construction work on the project, expectations about the future might well have been different, perhaps with important consequences for private investment. This, too, has been skipped over. The error, if any, would be in the direction of overstating the levels of regional activity and incomes in the hypothetical version of 1967.

1974 without the Project

The departures of the hypothetical exogenous variables from their actual values in that year are naturally more extensive. The most important of them is, of course, the level of paddy output. The small increase over the 1967 level largely reflects the improvement in yield of the main season crop, which would have occurred even if the project had not been undertaken. The sector's production technology was more advanced than in 1967, with greater mechanization and more intensive use of agrochemicals (mainly fertilizer). The small decline in the share in value added of nonproject farm households is the result of stagnant yields outside the project's command area over the 1967–74 period.

Since the region supplied about 45 percent of Malaysia's rice needs in 1974, one might ask whether the price of paddy would have risen even more if the project had not been undertaken. However, the country's source of marginal supplies was principally imports from Thailand and China, and it does not seem likely that the domestic price, which was close to the c.i.f. price of imports in 1974, would have risen further if there had been no Muda project.

In the agricultural complex, the gross output levels of the two rice milling sectors have been left at their 1967 levels. Since the output of paddy would have increased very modestly in the absence of the project and there were very small net imports of paddy into the region in 1967, this assumption seems sound. The gross output levels of sectors 3, 4, 6, 7, and 9 were set at their actual values in 1974, which amounts to assuming that, on balance, activity in these sectors was unaffected by the increased demand for domestic factors and goods generated by the project. In the case of smallholder rubber, however, some allowance was made for the fact that seasonal work in paddy production competes with rubber tapping. Net exports from sector 10 would have been somewhat lower in the absence of the project, because after 1970 the buffalo herd was run down rapidly in the face of mechanization, and the supply of animals on the hoof for export was increased. To reflect this, the sector's net exports were set $0.6 million lower than their actual level in 1974. As argued above, net export deliveries from sawmilling were left unchanged at their assumed value for 1974 (that is, their actual 1972 level). Both the exports of sector 14 and government outlays (29–31) were also left at their 1974 levels.

An inspection of the time series of buffalo livestock between 1967 and 1973 led to a guesstimate of zero deliveries from sector 10 to investment activities in 1974 in the absence of the project. Deliveries from sector 14 and government investment in housing were taken as given, but private investment in urban housing was reduced by about 50 percent to reflect the likely reduction in the demand for urban housing services had the project not been undertaken. Investment in rural housing was derived from the expenditures of single-cropping households in 1972. In the case of nonresidential construction, the tail end of Muda project work was cut out, together with a sizable chunk of the modest private demand for such output.

Following earlier practice, private transfers were left at their actual levels in 1974. Lump-sum taxes were altered to reflect guesstimates of what the stock of vehicles and the number of businesses would have been in the absence of the project. These changes affected mainly nonfarm households, reducing their tax burden by almost $4.5 million.

The only remaining issue is whether permanent out-migration from the region might have been higher if there had been no project. Although there is no direct evidence on which to base an answer, it seems plausible that the rapid rise in incomes between 1970 and 1974 induced more people to stay in the region than would otherwise have been the case. In rough and ready fashion, it has been assumed that the farm population would have been stationary after 1972, whereas the nonfarm population would have grown at 0.5 percent a year.

The Results

Columns A, B, and C of table 7-6 contain the necessary solutions for measuring the project's impact at maturity in 1972 prices. Asterisks placed against sector gross outputs indicate that their values were exogenously fixed as discussed above.

Total gross output was $836 million in 1974 when expressed in 1972 prices. This is $199 million larger (an increase of 31 percent) than it would have been had the project not been constructed, and $354.5 million larger than it would have been in 1967 had there been no project construction work in that year. According to expression (7.3), the project accounted for 56 percent of the total increase in regional gross output between 1967 and 1974 when measured in 1972 prices. Similar calculations for the other variables in table 7-6 lead to the numbers in the fourth column.

As one would expect, the project had its biggest impact on the paddy and rice milling sectors, and accounted for most of the increase in government expenditure on irrigation. The indirect effects were also substantial. The project accounts for at least one-third of the increase in output of all demand-driven sectors, and for a much higher proportion of the increase in output of such sectors as agricultural machinery services (95 percent), construction (70 percent), financial services (62–70 percent), and trade (57–59 percent).

In the aggregate, the project induced an increase of $110 million in regional value added. Of this, $60 million was due to the increase in paddy output, so that for every dollar of value added generated directly by the project at maturity, another 83 cents were generated in the form of downstream or indirect effects. Further, the rice milling sectors accounted for only $5.9 million of the increase in value added, which is equivalent to 10 cents of the 83 cents of indirect benefits. This still leaves 73 cents of indirect benefits generated in other sectors of the economy. To put it slightly differently, if we integrate the paddy and rice milling sectors—a common practice in project design—for every dollar of value added arising from the integrated project, 67 cents of value added were generated downstream.

The project accounted for about 80 percent of the increase in the real incomes of paddy farmers and landless paddy farm workers. It also accounted for about one-third of the increase in the incomes of nonagricultural households. Farm households outside the irrigation scheme area benefited only marginally from the project.

Table 7-7 shows the change in the distribution of per capita incomes brought about by the project. In the absence of the project in 1967, landless workers' per capita incomes would have been only 30 percent of the average income level for all households in the region (see the figures in parentheses). At the other extreme, nonagricultural households would have received per capita incomes which were 62 percent above the regional average. Without the project, the distribution of incomes would have been even worse in 1974. In fact, landless workers would have seen their per capita incomes fall to only 25 percent of the regional average, while nonagricultural households would have advanced to 72 percent above the average. The average income would have increased by 37 percent over this period, all households would have been better off in real terms, but the distribution of incomes would have become more skewed. The project was apparently successful in staving off this movement. Even though the average per capita income more than doubled in the presence of the project, the per capita incomes of landless farm workers increased to 34 percent of the regional average, while those of nonagricultural households fell to 57 percent above the average. Paddy farming households also improved their relative standing with the construction of the project, but farm households located outside the project's command area were made relatively worse off. Even so, they were still rather better off in absolute terms than they would have been without the project.

The direct and indirect income effects of the project can be separated. Table 7-7 shows the per capita income for each household type that was derived from paddy production. Landless workers received $149 of their total per capita income of $227 from the paddy sector in 1974, in this case through employment in paddy farming. The project brought about an

Table 7-6. *Results from the Regional Model, 1967–74*
(millions of Malaysian dollars)

Sector	$Z^o_{67}(p_{72})$ *(A)*	$Z^o_{74}(p_{72})$ *(B)*	$Z_{74}(p_{72})$ *(C)*
1. Commercial rice mills	73.08*	73.08*	107.00*
2. Small rice mills	41.46*	41.46*	68.00*
3. Food processing	5.76*	8.20*	8.20*
4. Fish processing	5.92*	7.33*	7.33*
5. Paddy production	84.15*	95.99*	170.56*
6. Fishing	24.48*	34.00*	34.00*
7. Estates rubber	2.80*	3.90*	3.90*
8. Smallholder rubber	7.71*	23.12*	22.30*
9. Rubber processing	1.11*	2.32*	2.32*
10. Other agriculture	13.67	20.43	23.66
11. Sawmilling	7.36*	8.45	9.62
12. Agricultural machinery services	1.24	1.34	3.11
13. Machinery repair	3.00	4.65	6.67
14. Manufacturing n.e.c.	16.51	19.99	22.57
15. Road transport	10.57	15.58	20.30
16. Rail transport	0.61	0.92	1.38
17. Hotels and restaurants	8.95	15.27	18.48
18. Entertainment	2.51	4.78	6.07
19. Services n.e.c.	0.80	1.22	1.79
20. Private health	2.66	3.50	4.27
21. Distributive trades	26.95	36.24	48.69
22. Petty trading	2.47	3.18	4.22
23. Electricity	3.53	4.99	5.73
24. Water	1.75	2.55	3.04
25. Posts and telecommunications	0.96	2.46	3.21
26. Residential construction	10.60	13.23	18.59
27. Other construction	17.37	18.66	21.67
28. Government, irrigation	3.70*	4.02*	8.37*
29. Government, education	25.22*	53.38*	53.38*
30. Government, health	7.50*	12.75*	12.82*
31. Government, other	30.81*	49.72*	49.88*
32. Traditional financial services	2.86	3.90	6.38
33. Modern financial services	1.45	2.00	2.92
34. Urban housing	11.52	18.72	22.61
35. Rural housing	20.43	25.71	32.96
Total gross output	481.46	637.04	836.00
Total value added	277.88	395.00	505.00
Household incomes			
Landless paddy workers	2.35	2.84	4.94
Labor-abundant paddy farms	17.84	21.55	35.10
Land-abundant paddy farms	52.62	63.20	105.71
Nonproject farms	33.80	50.78	54.77
Nonfarm	148.73	226.78	265.99

Note: An asterisk denotes exogenously fixed levels of gross output, except in connection with Z, where it denotes the absence of new varieties.

Project impact at 1972 prices $\frac{C-B}{C-A} \times 100$	$Z_{67}(p_{67})$ (D)	$Z^o_{67}(p_{67})$ (E)	$Z^o_{74}(p_{74})$ (F)	$Z^*_{74}(p_{72})$ (G)	$Z_{74}(p_{74})$ (H)
100	82.07*	82.07*	130.89*	107.00*	191.64*
100	46.56*	46.56*	74.25*	68.00*	121.79*
0	5.54*	5.54*	11.98*	5.76*	11.98*
0	3.42*	3.42*	10.73*	5.92*	10.73*
86.3	96.94*	96.94*	172.51*	170.56*	306.50*
0	14.13*	14.12*	49.78*	24.48*	49.78*
0	3.91*	3.91*	6.54*	2.80*	6.54*
−5.6	10.76*	10.76*	38.75*	7.71*	37.37*
0	1.42*	1.42*	4.45*	1.11*	4.45*
32.3	14.17	13.81	30.53	18.48	36.96
51.8	6.39*	6.39	12.70	7.36*	14.58
94.7	1.05	1.05	1.66	3.11	3.86
55.0	3.83	2.39	5.89	5.43	8.76
42.6	17.38	15.34	28.81	20.02	33.27
48.5	11.31	9.96	20.64	16.17	28.46
59.7	0.56	0.54	1.23	1.19	1.90
33.7	8.81	7.93	19.02	14.04	24.68
36.2	2.50	2.08	5.71	4.84	7.99
57.6	1.00	0.98	1.97	1.22	3.02
47.8	2.70	2.59	4.75	3.49	6.10
57.3	26.18	24.58	48.45	40.80	67.31
59.4	2.65	2.61	4.80	3.20	6.69
33.6	3.54	3.32	5.31	4.45	6.30
38.0	1.74	1.63	3.37	2.30	4.20
33.3	1.08	0.83	2.93	2.17	4.19
67.1	9.17	9.09	20.07	15.99	28.43
70.0	50.52	14.28	29.02	20.65	33.78
93.2	3.15*	3.15*	5.31*	8.34*	11.01*
0	20.83*	20.83*	53.86*	25.22*	53.86*
1.3	6.20*	6.20*	12.86*	7.57*	12.94*
0.8	25.81*	25.47*	50.18*	31.00*	50.36*
70.5	3.28	3.22	6.28	4.30	10.56
62.6	1.55	1.31	2.54	2.52	3.84
35.1	10.93	9.68	22.20	18.41	28.82
57.9	20.37	19.84	38.25	25.87	51.29
56.1	521.45	473.84	938.23	701.50	1283.84
48.4	292.60	269.98	556.40	406.84	752.03
81.1	2.51	2.48	4.60	4.70	8.46
78.5	20.23	20.09	36.82	33.92	61.88
80.1	60.24	59.95	109.71	102.99	188.45
19.0	40.94	40.65	84.18	39.96	91.80
33.4	138.65	125.71	275.62	191.23	341.22

Table 7-7. *The Level and Distribution of per Capita Household Incomes*
(Malaysian dollars in 1972 prices)

	$Z^o_{67}(p_{72})$		$Z^o_{74}(p_{72})$		$Z_{74}(p_{72})$	
Household	*Total*	*Paddy*	*Total*	*Paddy*	*Total*	*Paddy*
Landless paddy workers	115 (30)	59	133 (25)	65	227 (34)	149
Labor-abundant paddy farms	181 (47)	124	209 (39)	138	333 (50)	248
Land-abundant paddy farms	298 (77)	226	342 (65)	250	561 (85)	450
Nonproject farms	268 (69)	121	385 (73)	116	407 (62)	127
Nonfarm	627 (162)	16	913 (172)	17	1039 (157)	39
All households	387 (100)	110	530 (100)	118	661 (100)	200

Note: Figures in parentheses are income indices with the average per capita income over all household types as the base.

increase of 129 percent in the earnings from paddy of landless workers, whereas it increased the incomes from paddy of farm households in the scheme area by only 80 percent. The percentage increases in total per capita incomes, however, were much more even—71 percent, 59 percent, and 64 percent for landless workers, labor-abundant, and land-abundant paddy farms, respectively—because paddy farm households gained relatively more from increases in income from other sources than did the landless workers. The paddy earnings of nonfarm households and farms outside of the command area changed very little in absolute terms with the advent of the project, though the former gained substantial increases in their incomes from other activities.

To sum up, the increases in incomes from sources other than paddy, and hence the indirect effects of the project, were skewed very much in favor of the nonfarm households in the region, though paddy farm households benefited more from the indirect effects than did the landless workers. These results carry an important policy implication; namely, that project analysts intent on improving the distribution as well as the absolute levels of incomes in project areas may well find that their projects have awkward indirect effects, even though the direct benefits might, with some care, be directed toward the poorer households in a region. In the absence of suitable complementary policy interventions, the indirect benefits of the project are likely to work in favor of the richer, nonagricultural households, the members of which are generally better placed to participate in non-farm, entrepreneurial activities. In the Muda case, the distribution of the direct benefits of the project more than offset the adverse distribution of the indirect benefits, so that the overall distribution of total incomes did, in fact, improve. However, this happy result is by no means necessarily the general rule.

Production and Consumption Linkages

Given the magnitude of the indirect or downstream effects of the project, it is important to ascertain the relative strengths of the project's production and consumption linkages in inducing these effects at project maturity. Production linkages between agriculture and the rest of the economy have generally been thought to be weak (see, for example, Hirschman, 1959). If this is so, an agricultural investment project is unlikely to have strong production linkages in its surrounding region. In contrast, Mellor (1976) has stressed the role of households' consumption linkages in generating important secondary effects after an initial increase in farm incomes. To test these ideas in a quantitative way, we have solved a variant of the semi-input-output model in which household incomes (and hence consumers' expenditures) are fixed exogenously, so that only the production linkages are free to work. More specifically, household incomes have been set at the levels obtained from the $Z^o_{74}(p_{72})$ solution, and a new solution has been obtained for the 1974 "with project" situation, $Z_{74}(p_{72})$.

Under these circumstances, the region's gross output and value added would have been \$796.6 million and \$474.6 million, respectively. Value added in the absence of the project would have been \$395 million, and the value added generated directly by the project stays at \$60 million. In this case, the downstream value added generated by production linkages alone is \$19.6 million. Hence, for every dollar of value added generated directly by the project, 33 cents of value added were generated in the form of downstream effects through the operation of production linkages alone, compared with 83 cents when both production and expenditure linkages are allowed for. In the case of this project, then, the omission of consumers' expenditure linkages, which interact with the system's interindustry linkages, would lead to a gross underestimate of the project's downstream effects.

Balancing Investments

To complete this analysis of the project's impact on the region, it is necessary to form some estimate of the investment needed to realize the downstream effects. At one extreme, it could be assumed that all sectors other than that producing paddy were suffering from excess capacity to such an extent that no additional investment was needed to complement the project itself. But this is hardly realistic in the light of the investments in buildings, housing, vehicles, and rice milling that accompanied the surge in incomes after 1970. Unfortunately, the available data provide only a sketchy investment series for the period, so the following estimates of investments related to downstream effects are inevitably somewhat tentative.

For present purposes, the relevant investment deliveries were made by sectors 14, 26, 27, and imports. An estimate of the investment needed to realize both autonomous and downstream increases in regional value added is obtained by omitting construction deliveries to the Muda project. Without discounting, this investment reached a cumulative total of $240 million in 1972 prices. Aside from additional investment in the rice milling sectors, which was mainly for extra drying capacity to handle a second crop and did not exceed $5 million (FAO, 1975), it seems plausible that the composition of the investments needed to support autonomous growth and downstream effects would not be very different. In this case the autonomous increase in value added between 1967 and 1974 was $117.1 million, while the downstream value added attributable to the project at maturity in 1974 was $50 million. Apportioning the $240 million additional investment in the same ratio as that for value added, we arrive at an estimate of $72 million for the cumulative, undiscounted total investment associated with the steady-state, downstream increase in value added.

This estimate is, however, almost certainly on the high side, for much of the residential and nonresidential construction was undertaken by the state or federal governments. Moreover, it is unlikely that extra investments were made much in advance of the appearance of the project's downstream effects in 1970. Fragmentary evidence suggests that annual private investment in housing and other buildings would have been about $7 million lower in the absence of the project from 1970 onward. Accumulated over four years, downstream related investments are estimated at $38 million. In the light of these two estimates, it seems fairly probable that for each dollar of downstream value added associated with the project, between $0.75 and $1.50 of complementary investment was needed, appropriately distributed over all other sectors.

Decomposition of the Sources of Change, 1967–74

When the project's impact is measured with expression (7.3), the project is credited with some of the benefits arising from autonomous changes in the economy, particularly those from the simultaneous occurrence of the project with other autonomous changes. Such interaction effects represent a windfall gain to the project from fortunate timing, and although they can rightly be attributed to the project in an ex post evaluation, it is instructive to explore what might have happened had the project not been accompanied by such changes.

To this end, we turn to a decomposition procedure similar to that used in the analysis of the paddy sector in chapter 4 in order to separate the contributions of the various sources of growth in the regional economy between 1967 and 1974. The appropriate identity is:

$$(7.4) \quad Z_{74}(p_{74}) - Z^{o}_{67}(p_{67}) \equiv [Z^{*}_{74}(p_{72}) - Z^{o}_{67}(p_{72})] + [Z^{o}_{74}(p_{72}) - Z^{o}_{67}(p_{72})]$$
$$+ [Z_{74}(p_{72}) - Z^{*}_{74}(p_{72}) + Z^{o}_{67}(p_{72}) - Z^{o}_{74}(p_{72})]$$
$$+ [Z_{74}(p_{74}) - Z_{74}(p_{72}) + Z^{o}_{67}(p_{72}) - Z^{o}_{67}(p_{67})]$$

[Total change] ≡ [Project effect] + [Autonomous growth]
+ [Project-autonomous interactions]
+ [Price and price-interaction effects]

The identity (7.4) differs from (4.1) in that we have used 1972 prices rather than 1967 prices to measure the nonprice effects. Also, since there was some initial project construction work in 1967, it is necessary to use $Z^{o}_{67}(p_{67})$ as the base. The treatment of the autonomous sources of growth closely parallels the treatment of the introduction of improved varieties in chapter 4, the absence of any autonomous growth being denoted by an asterisk.

All the solutions necessary to implement (7.4) are to be found in table 7-6. The only specification that has not been discussed so far is that for $Z^{*}_{74}(p_{72})$. The assumptions underlying this solution are as follows. We are interested in simulating the economy in the presence of the project but the absence of autonomous changes. Thus, it is natural to draw the gross outputs for paddy and rice milling from $Z_{74}(p_{72})$, while (conservatively) leaving all other exogenously determined gross outputs at their levels in $Z^{o}_{67}(p_{72})$. Since "other agriculture" is heavily influenced by its livestock subsector, export and investment deliveries have been drawn from $Z_{74}(p_{72})$, for the technology for producing paddy is largely mechanized in $Z_{74}(p_{72})$. Export and investment deliveries by sector 14 have been taken from $Z^{o}_{67}(p_{72})$. Investment deliveries by the construction sectors and the fixed taxes $\{\bar{T}_k\}$ have been defined as their values in 1967 in the absence of the project plus the difference between those in 1974 in the presence and absence of the project, respectively. In effect, this is a rough and ready way of estimating the effects of the project alone on variables which, while set exogenously, are clearly influenced by its presence. Finally, the farm

Table 7-8. *An Analysis of the Components of Change, 1967–74*
(millions of Malaysian dollars)

Variable	*1967 value (without project)*	+	*Pure project effect*	+	*Autonomous change*	+	*Project-autonomous interaction*	+	*Price and price-interaction effects*	=	*1974 value (with project)*
Total gross output	473.84		220.04		155.58		−21.08		455.56		1283.94
			(27.16)		(19.21)		(−2.61)		(56.24)		
Total value added	269.98		128.96		117.12		−18.96		254.93		752.03
			(26.75)		(24.30)		(−3.93)		(52.88)		
Household incomes											
Landless paddy workers	2.48		2.35		0.49		−0.25		3.39		8.46
			(39.30)		(8.19)		(−4.18)		(56.69)		
Labor-abundant paddy farms	20.09		16.08		3.71		−2.53		24.53		61.88
			(38.48)		(8.88)		(−6.06)		(58.70)		
Land-abundant paddy farms	59.95		52.62		10.58		−10.11		75.41		188.45
			(40.95)		(8.23)		(−7.86)		(58.68)		
Nonproject farms	40.65		6.16		16.98		−2.17		30.18		91.80
			(12.04)		(33.20)		(−4.24)		(59.00)		
Nonfarm	125.71		42.50		78.05		−3.29		98.25		341.22
			(19.72)		(36.22)		(−1.53)		(45.59)		
Total household income	248.88		119.71		109.81		−16.10		231.76		691.81
			(27.03)		(24.79)		(−4.14)		(52.32)		

Note: Figures in parentheses are percentages of total change.

populations were drawn from $Z_{74}(p_{72})$, while the increase in the nonfarm population has been halved to allow for more modest levels of nonfarm incomes in the absence of autonomous changes.

The results of this decomposition are summarized in table 7-8. The "pure" project effect is defined as the change in key variables that would have been observed between 1967 and 1974 had prices remained at their 1972 levels throughout, and had there been no autonomous changes. A comparison of expressions (7.1) and (7.4) reveals that the "pure" project effect differs from the measure of project impact defined in expression (7.3) in that the latter is the sum of the pure project effect and the interaction effect of the project with autonomous changes. Specifically,

$$(7.5) \qquad Z_{74}(p_{72}) - Z^{o}_{74}(p_{72}) \equiv [Z^{*}_{74}(p_{72}) - Z^{o}_{67}(p_{72})] + [Z_{74}(p_{72}) - Z^{*}_{74}(p_{72}) + Z^{o}_{67}(p_{72}) - Z^{o}_{74}(p_{72})].$$

The pure project impact should therefore be smaller than the effects of the project discussed in the previous section by the amount of the interaction term (or larger in those cases where the interaction term is negative).

The pure project effect accounts for a little over a quarter of the increase in the nominal value of regional gross output and value added between 1967 and 1974. As expected, it accounts for a much larger proportion of the increase in the incomes of farm households in the project's command area than in those of farm households outside and of nonfarm households. Of the outsiders, nonfarm households do the better. As absentees, they receive extra rents; as rice millers and suppliers of goods and services to agricultural production, they gain extra business; and as producers of the goods and services on which farm households spend their extra incomes, they benefit from additional demand. Farm households outside the project get a few crumbs: more casual employment on paddy farms inside the project, a few new jobs in the nonfarm sectors of the economy, and some extra income from petty trading.

With regard to the effects of "autonomous" changes, however, the position is reversed. Farm households outside the project benefit from the rise in the area under smallholder rubber and the yield of the trees, while nonfarm households gain most of the additional incomes generated (directly and indirectly) by the increase in government expenditures and the expansion of other sectors producing traded goods. Households engaged in paddy farming participate only marginally in these activities. In the aggregate, the effects of the project alone are slightly larger than those of the autonomous forces at work in that period.

The interaction effects of the project and autonomous changes are negative, but relatively small. That they are negative is partly a reflection of the fact that the supplies of resources, especially farm labor, are not perfectly elastic. The structure of direct taxation is also very mildly progressive. Hence, when the project and autonomous changes appear together, "fiscal drag," enhanced in the face of inelastic labor supplies, works to moderate their combined effects, because the government's outlays are fixed independently of its revenues.

This result permits us to say somewhat more about the multiplier effects estimated above. The pure project effect accounts for an increase of $128.16 million in regional value added (at 1972 prices). Of this, $63.05 million is due to the increase in the output of paddy. Hence, for each dollar of value added generated directly by the pure project effect, another $1.03 of value added is generated downstream. This value added multiplier is 20 cents larger than the one reported above (the subsection "The Results") the difference being due to the negative interaction effects between the project and autonomous sources of growth (see identity [7.5]). Put another way, if there had been no competition for scarce resources arising from the demands induced by the project and the autonomous sources of growth, the value added multiplier of the project at maturity would have been 2.03 instead of 1.83. At the margin, therefore, the interaction effects are quite important.

As in chapter 4, changes in prices and the effects of their interaction with the project and other changes account for the lion's share of the increases in nominal output and incomes between 1967 and 1974. To judge from households' incomes, however, it appears that agricultural households fared relatively better from the movements in prices than did nonfarm households. Price movements accounted for only 46 percent of the increase in the incomes of the latter, compared with 57 to 59 percent for the former, which reflects the facts that relative prices moved in favor of agricultural producers between 1967 and 1974 and agricultural output expanded rapidly.

To go into this matter more deeply, two aspects of changes in prices concern us here. First, changes in absolute prices affect the nominal values of output and factor rewards. Second, changes in relative prices, without an accompanying change in the general level of nominal prices, may have important consequences for production and income distribution.

Beginning with the movements in nominal prices, the price and price-interaction effects in table 7-8 account for an increase of $254.9 million in the nominal value of the region's income between 1967 and 1974, or an increase of 94 percent. Given that the local cost of living rose by about 50 to 60 percent over this period (table 7-1), the general upward movement in the prices facing producers more than kept pace. Nominal and real incomes rose even more than this, of course, thanks to the "real" contributions of the project and other autonomous sources of growth.

With regard to the terms of trade effects, table 7-9 contains the model's estimates of sectoral gross outputs (in 1972 prices) in 1967 and 1974 at both

Table 7-9. *Real Gross Output, 1967 and 1974*
(millions of Malaysian dollars in 1972 prices)

Sector	*1967*		*1974*	
	1967 terms of trade	*1972 terms of trade*	*1974 terms of trade*	*1972 terms of trade*
1–9. Agriculture	246.47*	246.47*	423.61*	423.61*
10. Other agriculture	15.47	14.10	24.72	23.66
11. Sawmilling	7.36*	7.36*	9.76	9.62
12. Agricultural machinery services	1.24	1.24	3.12	3.11
13. Machinery repair	4.87	4.74	6.90	6.67
14. Manufacturing n.e.c.	18.43	18.72	21.56	22.57
15. Road transport	12.95	12.14	22.71	20.30
16. Rail transport	0.66	0.63	1.45	1.38
17. Hotels and restaurants	9.68	10.00	17.13	18.48
18. Entertainment	2.82	3.02	6.50	6.07
19. Services n.e.c.	1.18	0.82	2.07	1.79
20. Private health	2.85	2.77	4.92	4.28
21. Distributive trades	29.28	28.80	51.50	48.69
22. Petty trading	2.99	2.52	5.49	4.22
23. Electricity	3.77	3.78	6.13	5.73
24. Water	2.05	1.89	3.37	3.04
25. Posts, telecommunications	1.25	1.27	3.17	3.21
26. Residential construction	10.80	10.69	18.56	18.59
27. Other construction	61.53	61.51	21.71	21.67
28–31. Government	67.67*	67.67*	124.45*	124.45*
32. Traditional financial services	3.69	2.92	8.67	6.38
33. Modern financial services	1.71	1.72	2.95	2.92
34. Urban housing	12.20	13.02	20.90	22.61
35. Rural housing	23.96	21.06	33.70	32.96
Total gross output	544.86	538.84	845.06	836.00

Note: An asterisk denotes exogenously fixed levels of gross output.

the actual terms of trade for each of those years as well as those for 1972. Although the movements in prices away from their base year (1972) values had little effect, in the aggregate, on the level of endogenously determined output,[5] the composition of such output underwent significant changes. Since the values of the demand parameters vary considerably across household classes, such changes in composition were the result of shifts in the distribution of incomes as well as substitution effects, total income being virtually unchanged. The sectors most favorably affected in both 1967 and

5. An analysis of the effects of devaluation that has something in common with our analysis of the effects of changes in the barter terms of trade may be found in Krugman and Taylor (1978).

Table 7-10. *The Level and Distribution of per Capita Household Incomes*
(Malaysian dollars in 1972 prices)

	1967				1974			
	1967 terms of trade		*1972 terms of trade*		*1974 terms of trade*		*1972 terms of trade*	
Household	*Total*	*Paddy*	*Total*	*Paddy*	*Total*	*Paddy*	*Total*	*Paddy*
Landless paddy workers	135 (31)	77	116 (28)	59	266 (38)	188	227 (34)	149
Labor-abundant paddy farms	221 (51)	159	181 (44)	123	398 (58)	311	333 (50)	248
Land-abundant paddy farms	373 (86)	292	298 (72)	223	690 (100)	573	561 (85)	450
Nonproject farms	355 (82)	157	269 (65)	120	469 (68)	161	407 (62)	127
Nonfarm	639 (147)	20	700 (170)	16	970 (140)	72	1039 (157)	39
All households	435 (100)	143	412 (100)	109	692 (100)	261	661 (100)	200

Note: The figures in parentheses are income indices with the average per capita income over all household types as the base.

1974 were services not elsewhere classified, private health, petty trading, and traditional financial services. Those most adversely affected were hotels and restaurants and urban housing services. This pattern suggests that (redistributive) income effects were very much at work in these cases.

Table 7-10 shows the corresponding levels of per capita household incomes in 1972 prices. Per capita incomes are apparently much more sensitive to changes in the terms of trade than are sectoral outputs. The average per capita income was 6 percent larger in 1967 than it would have been had the terms of trade in 1972 prevailed. Similarly, in 1974 the average per capita income was 5 percent higher with the terms of trade actually ruling than with those ruling in 1972.

The movements in the terms of trade between 1967 and 1972 favored nonfarm households; their per capita incomes would have been almost 10 percent higher if the terms of trade in 1972 had ruled in 1967. The incomes of all other household types would have been much lower, however, particularly those of farm households outside of the command area. In contrast, the movements in the terms of trade between 1972 and 1974 worked strongly in favor of agricultural households, whereas nonfarm households suffered a modest reverse.

Conclusions

It must be emphasized that the frailties of the data base, and the many assumptions thus entailed in deriving results, demand some caution in drawing conclusions from the empirical analysis. Nevertheless, the broad orders of magnitude of certain key variables should be sufficiently solid to warrant some confidence in our main findings. In the aggregate, the Muda project's downstream effects were of the same order as its direct effects. For every additional dollar of value added in paddy production generated by the project at maturity, about 83 cents of value added were generated by downstream effects, partly in rice milling. Also, each dollar of downstream value added was probably supported by just over a dollar of additional investment in plant and equipment, spread appropriately over the sectors that expanded in response to the project. The direct effects of the project did not worsen the distribution of income among farm households, but its downstream value added accrued mainly to the nonfarm households engaged in paddy milling and the production of nontradables. Although the spread in nonfarm incomes was wide, the lion's share of downstream income went almost certainly to households which were better-off than those engaged in paddy farming. Thus, while the project's downstream effects did much to boost the aggregate income of this relatively poor region, they kept the intraregional distribution of income from improving as much as the project's direct effects would, at first sight, suggest. Lastly, it is clear that, new technology in paddy production notwithstanding, the project's production linkages alone were much weaker than its combined production and consumption linkages, even when the expansion of paddy milling, agricultural machinery services, and irrigation activities are taken into account.

Although the movements in the terms of trade away from their base values were quite strong in 1967–74, the results from the model suggest that in the aggregate their net contractionary or expansionary effects were rather small. Such movements in prices did, however, produce strong shifts in the distribution of household incomes. When prices favored agriculture as a whole (and paddy producers in particular), the gap between farm and nonfarm household incomes was certainly narrower, other things being equal. However, this structure also made for a somewhat more unequal income distribution within agriculture itself.

Shifts in the absolute prices of tradables also produce changes in those prices relative to nontraded goods, and variations in the output of nontradables constitute the main part of expansionary and contractionary movements overall. Hence, both substitution and income distribution effects were at work in maintaining an apparently fairly stable level of effective demand. When prices shifted in favor of agriculture, it appears that the gains in real agricultural incomes and the structure of farm households' tastes as consumers generated just enough extra demand for nontraded goods to offset the decline in such demand resulting from the associated fall in the real incomes of nonfarm households. To put it slightly differently, the employment of factors in the nonfarm sectors of the economy did not change much, but the real rewards accruing to them did.

All in all, it would appear that attempts to shift prices in favor of this set of peasant farmers will improve their incomes without changing appreciably the level of household income for the region as a whole. Nevertheless, the attendant effects on the distribution of income are not unambiguously equalizing. In chapter 9, we will attempt a more extensive analysis of the role of price policy in raising farm incomes in the future.

8

A Social Cost-Benefit Analysis of the Project

CHAPTERS 6 AND 7 were concerned solely with the effects of the Muda project on output and incomes in the surrounding regional economy. Naturally, the project has also had an impact on the national economy, principally through its effects on the uses of labor, on tradable goods, and on the level of government revenues. Measuring this impact, a task that lies in the province of social cost-benefit analysis, is the subject of this chapter.

Some of the recent literature in this field (Blitzer, Little, and Squire, 1978; Scott, 1976) has been concerned with the multiplier or downstream effects of projects, and especially with the derivation of shadow prices that capture all such effects in full. If these shadow prices are correctly calculated, so it is asserted, then valuing a project's *direct* inputs and outputs at these prices yields the right measure of its social profitability. This approach is in the spirit of, and consistent with, that of the various manuals on social cost-benefit analysis (UNIDO, 1972; Little and Mirrlees, 1974; Squire and van der Tak, 1975).

This chapter also deals with the problem of allowing fully for multiplier effects in project evaluation, but approaches it on a different tack. The total (direct and indirect) impact of a project on outputs, incomes, savings, and expenditures, which are the elements entering into the calculation of social profits, is derived with the use of a linear model of the economy. In other words, we follow the "primal" route of working solely with quantity and income flows rather than the "dual" route of shadow prices. The notion of "primal" and "dual" is more than suggestive, for there is an intimate connection between the model developed in this chapter and that which underlies the manuals referred to above. Although this connection will not be analyzed in great detail here, the concepts and formulations shared by the two approaches will be discussed as they appear. Readers who are more

Clive Bell and Shantayanan Devarajan are solely responsible for this chapter.

at home with the approach in the manuals should therefore find at least a few familiar points of reference as the chapter unwinds, while those who were nurtured on semi-input-output analysis may find it insightful to have this extension laid out explicitly in its context of social cost-benefit analysis.[1] The relation between the primal and dual approaches is illustrated in a formal and rigorous way in the appendix to this chapter. The model employed there is highly simplified, since our aim is to demonstrate that the two approaches are formally equivalent rather than to derive the complete set of shadow prices corresponding to the primal model developed below. Simple though it is, the model in the appendix has sufficient features in common with the full version that extending the analysis in the appendix to cover more elaborate cases is intuitively straightforward.

The Model

This section presents the linear model used to estimate the values of all the variables of interest associated with the project's direct and indirect effects. After a brief description of the sectoral classification employed, the various ingredients of the model, that is, the accounts on which it is built, are taken up. At this point, the model is best thought of as one of a national economy, although its application to a regional context is discussed in the next section. Since the model is a good deal more elaborate than its cousins in chapters 6 and 8, there are, unavoidably, some changes in notation.

Like its simpler cousins, the model is a variant of Tinbergen's (1966) semi-input-output framework. The *M* production activities in the economy are divided into two groups: those producing tradables and those producing nontradables.[2] The tradables sectors have parametrically fixed output levels, that is, they operate at full capacity, so that variations in demand are met entirely by changes in net exports. The nontradables sectors, on the other hand, respond to changes in demand solely by adjusting output. Either they have excess capacity, in which case they are producing at constant marginal costs, or they add new plant and equipment to satisfy demand, in which case production takes place at constant average costs.

1. Wanhill (1974) and Kuyvenhoven (1975) have tried to synthesize the Little-Mirrlees and semi-input-output approaches without alluding to the primal-dual distinction. Elsewhere (Bell and Devarajan, 1980*a*), we show why their approach departs from the spirit of Little and Mirrlees. The only other empirical application of a similar nature known to us is by Setor (1973), who attempts to calculate shadow prices for Chile, allowing for output- and expenditure-linkage effects. Unfortunately, as noted in Bell and Devarajan, Seton's otherwise commendable paper is marred by a theoretical flaw.

2. Tinbergen calls these two kinds of sectors "international" and "national." To highlight the relation of our approach to that of Little and Mirrlees, we have adopted the terminology used by the latter.

The classification into tradables and nontradables is not necessarily a statement about whether these sectors are, in fact, trading. Rather, the distinction refers to the source of supply response—whether it is net exports or domestic output. As before, to keep the national income accounts consistent, it is necessary to introduce a third type of sector—for distribution and transport services—since export activities, both exogenous and endogenous alike, purchase those services domestically. The full rationale for this modification is given in chapter 6.

Material Balances

The subscripts T, D, and N denote the set of tradables, distribution and transport, and nontradables sectors, respectively. The rows and columns of the fixed coefficient technology matrix, A, are ordered so that it can be partitioned as follows:

$$A = \left[\begin{array}{c|c|c} A_{TT} & A_{TD} & A_{TN} \\ \hline A_{DT} & A_{DD} & A_{DN} \\ \hline A_{NT} & A_{ND} & A_{NN} \end{array}\right].$$

The submatrix A_{TN}, for example, represents the inputs of tradables required by the nontradables sectors when the latter are operating at unit activity levels.

The material balance equations for the economy are:

$$(8.1) \quad X_T = A_{TT}X_T + A_{TD}X_D + A_{TN}X_N + C_T + G_T + J_T + E_T$$

$$(8.2) \quad X_D = A_{DT}X_T + A_{DD}X_D + A_{DN}X_N + C_D + G_D + J_D + E_D$$

$$(8.3) \quad X_N = A_{NT}X_T + A_{ND}X_D + A_{NN}X_N + C_N + G_N + J_N + E_N$$

where $X = [X_T, X_D, X_N]$ are gross outputs
$C = [C_T, C_D, C_N]$ are consumption demands
$G = [G_T, G_D, G_N]$ are government demands
$J = [J_T, J_D, J_N]$ are private investment demands
$E = [E_T, E_D, E_N]$ are net export demands.

Note that

$$(8.4) \quad E_D = \mu[E_T, E_N]$$

where μ is the matrix of trade and transport margins on net commodity exports accruing to the domestic economy.[3]

3. Since these margins are earned on gross rather than net flows, the above formulation is strictly correct only when there are no competitive imports of characteristic commodities which are exported. In the present application, that is a fair approximation to the observed

(*Note continues on the following page.*)

Now equations (8.1)–(8.3) are expressed in physical units. By choosing units of physical measure such that all domestic prices are unity, the same equations result if everything were measured in value terms, provided the user price of each commodity (*net* of transport and distribution margins, which may vary across users) is the same for all users.

Incomes

The vector of incomes $Y = [Y_1, \ldots, Y_L]$ earned by the L household types in the economy is determined by direct earnings in production and distributed profits. Let v_{kj} and ω_{kj} be the direct earnings and dividend payments, respectively, accruing to the kth household type for each unit of output produced by industry j. Then,

$$Y = (V + \Omega)X \tag{8.5}$$

where $$V = \| v_{kj} \| , \quad \Omega = \| \omega_{kj} \| .$$

Consumption

The gross income of household class k is used to pay lump-sum taxes, $\overline{T}_k$, and income taxes, $\tau_k Y_k$. A constant share, s_k, of posttax income is saved. Also, a fixed proportion of its marginal income is spent on the consumption of each good. These proportions are given by the matrix B, where β_{ik} is household k's marginal propensity to consume good i. By denoting the intercept terms of these consumption functions by γ_{ik}, and noting that domestic prices are normalized to unity, we have

$$C = \Gamma u + B(I - s)[(I - \tau)Y - \overline{T}] \tag{8.6}$$

where $\Gamma = \|\gamma_{ik}\|$; $(I - s)$ and $(I - \tau)$ are diagonal matrices whose k,kth elements are $1 - s_k$ and $1 - \tau_k$, respectively; $\overline{T} = (\overline{T}_1, \ldots, \overline{T}_L)$; and u is a column vector of ones.

Government

Government derives its revenue from direct taxes, tariff collections, and ownership of state corporations. In addition to income taxes, the government collects corporate taxes, which are denoted by $\tau_F \zeta' X$, where τ_F is the corporate tax rate and ζ' is the vector of profit margins on gross outputs. If

---

trade patterns. The transport and distribution margins on noncompetitive intermediate and consumer goods are included in A_{DT}, A_{DD}, A_{DT}, and C_D, respectively. Noncompetitive imports may be placed in category T, the corresponding element of X_T being zero.

δ'_G is the vector of profit margins on state-owned corporations, the government's receipts from these enterprises are $\delta'_G X$. Finally, let P_i be the ratio of the world price to the domestic producer price of good i, and invoking the "small country" assumption, the P_i's are parametrically given. Then government revenue from tariff collections is $[P' - (u' + \bar{\mu}')]E$,[4] where u' is a row vector of ones, (1, 1, . . . , 1), and $\bar{\mu}$ is the vector of the combined trade and transport margins on net exports which accrue to domestic producers:

$$\bar{\mu}_j = \begin{cases} \sum_i \mu_{ij}, & i \,\varepsilon\, D, j \,\varepsilon\, T,N. \\ 0, & j \,\varepsilon\, D. \end{cases}$$

We now assume that the government's outlays are exactly equal to its entire revenue. The bundle (G) of government purchases is given by: a vector (C_G) of obligatory purchases for current consumption (education, health, and so on), a vector (K) of exogenous outlays on investment goods (project-related expenditure), and a residual whose composition is fixed. The latter is given by the vector Φ, where ϕ_i is the government's outlay on goods from sector i out of each dollar of the revenue remaining after current consumption and project-related outlays. We have, therefore,

$$G = K + C_G + \Phi u' \; (G - K - C_G). \tag{8.7}$$

Equating the government's purchases to its revenue yields

$$u'G = t'Y + u'\bar{T} + (\tau_F \zeta' + \delta')X + [P' - (u' + \bar{\mu}')]E \tag{8.8}$$

where $t' = (\tau_1, \ldots, \tau_L)$.

It is now necessary to relate government revenues (and purchases) as defined here to the concept of "uncommitted social income, measured in terms of convertible foreign exchange," which is the numéraire advocated by Little and Mirrlees (1974, pp. 145–51). In our model, the impact of a project on government revenue in a particular period is simply the change in $u'G$ associated with the project, that is, the difference between $u'G$ in equilibrium with the project and $u'G$ in equilibrium without the project. Since K is the set of government purchases required to put the project into effect, the change in $u'(G - K - C_G)$ arising from the project will measure the change in social income available for commitment to other uses only if there is no change in public consumption, C_G. However, this change in "uncommitted social income" will indeed take the form of a change in the holdings of foreign exchange reserves only under special conditions. Since the formulation of the international trade regime is explicitly that of the

4. When the sum of the trade and distribution margins ($\bar{\mu}_j$) on net exports of a commodity accrue to domestic producers, the ex tariff f.o.b. price of the exports is $(1 + \bar{\mu}_j)$.

"small country" case, the government can exchange one tradable good against another in any quantity it desires at fixed world prices. Therefore, a necessary condition for its marginal income to be in effect "uncommitted," in the sense that the composition of its marginal purchases (Φ) does not matter, is that there are no allocations for the purchase of nontradables, that is, Φ_D and Φ_N are both zero. In regard to tradables, only if the government's marginal purchases fall solely on goods bearing a zero tariff will the change in social income be exactly equal to its command over foreign exchange, for in the presence of a tariff, outlays on the good in question will affect revenues. In the present case, all tradables carry a positive tariff (or subsidy), a difficulty we skirt by setting all Φ_i's in equation (8.6) equal to zero, except that for noncompetitive imports, which is set at unity. Since C_G is kept invariant with respect to K, the project's effects on uncommitted social income is simply the associated change in $u'(G - K - C_G)$, all of which is spent on the numéraire good, namely, noncompetitive imports. Measured in terms of convertible foreign exchange, the change in uncommitted social income is, therefore, the change in $u'(G - K - C_G)$ times the ratio of the world to the domestic price of the numéraire good.[5]

Savings and Investment

A crucial feature of the model developed here is that private investment is assumed to be financed solely by private (household and corporate) savings. Over the long run, this is a plausible assumption in a national setting; but it introduces certain difficulties in the context of a regional economy, which will be discussed in a later section.

Taking corporate savings to be equal to undistributed, net-of-tax corporate profits, we have:

$$u'J = u'\hat{S} + [(1 - \tau_F)\zeta' - u'\Omega]X \tag{8.9}$$

where household savings are given by

$$\hat{S} = s[(I - \tau)Y - \overline{T}]. \tag{8.10}$$

Investment demand in this model has two components. First, "balancing" investment may be needed to release capacity bottlenecks in the nontradables sectors as the economy moves from one steady state to another; this is denoted by the vector H. Second, there is "voluntary"

5. This is a consequence of the fact that the economy is in balance of payments equilibrium, as is shown below in this section. For an extensive discussion of these issues, see Bell and Devarajan (1980*b*).

investment, which is taken to be the residual investible funds allocated in fixed proportions (ψ_i) across the various sectors. Thus,

$$J = H + \Psi u' [J - H]. \tag{8.11}$$

To value private investment in the social cost-benefit calculation, we must know the distribution of investment demands among households (since, presumably, investment by poor households will be socially more valuable than that by the rich). Since our data do not permit this level of disaggregation, however, we are forced to represent private investment by a single vector, implicitly assuming that the composition of investment demand is the same across households.

The particular set of assumptions we have made also guarantees balance of trade equilibrium for the economy, that is, $P'E = 0$. This is readily seen by adding equations (8.1)–(8.11) and setting prices equal to costs at domestic prices.

Labor Supply

When a project is undertaken, one of its direct effects is to increase the demand for labor, both during the construction phase and when the project is producing output. The laborers thus employed would otherwise have been idle or engaged in some other activity, depending on whether there was full employment to start with. (If, initially, some mobile labor were unemployed and the project were big enough, the project would draw off labor from other sectors as well as exhaust the pool of unemployed.) Since we are dealing with a general equilibrium system, the total effects of introducing a project will, of course, usually differ from its direct effects. Yet the central question remains: Where is the labor coming from? Since the output of nontradables is driven by the level of domestic demand, the ultimate source of labor to the economy, in the sense that all adjustments in the total demand for labor elsewhere in the economy fall upon it, is a subset of sectors producing tradables or the pool of unemployed, or both. (There may be more than one such pool if labor is not perfectly mobile and homogeneous.)

If labor is perfectly mobile and homogeneous and its supply is perfectly inelastic, we have

$$\ell' X + L_p = \bar{L}$$

where $\ell' = (\ell_1, \ldots, \ell_M)$ is the vector of labor-output ratios, L_p is the pool of unemployed workers, and $\bar{L}$ is the total labor force. This equation implies nothing about which classes of households reallocate their labor in response to changes in output, however, and it will be consistent with (8.5) only if there is but one class of households.

Where there is more than one such class, there is, in effect, more than one sort of labor, or "characteristic factor." The appropriate generalization to the case of L household classes is immediate:

$$\Lambda X + L_p = \bar{L} \tag{8.12}$$

where $\Lambda = \|\ell_{kj}\|$ is the matrix of labor-output ratios, and L_p and $\bar{L}$ now denote vectors whose elements are the pools of unemployed workers and total work forces, respectively.

Solving the Model

The linear system (8.1)–(8.12) has $4M + 2L$ linearly independent equations in $5M + 2L$ variables. Hence, to solve the system, M of the variables must be specified exogenously. There are two cases of interest.

First, if there is unemployment of all kinds of labor ($L_p > 0$), (8.12) can be ignored. In this case, we fix the levels of outputs of tradable goods, X_T, and the levels of (net) exports by sectors in the subset N, so that the system is solved for E_T, E_D, X_D, X_N, and Y, as in chapter 6. The system may be written compactly:

$$\begin{bmatrix} R_{11} & R_{12} \\ -(V+\Omega) & I \end{bmatrix} \begin{bmatrix} X \\ Y \end{bmatrix} = \begin{bmatrix} R_{13}E + R_{14} \\ 0 \end{bmatrix} \tag{8.13}$$

where

$$R_{11} = I - A - \Phi(\tau_F\zeta' + \delta') - \Psi[(1-\tau_F)\zeta' - u'\Omega]$$

$$R_{12} = -[B(I-s)(I-\tau) + \Phi t' + \Psi u's(I-\tau)]$$

$$R_{13} = \Phi[P' - (u' + \bar{\mu}')] + I$$

$$R_{14} = \Gamma u + B[(I-s) + (\Phi - \Psi)]\bar{T} + (I - \Phi u')H + (I - \Psi u')(C_G + K).$$

A comparison of (8.13) with (6.11), its counterpart in chapter 6, indicates at once the changes in structure resulting from treating private investment and government expenditure endogenously.

Second, for each instance in which there is full employment of one kind of labor, the output of a traded good (or a bundle of traded goods) becomes endogenous, as well as the exports of that good (or goods). This restores the degree of freedom lost whenever an element of L_p is zero, and so (8.12) becomes an effective part of the system, which may be written as:

$$\begin{bmatrix} R_{11} & R_{12} & 0 \\ -(V+\Omega) & I & 0 \\ \Lambda & 0 & I \end{bmatrix} \begin{bmatrix} X \\ Y \\ L_p \end{bmatrix} = \begin{bmatrix} R_{13}E + R_{14} \\ 0 \\ \bar{L} \end{bmatrix}. \tag{8.14}$$

After suitable transpositions of columns in the manner described in chapter 7 and simple rearrangement of variables, both variants of the above model can be represented in the following schematic form:

$$Z = A^{*}Q + A^{**}u \tag{8.15}$$

where Q and Z are the vectors of exogenous and endogenous variables, respectively, and the system's parameters enter into the elements of the matrices A^* and A^{**}. The precise form of the scheme will, of course, depend on whether there is full employment of each and every kind of labor.

In all the above cases, the levels of the endogenous variables represent their steady-state values as responses to the exogenous variables. We assume the system adjusts instantaneously to its steady state when perturbed by an exogenous "shock." Since this assumption has been made in chapters 4 and 7, it hardly needs elaboration here. It should be emphasized, however, that this use of the term "steady state" is different from that later in the chapter, where we characterize the steady-state trajectory of the economy after it has adjusted fully to the impact of the project at maturity. The latter is the notion of the annual values of the endogenous variables (which are themselves steady-state values in the former sense) eventually reaching a steady state. The adjustment process to this steady state, though, takes place over time (that is, between 1967 and 1975) rather than instantaneously.

Social Cost-Benefit Analysis

To determine the impact of a project, we must specify the values taken by the exogenous variables, Q, both in the presence of the project and in its absence, in each case solving for the endogenous variables, Z. If, as is usual, the project has an effect on the technology employed in its own sector, or even elsewhere in the economy, one or both of the matrices A^* and A^{**} will be different in the two cases. With the superscript ^{o} denoting the case in which the project is not undertaken, the difference between the ordered pairs of vectors (Q, Z) and (Q^o, Z^o) represents the project's impact on the economy. Recalling the discussion in chapter 4, we are therefore assuming that the autonomous changes in the economy are anticipated in full.

How can the project's impact on outputs, incomes, savings, and so forth be used to determine the contribution of the project to social welfare in a given year? Following well-established tradition, the arguments of the social welfare function are: private consumption, private savings over and above those needed to finance the balancing investments in the nontradables sectors, and uncommitted social income. Here, it will be assumed that the weights for private savings and consumption relative to uncommitted social income (the numéraire) are estimated so as to remain constant over the relevant ranges of these variables, ranges which may be quite large. Hence, the social profit, ΔU, associated with the project for a unit of time is given by

$$\Delta U = (G - G^o - K) + \lambda'(\hat{C} - \hat{C}^o) + \xi'(\hat{S} - \hat{S}^o) + \xi_F[(1 - \tau_F)\zeta' - u'\Omega](X - X^o) - \xi_H H \tag{8.16}$$

where $\hat{C} = (I - s)[(I - \tau)Y - \overline{T}]$ and $\hat{S} = s[(I - \tau)Y - \overline{T}]$. The vectors λ' and ξ' denote, respectively, the social valuations of a unit increase in the levels of consumption and savings for each of the household classes; ξ_F is the social valuation of a unit of retained corporate profits; and ξ_H is the social valuation of the savings that finance the balancing investment, H, whose composition by source may differ from that for savings as a whole. Thus defined, ΔU may be thought of as the social valuation of the change in national income that the project brings about.

Equation (8.16) departs from Little and Mirrlees in that the weights λ, ξ, ξ_F, and ξ_H are applied to private consumption and savings measured at market prices as opposed to accounting prices. In their discussion of this choice, Little and Mirrlees (1974, pp. 236–37) seem to be concerned principally with the fact that different consumers may pay different prices for the same good. Faced with this difficulty, they suggest as a shortcut that consumers' expenditures be valued at accounting prices. Although we do not find this suggestion a very appealing way to make "baskets" of goods purchased at different market prices comparable with one another, it is the only strictly correct way of comparing consumers' expenditures with uncommitted social income unless the consumers' consumption conversion factors happen to be the same. As Little and Mirrlees recognize, however, an assessment of *equity* among consumers requires comparisons of their expenditures (or incomes) at market prices, even though this may mean that incomes of two people which differ greatly at market prices may differ little at accounting prices—and hence be of roughly the same social value if their shortcut method is adopted.

We have formulated exactly this issue of equity among consumers at the possible cost of introducing some errors into the valuation of their expenditure relative to the numéraire. Indeed, since the method advanced here is designed to finesse the task of estimating accounting prices for nontradables and shadow wage rates, we have no choice in the matter. As for the possibility that different people may pay different prices for the same good, the region under study is blessed with a good transport system. Moreover, the provincial capital is both modest in scale and readily accessible to all rural households, so that urban and rural cost of living indexes probably differ little.

To complete the evaluation of the project, the stream of social profits, as given by equation (8.16), generated by the project throughout its lifetime is discounted at the accounting rate of interest.

Applying the Model to the Muda Irrigation Project

The model set out in the previous section describes the flows within a national economy, and those to and from the rest of the world. However, the social accounts matrix that provides the primary data base for the model has been estimated for the Muda region. Although the basic structure of the model presented above is preserved if the data pertain to a region, some additional assumptions must be made if the results are to be interpreted correctly.

First, some primary commodities do not normally enter international trade until they are processed, but they are shipped in raw form from the producing region to other regions for processing. In the present case, there are two such goods: paddy and smoked rubber sheet. Until the advent of irrigation in 1970, the region's net exports of paddy were roughly zero. Thereafter, the capacity of the region's milling sector did not expand rapidly enough to keep pace with the surge in paddy output, and as a result exports of paddy to rice mills in southern Kedah and Province Wellesley were substantial. Only in the late 1970s was the region's capacity sufficient to mill the whole of the region's paddy crop. As for rubber, less than 15 percent of the output of the region's smallholders and estate sectors is processed within the region itself. Strictly speaking, therefore, these two commodities should be treated as tradables with respect to the region. Yet, the processing of paddy and rubber generates value added within the Malaysian economy, and the multiplier effects of this value added would go unaccounted for if the model were applied literally. To correct for this, processing activities outside the region are treated in the model as if they took place inside the region, which entails the assumption that technology, the distribution of value added among households, and the levels of household incomes do not vary across regions. Since this comparison involves northern Kedah on the one hand and southern Kedah and the mainland section of Pinang on the other, all of them rather poor regions, the assumption seems defensible.

Second, as indicated in chapters 5 and 7, by no means is the whole of private savings invested in the region itself. The outflow of private capital from the region—much of it through the banking system—was especially large in 1972 and for some years thereafter. Very little is known about the uses to which these savings are put, and for want of anything better, it has been assumed that they are used to finance the purchase of a bundle of investment goods with the same composition as that of the region's own investment activities. Since, in the model presented above, all private savings are invested, the procedure adopted here is tantamount to incorporating within the region those activities making deliveries to meet investment demands outside the region which are also financed by the region's savings.

Both these enlargements of the region's activities, whether by raising output, and hence income, in the tradable sectors, or by increasing the demand for investment goods from the nontradable sectors, will lead to further increases in the steady-state values of incomes and the output of nontradables. Thus, the results from the model will not square with those for the Muda region reported in chapter 7, which are derived from a closely related semi-input-output model featuring *exports* of paddy and smoked rubber sheet as well as exogenously fixed investment. In particular, value added will be higher in the extended region if, initially, there was any unemployment in the peripheral areas thus incorporated. Moreover, since output increases in all the sectors producing nontradables in the other regions of the economy thus affected, the earlier assumption of homogeneous technology, distribution of value added among households, and household income levels is not only vital, but also far stronger than it appears to be at first sight. Be that as it may, the data admit of nothing better. Certainly, such assumptions of homogeneity are every bit in keeping with those employed in estimating "national" shadow prices in the literature on social cost-benefit analysis.

A third consideration in the interpretation of the model's results stems from the stream of seasonal workers into the Muda region during the paddy harvest. Some of these agricultural workers come from southern Kedah and Kelantan, but others come from Thailand. The former's household incomes are, in all probability, at best equal to those of landless paddy farm worker households in the Muda region. At one extreme, the migrants' seasonal earnings could be simply lumped with those of Muda's landless households, on the assumption that the output forgone in their home regions is zero. Alternatively, it could be assumed that their seasonal earnings in Muda are exactly equal to the output forgone elsewhere in the economy, both valued at market prices. The assumption underlying the results presented below rests somewhat uneasily in between, inasmuch as all payments to seasonal workers, Malay and Thai alike, are treated as imports used in paddy production. This is undeniably rough and ready, but the sums involved are small and hardly worthy of more elaborate treatment. In one respect, however, payments to Malay and Thai seasonal workers must be distinguished. Presumably the Malaysian government attaches no social value to the extra income going to Thai workers, whereas it ought to be gratified by the additional income to poor Malay agricultural workers. For this reason, payments to Malay seasonal workers must feature in an appropriate way in the calculation of the project's stream of social profits.

The next step in the calculation of the project's social profitability is to set up a pair of vectors of exogenous variables for each year of the project's expected lifetime, one denoting the estimated actual values of such variables and the other their estimated values in the hypothetical absence of the project. In brief, construction work on the project began in a small way in 1965, reached a peak in 1967 and 1968, and then tailed off steadily to completion in 1974. Irrigation commenced on a small scale in 1970, about three-quarters of the project command area was served in 1972, and the project attained full maturity in 1974. Since the ratio of the domestic to the world price of rice was abnormal in 1974, however, the 1975 ratio seems more representative of the conditions likely to hold over the project's lifetime. On the assumption that a steady state ruled after 1975, this chronicle of events thus appears to entail the estimation of no fewer than twenty-two separate vectors of exogenous variables—a daunting task indeed.

Fortunately, during that part of the project construction period when no output is produced, it suffices to estimate the two sets of exogenous variables for just one year. From equation (8.5), so long as A^* and A^{**} are unchanged, that is, there are no changes in technology and expenditure propensities and no shift from unemployment to full employment, the change in the levels of the endogenous variables is given by:

$$Z - Z^o = A^*(Q - Q^o). \qquad (8.15)'$$

If, during this phase of the project, the members of the sequence of exogenous vectors $(Q - Q^o)$ are scalar multiples of one another—in this case, the vector $Q - Q^o$ is the project's demands for investment deliveries from the construction sector—the members of the sequence $(Z - Z^o)$ are also scalar multiples of one another. Hence, once the sequence of $(Q - Q^o)$ has been estimated, it is enough to estimate one member of the sequence of $(Z - Z^o)$ to arrive at the rest. The year actually chosen for this purpose is 1967, in keeping with the discussion in chapter 7.

Once the project begins to produce a stream of output, however, A^* will change continuously until the project reaches maturity. Hence, in principle, it is necessary to estimate

$$Z - Z^o = A^*Q - A^{*o}Q^o \qquad (8.15)''$$

for all years after 1970, the matrix A^{**} being unaltered by the project.

In the present case, the pair of vectors (Q, Q^o) has already been estimated for the years 1967, 1972, and 1974 in connection with the model in chapter 6. To keep the amount of additional work within reasonable bounds, it was decided to estimate (A^*, A^{*o}) and (Q, Q^o) for 1970 alone and to use linear interpolation to arrive at the values of relevant variables for the intervening transitional years, 1971 and 1973. Naturally, the estimation procedure for these sets of variables for 1970 followed closely that in chapter 7. In brief, the technologies for paddy production entering into A^* and A^{*o} were taken to be output-weighted averages of the corresponding technologies for 1967 and 1972. Apart from the obvious difference of deliveries from the construction sector in the presence of the project, Q departs from Q^o by virtue of the additional paddy output thanks to irrigation and extra throughput in the region's rice mills. There is a fall in rubber output and processing as labor is drawn into paddy harvesting. Also, exogenous taxes change somewhat as farmers benefit from irrigation subsidies and rising regional activities swell tax receipts from vehicle and business registrations.

In treating private investment, government outlays, and labor demand endogenously, the model employed here features three sets of parameters, Φ, Ψ, and ℓ, which are not found in chapter 6. As already explained, residual government purchases are allocated entirely to noncompetitive imports. The vector Ψ denotes the composition of demand for investment goods by the private sector after all necessary purchases of balancing investments have been made. A critical assumption underlying the estimation of Ψ is that the share of noncompetitive imports in total outlays on such investments stayed constant at its 1972 level. That being so, Ψ varied little up to 1970, thereby keeping A^* invariant to all intents and purposes. It showed only slight year-to-year variation thereafter, so it seems reasonable to assume that there were no fluctuations after 1974, when the project attained its steady-state configuration.

Turning, at last, to labor supply and demand, we begin by recalling that virtually all the people engaged in farming are ethnic Malays, whereas most of the workers employed in manufacturing and services are Chinese and many of the firms in those sectors are family businesses. Thus, the mobility of labor between the farm and nonfarm sectors is somewhat imperfect, so that the labor market is segmented—at least, to a degree.

In keeping with the evidence in earlier chapters, the endowments of labor of Muda's paddy farming households are treated as if they were specific to the paddy farming sector. However, the project brings about such a large increase in the demand for labor by this sector that family labor supplies cannot meet it, and so labor is drawn in, first, from the rubber smallholdings on the periphery of the scheme and then from southern Kedah, Kelantan, and Thailand. The losses of output that result elsewhere in the economy (discussed in chapter 7 and earlier in this section) capture the effects of the project on the allocation of labor in the farming sectors.

To complete the story, it is necessary to specify the source of labor to the sectors producing nontraded goods, the output of which will, in general, change in response to the project. If the labor force available to these sectors is not fully employed, even after the project has reached maturity, there are no effects on output elsewhere in the economy, and the appropriate model is given by equation (8.13). Conversely, if there is always full employment, labor is drawn off from one or more sectors producing traded goods, and the appropriate model is furnished by equation (8.14).

In the present case, the real state of affairs lay somewhere in between these two extremes. From what we know of the region's economy in the late 1960s, it is hard to avoid the conclusion that there was some underemployment of family labor in the sectors producing nontraded goods. At the same time, employment in the free trade zones in Penang began to boom, drawing in large numbers of young women from Kedah and Pinang and raising real wage rates. Thus, in the recent past, the Muda region and Penang's free trade zones were probably competing for labor. The evidence is far too shaky to warrant simulations in which there is a switch, at some point, from equation (8.13) to (8.14). Rather, we shall estimate the effects of the project using first the one and then the other. This should suffice to bracket what actually happened.

It is also apparent that we are treating the industries in Penang's free trade zone as the source of additional labor to the Muda region's economy, at least for the production of nontraded goods. At first sight, this may appear somewhat odd to many hardened practitioners of social cost-benefit analysis, who naturally think of subsistence agriculture as the source of labor to the rest of the economy. Nothing in the celebrated manuals, however, commands subsistence agriculture to fulfill this role; on the contrary, by examining carefully how the labor market actually works, we have adhered closely to the spirit of their recommendations.

For simplicity, we assume that only one sector yields its labor in response to increased demand for workers by other activities. Without straining anyone's sense of the plausible, the sector in question is taken to be electronics, this being far and away the biggest employer of labor in Penang's free trade zone. Given the tax and commercial policies toward these zones and the nature of the manufacturing technology in electronics, this sector affects the national economy solely through its demand for domestic labor. Imports of intermediate inputs, final output, and profits all go untaxed; foreigners provide all finance; and the requirements for nontraded goods, such as utilities, are negligible.

In the case of full employment with competitive labor markets, any change in the wage bill for this sector must be accompanied by an equal and offsetting change in the incomes accruing to nonfarm households from sectors producing nontraded goods.[6] Thus, we have the following special case of (8.12) for nonfarm households:

$$(8.12)' \qquad v_5' (X - X^o) = 0$$

where v_5' is the row vector $(v_{50}, v_{51}, \ldots, v_{5M})$, the electronics sector being labeled zero in the sectoral ordering. In this case, (8.14) becomes

$$(8.14)' \qquad \begin{bmatrix} R_{11} & R_{12} \\ -(V+\Omega) & I \end{bmatrix} \begin{bmatrix} X \\ Y \end{bmatrix} = \begin{bmatrix} R_{13}E + R_{14} \\ R_{21} \end{bmatrix}$$

where $R_{21} = [0, 0, 0, 0, v_5' X^o]$. Since we are interested in the differences resulting from the introduction of the project, the output of electronics may be set at zero in the absence of the project. With the output of electronics thus fixed, the values of all the endogenous variables in the absence of the project are determined by (8.13), and with Y_5^o determined, (8.14)′ can be solved, electronics output now being an endogenous variable and nonfarm incomes fixed at Y_5^o. (v_{50} equals unity and is the only parameter needed for the electronics sector.)

Although we assume no change in the overall wage bill for this full employment case, we do allow for a change in the level of corporate profits. The alert reader may question this discrepancy. Why, when assuming perfectly competitive labor markets, do we allow for what seems to be an imperfection in the capital market? It turns out that an increase in corporate profits from a reallocation of outputs (from X^o to X) is consistent with a uniform rate of profit across the board. Put simply, if the capital-labor ratios in two industries are different, a transfer in output from the labor-intensive to the capital-intensive industry which keeps the total payment to labor constant *must* increase the total payments to capital. This is true even if the rate of return to capital (as well as the wage rate) is the same in both industries.

In fact, the electronics sector from which the labor is drawn uses no domestically supplied capital, so that *any* transfer of output to the other industries is a shift to a capital-intensive sector and would result in higher profits. The reason for this state of affairs, however, goes a bit deeper. In an economy with only one factor, the assumption of full employment implies the economy will specialize in producing only one good. Since we have constrained the economy to produce more than one tradable, the economy is clearly not at its free trade specialization point. Hence, a shift in output that increases earnings from capital while leaving labor payments unchanged (in a fixed-price world) is tantamount to moving the economy closer to its free trade specialization point.

6. The supply of labor to each sector is assumed to be perfectly elastic, and there are constant returns to scale with no joint production—and no fixed factors outside agriculture.

Estimation of National Parameters for Social Cost-Benefit Analysis

The estimation of most of the parameters of the model (equations [8.1]–[8.16]) has been described in chapters 5 and 7. In making private investment and government outlays endogenous, we have added the vectors Φ and Ψ, which were discussed in the preceding section. In this section, we derive the remaining national parameters needed to yield a calculation of the project's social costs and benefits.

Four such sets of national parameters are used in our calculus: the accounting ratios of traded goods, that is, the ratio of the world price to the domestic price of each traded good; the social valuation, in terms of uncommitted social income in the hands of the government, of a unit of private consumption enjoyed by each household class (the vector λ); the social valuation of private savings in terms of the same numéraire (ξ); and the accounting rate of interest.

Veitch (1976) has estimated the accounting ratios of traded goods for 1974. Except for the two rice milling sectors, we assume these ratios remain the same throughout the period under study. Since the ratio of the world to the domestic price of rice changes considerably over the 1967–75 period, and rice is the sole direct output of the project, the accounting ratio for rice was estimated separately for each year in this period (Goldman, 1975; and Malaysia, 1974):

1967	*1968*	*1969*	*1970*	*1971*	*1972*	*1973*	*1974*	*1975*
0.867	0.914	0.905	0.895	0.776	0.732	0.716	1.110	0.901

The two sectors producing nontraded goods in this economy that also export significant amounts (albeit in exogenously fixed quantities) are "other agriculture" (10) and manufacturing not elsewhere classified (14). Their accounting ratios were estimated by using the accounting ratios of the categories of goods defined by Veitch that correspond most closely to the principal commodities of the two sectors. Finally, the accounting ratio for noncompetitive imports was derived by taking a weighted average of the different imports, the weights being obtained from the original SAM tableau. The complete set of accounting ratios for 1974 is (Veitch, 1976):

	Accounting ratio
1. Commercial rice mills	1.11
2. Small rice mills	1.11
3. Food processing	0.91
4. Fish processing	1.19
5. Paddy production	1.00

		Accounting ratio
6.	Fishing	1.19
7.	Estates rubber	1.22
8.	Smallholder rubber	1.22
9.	Rubber processing	1.22
10.	Other agriculture	0.91
11.	Sawmilling	1.32
14.	Manufacturing n.e.c.	0.80
	Noncompetitive imports	0.90

(No paddy is in fact exported, and the value of unity is used solely for accounting requirements.)

To estimate λ, the set of social weights for household consumption, two parameters are needed: $\bar{c}$, the so-called critical per capita consumption level, at which a unit of additional consumption (valued at market prices) is just as valuable socially as a unit of uncommitted government income; and η, the elasticity of the marginal utility of consumption. For $\bar{c}$, it is reasonable to take the official poverty line, on the grounds that a prime objective of government policy is to bring all individuals to that level of living as soon as possible. At the time of writing, the value of $\bar{c}$ in 1972 prices is $375.50 a year (Visaria, 1979, p. 29), which is about one-third of per capita GDP. It is difficult to be so definite about η. Following a well-established tradition in this literature, two values will be used: unity, which arises from a logarithmic utility function and is mildly egalitarian; and two, which corresponds to Atkinson's (1970) equally distributed equivalent income being the harmonic mean of all incomes and which is quite strongly egalitarian.

As shown in chapter 7, the project brought about large changes in the per capita consumption levels of the five classes of households. Hence, the concavity of the utility function implies that potentially serious errors will arise if λ is estimated at the steady-state values of household consumption when the project reaches maturity. Instead, we proceed as follows. In view of equation (8.16), two additional parameters of the utility function may be fixed. First, the marginal social utility of a unit of private consumption at $\bar{c}$ should be unity. Second, the absolute social utility attached to a private consumption level of $\bar{c}$ should be $\bar{c}$. Hence, we have, for $\eta = 1$ and $\eta = 2$, respectively,

(8.17a) $$V(c) = \bar{c}[1 + \log_e (c/\bar{c})]$$

(8.17b) $$V(c) = \bar{c}[2 - (\bar{c}/c)].$$

The value of λ arising from a movement from c^o to c is, therefore,

(8.18) $$\lambda = [V(c) - V(c^o)]/[c - c^o].$$

The per capita consumption levels of the various classes of household for 1967 and 1974, both with and without the project, may be derived from chapter 7, while those for 1972 with the project have been estimated directly in the course of constructing the SAM for that year. Those for 1970 and 1972 (without the project) were derived by linear interpolation. The resulting values of λ are set out in table 8-1. There is no call for the (spurious) accuracy of similar calculations for the intervening years. Up to 1970, the values of λ for 1967 will do; those for 1972 will be applied to 1971; and those for 1974 are assumed to hold for 1973 and all years thereafter.

To place a social value on private savings, it is necessary to ascertain not only to what uses they will be put, but also what their effects will be on all incomes (including taxes) in the future. In general, this is a stiff task indeed. To tackle it rigorously, an intertemporal model of resource allocation that pays close attention to the working of the capital market is needed. The only simple case is that in which income distribution and savings are already optimal and the capital market is perfect, for the social values of an additional unit of private consumption, household savings, retained corporate profits, and uncommitted government income will then be equal at unity. In all other cases, the only alternative is to employ short-cut methods, and that is what we shall do here without apology.

Beginning with household savings, Little and Mirrlees (1974, p. 243) make the suggestion that, as a shortcut, the weight attached to an additional dollar of savings by household class k, ξ_k, be taken as halfway between unity, that is, the weight for uncommitted government income, and λ_k, the weight for an additional dollar of consumption by that household. This rule of thumb appears to be prompted by the consideration that while private consumption is usually deemed to be socially less valuable at the margin than government income, private savings are also suboptimal. Of course, if the social weight placed on additional private consumption were greater than unity, the rule of thumb yields the rather paradoxical result that private consumption is more valuable at the margin than private savings. A glance at table 8-1 reveals that this is exactly what happens in the case of all agricultural households in all years. That is to say, the households

Table 8-1. *Social Weights for Household Consumption at Market Prices*

Household	*1967* $\lambda(1)$	*1967* $\lambda(2)$	*1970* $\lambda(1)$	*1970* $\lambda(2)$	*1972* $\lambda(1)$	*1972* $\lambda(2)$	*1974* $\lambda(1)$	*1974* $\lambda(2)$
Landless paddy worker	3.25	10.53	2.62	6.90	2.29	5.39	2.14	4.70
Labor-abundant paddy farm	2.37	5.63	2.11	4.45	1.83	3.41	1.61	2.63
Land-abundant paddy farm	1.54	2.36	1.38	1.92	1.19	1.43	1.03	1.08
Nonproject farm	1.75	3.09	1.59	2.55	1.39	1.94	1.25	1.42
Nonfarm	0.78	0.61	0.73	0.53	0.63	0.41	0.53	0.29

Note: $\lambda(\eta)$ denotes the value of λ evaluated at $\eta = 1,2$.

concerned are deemed to be saving "too much," given their relative poverty. If this specific instance of the overriding of consumer sovereignty is thought to be disquieting, it should be added that these households do not have access to all the investment opportunities open to nonfarm households and incorporated businesses. In particular, since the poorer households in the Muda region place much of their savings in bank accounts yielding zero or negative real rates of interest, to the advantage of affluent borrowers who have access to highly profitable investment opportunities, then the social value of additional savings by farm households will be relatively low.

In regard to retained posttax profits, the investment of which generates future dividends for shareholders, why should they be valued differently from the savings made by the shareholders' households? There are two necessary conditions for the valuations to be the same: First, shareholders must control fully the dividend policies of the firms in which they hold equity; and second, shareholders must be representative of the class of households to which they are assigned. Neither is especially plausible. Control by shareholders is usually tenuous at best, and managerial decisions are likely to result in higher total savings by firms and shareholders combined than under shareholders' control, though it is unclear that the difference would be very large. In any event, the meager market for securities in Malaysia would make it difficult for investors to arrange their portfolios optimally, so that some divergence between stockholder and managerial preferences is to be expected. Furthermore, shareholders are extremely unlikely to be representative of nonfarm households as a group. Since they are either very affluent or foreigners, or both, the social value of the dividends they receive as a result of current investment out of profits should be very low, if not zero. In the light of the social weight attached to additional savings by nonfarm households as a group (see table 8-1), a social value of 0.2 for each additional dollar of retained, posttax corporate profits (ξ_F) does not seem particularly objectionable.

Finally, there is the matter of arriving at the social valuation of the savings needed to finance the set of balancing investments. Most of these investments were undertaken in the nonfarm sector of the economy between 1970 and 1974, by incorporated and unincorporated enterprises alike, while others took the form of construction or improvement of housing by farm households. Accordingly, the social value of a unit of such investment should be a properly weighted average of ξ and ξ_F. In rather rough and ready fashion, the value of ξ_H is taken to be 0.7, which is roughly the value of ξ_5 when the project has reached maturity.

By and large, the argument advanced above rests on the notion that all households and firms have access to a perfect capital market and so receive the same rate of return on their savings—an unlikely situation for the poorer households in the Muda region. To put a lower bound on the social value of private savings, suppose that all savings by farm households were placed in the banking system and then lent to the government, to the richest class of households, or to private corporations. The social weight attached to all savings, irrespective of their source, would then lie somewhere in the range of the social weights for savings made by these three institutional categories and the social weights for consumption by farm households. Very little is known about these matters, and it is profitless to attempt a refined calculation of an appropriately weighted average. Moreover, since the purpose of setting up this extreme case is to establish a parameter value for sensitivity analysis, a social weight of 0.7 for all private savings should provide a plausible lower bound on the valuation of this stream of outcomes flowing from the project.

The accounting rate of interest (ARI) for Malaysia is estimated by Anand (1977) to be 10 percent. An alternative, crude approach is as follows. A nominal rate of 10 percent is fairly close to the rate at which Malaysia was able to borrow on the world market, since the country's balance of payments position was quite strong in 1967–74 and foreign reserves remained high thereafter. However, the model set out here is formulated in constant (domestic) prices. Adjusting for an average annual inflation rate of 5 percent and assuming exchange rate stability against an appropriate basket of currencies, we estimate the ARI to be around 5 percent. Since it may be felt that public sector projects should have a higher rate of return than the marginal cost of foreign borrowing, this figure of 5 percent is probably a lower bound. To reflect this consideration, we perform sensitivity experiments using 5, 10, and 20 percent for the ARI.

Social Cost-Benefit Calculations

As indicated by equation (8.16), the stream of social profits generated by the project is made up of the project's effects on household consumption and savings, on retained corporate profits, and on government income. Setting out the individual trajectories of these elements, before weighting them together to yield the time path of social profits, makes clear the positive basis of the subsequent normative aggregation.

The trajectories for the two extreme cases of labor supply conditions facing the nonfarm sectors of the economy, namely, unemployment and full employment, are laid out in tables 8-2 and 8-3, respectively. Qualitatively, there is nothing very startling about them, given the nature of the project. In the construction phase, the government made large outlays on construction, which accrued, in the first instance, as incomes to nonfarm households and corporations, both domestic and foreign. (In the full-

Table 8-2. *Changes in Incomes Owing to the Project: Unemployed Nonfarm Labor*
(thousands of Malaysian dollars in 1972 prices)

Institution	*1965*	*1966*	*1967*	*1968*	*1969*	*1970*	*1971*	*1972*	*1973*	*1974*	*1975*
Households											
Landless paddy worker											
$C_1 - C_1^o$	4	16	44	56	40	648	1,120	1,591	1,858	2,124	2,124
$S_1 - S_1^o$	0	0	0	0	0	2	3	4	5	5	5
Labor-abundant paddy farm											
$C_2 - C_2^o$	9	36	102	129	94	2,875	5,345	7,814	10,013	12,211	12,211
$S_2 - S_2^o$	1	5	13	16	12	359	667	975	1,249	1,523	1,523
Land-abundant paddy farm											
$C_3 - C_3^o$	6	23	64	81	59	8,091	15,383	22,674	28,989	35,304	35,304
$S_3 - S_3^o$	1	5	14	18	13	1,717	3,264	4,811	6,151	7,490	7,490
Nonproject farm											
$C_4 - C_4^o$	−6	−22	−62	−78	−57	−96	405	906	2,112	3,317	3,317
$S_4 - S_4^o$	−1	−5	−13	−16	−12	−20	85	190	442	694	694
Nonfarm											
$C_5 - C_5^o$	1,527	5,834	16,468	20,832	15,121	13,904	20,343	26,782	32,271	37,760	37,760
$S_5 - S_5^o$	289	1,105	3,123	3,946	2,864	2,633	3,853	5,073	6,113	7,152	7,152
Change in retained corporate profits	617	2,358	6,664	8,421	6,112	3,911	4,460	5,008	5,354	5,700	5,700
Change in uncommitted government income	−6,539	−16,920	−42,483	−51,401	−40,540	−14,874	−29,713	−16,259	−2,861	16,227	11,350
Balancing investments	−720	−2,160	−5,040	−5,760	−5,040	−5,760	−10,080	−12,240	−14,400	−10,800	0
Total	−4,812	−9,725	−21,088	−23,756	−21,334	13,390	15,135	47,329	77,296	118,707	124,630

employment case, of course, output and incomes fell in the sector that is the source of labor to the rest of the economy, so that these households gained little on balance.) Agricultural households participated only marginally in construction and the extra activity in other domestic sectors induced in its wake. Indeed, the incomes of farm households on the periphery of the scheme fell slightly, since they earned much of their livelihood from rearing livestock, which declined in number in the face of mechanization induced by the irrigation project. After the project came on stream in 1970, the incomes of agricultural households rose sharply. In the case of unemployed labor in the nonfarm sectors, the additional spending by paddy farming households boosted nonfarm incomes and corporate profits as construction work on the project wound down. Uncommitted government income recovered too, as revenues from income and indirect taxes increased by more than enough to offset the losses arising out of the effective protection afforded to the paddy sector in all years. (In 1974, an almost mature project and a very high world price of rice combined to yield an exceptionally high payoff for the government.) In the case of fully employed nonfarm labor, the increase in incomes accruing to nonfarm households are naturally rather modest, with attendant repercussions on farm households, which do earn small incomes from the production of nontraded goods. The fall in the output of traded goods as labor is drawn off from the electronics sector results in a fall in uncommitted government income, and a substantial one it is when the project has attained maturity. The outlays on balancing investments are the higher of the two (rather rough) estimates discussed in chapter 7, just to be on the safe side. The flows for 1975 are assumed to be the steady-state values of relevant variables for the rest of the project's life. In keeping with the procedures used in the final appraisal by the World Bank, the project is assumed to have a salvage value of $100 million in 2000.

The streams of social profits associated with the flows in tables 8-2 and 8-3 have been calculated under five sets of assumptions about the social valuation of household consumption, savings, and corporate profits. These correspond to three values of the elasticity of the marginal utility of consumption and two methods of placing a social valuation on household savings and retained corporate profits, as already discussed in the previous section. In particular, to satisfy those who would have no truck with distributional weights to value private consumption and savings in terms of government income, the streams of social profits have been calculated with all such weights set to unity. In this case, the streams are then simply the algebraic sums of the elements making up the individual columns of tables 8-2 and 8-3.

These alternative streams are set out in tables 8-4 and 8-5. In all cases, the project's net present social value is handsomely positive at an account-

Table 8-3. *Changes in Incomes Owing to the Project: Fully Employed Nonfarm Labor*
(thousands of Malaysian dollars in 1972 prices)

Institution	*1965*	*1966*	*1967*	*1968*	*1969*	*1970*	*1971*	*1972*	*1973*	*1974*	*1975*
Households											
Landless paddy worker											
$C_1 - C_1^o$	2	6	17	21	16	630	1,094	1,558	1,815	2,071	2,071
$S_1 - S_1^o$	0	0	0	0	0	2	3	4	5	5	5
Labor-abundant paddy farm											
$C_2 - C_2^o$	−3	−11	−31	−39	−28	2,787	5,220	7,652	9,800	11,947	11,947
$S_2 - S_2^o$	0	−1	−4	−5	−4	348	651	954	1,222	1,490	1,490
Land-abundant paddy farm											
$C_3 - C_3^o$	−20	−76	−216	−273	−198	7,904	15,118	22,332	28,541	34,749	34,749
$S_3 - S_3^o$	−4	−16	−46	−58	−42	1,677	3,208	4,738	6,056	7,373	7,373
Nonproject farm											
$C_4 - C_4^o$	−28	−106	−300	−379	−275	−255	181	617	1,732	2,847	2,847
$S_4 - S_4^o$	−6	−22	−63	−80	−58	−53	38	129	363	596	596
Nonfarm											
$C_5 - C_5^o$	396	1,513	4,275	5,402	3,921	5,563	8,743	11,923	12,806	13,688	13,688
$S_5 - S_5^o$	75	287	810	1,024	743	1,054	1,656	2,258	2,426	2,593	2,593
Change in retained corporate profits	561	2,143	6,054	7,650	5,553	3,489	3,878	4,266	4,383	4,499	4,499
Change in uncommitted government income	−6,787	−17,867	−45,161	−54,785	−42,996	−20,942	−37,080	−24,924	−11,801	7,011	2,060
Balancing investments	−720	−2,160	−5,040	−5,760	−5,040	−5,760	−10,080	−12,240	−14,400	−10,800	0
Total	−6,534	−16,130	−39,705	−47,282	−38,408	−3,556	−7,370	19,267	42,948	78,069	83,918

Table 8-4. *Social Profits from the Project: Unemployed Nonfarm Labor*
(thousands of Malaysian dollars in 1972 prices)

		$\eta = 1$		$\eta = 2$	
Year	$\eta = 0$	*A*	*B*	*A*	*B*
		Stream of undiscounted social profits			
1965	−4,812	−5,332	−5,388	−5,556	−5,590
1966	−9,725	−11,857	−12,071	−12,717	−12,843
1967	−21,088	−27,482	−28,087	−29,910	−30,268
1968	−23,756	−32,006	−32,770	−35,074	−35,526
1969	−21,334	−27,082	−27,637	−29,309	−29,637
1970	13,390	15,016	13,444	26,625	24,439
1971	15,135	18,971	16,546	33,862	30,707
1972	47,329	52,730	49,452	70,904	66,779
1973	77,296	77,528	73,867	92,290	88,109
1974	118,707	112,048	108,002	123,378	119,161
1975	124,630	114,731	110,686	126,061	121,844
Accounting rate of interest (percent)		*Net present social value of the project*			
5	1,260,000	1,160,000	1,110,000	1,300,000	1,250,000
10	547,375	494,632	469,906	568,036	541,805
20	132,719	111,152	101,915	136,891	127,167

Note: η is the elasticity of the marginal social utility of private consumption.

In case A the social weights for private savings (ξ_k) are drawn from table 8-1 by use of the rule $\xi_k = [1 + \lambda_k(\eta)]/2$. In case B a uniform weight of 0.7 applies to all private savings.

ing rate of interest of 10 percent, and it can get by even at 20 percent. Not surprisingly, the project turns out to be more profitable if nonfarm labor is not fully employed. Thus, the great improvements the project has made in the material standards of people in the Muda region—improvements at once evident to the casual observer—are reflected in, and consistent with, the project's high social profitability.

Interestingly enough, the value of η makes relatively little difference to the project's net present social value, even at the extreme ends of its range. This has come about by chance: the poverty line, $\bar{c}$, at which a marginal change in private consumption and uncommitted government income are equally valuable, happens to be roughly equal to the mean social value of additional household consumption as actually distributed over the five classes of households. If, for example, $\bar{c}$ had been set at the level at which households are just drawn into the income tax net, which is much higher than the official poverty line, the project would show much higher social profits for $\eta = 2$ than $\eta = 0$. As it is, in the case of unemployed nonfarm labor, the results for these cases are almost identical (given the ARI), but

Table 8-5. *Social Profits from the Project: Fully Employed Nonfarm Labor*
(thousands of Malaysian dollars in 1972 prices)

		$\eta = 1$		$\eta = 2$	
Year	$\eta = 0$	*A*	*B*	*A*	*B*
		Stream of undiscounted social profits			
1965	−6,534	−6,796	−6,803	−6,927	−6,922
1966	−16,130	−17,453	−17,481	−17,956	−17,936
1967	−39,705	−43,292	−43,374	−44,714	−44,660
1968	−47,282	−51,984	−52,087	−53,780	−53,712
1969	−38,408	−41,583	−41,658	−42,887	−42,837
1970	−3,556	559	−704	13,423	11,428
1971	−7,370	826	−1,246	17,815	14,819
1972	19,267	30,897	28,016	52,011	48,014
1973	42,948	52,773	49,508	71,887	67,718
1974	78,069	84,371	80,721	101,484	97,144
1975	83,918	86,980	83,330	104,094	99,754
Accounting rate of interest (percent)		*Net present social value of the project*			
5	734,979	779,666	737,101	992,354	940,839
10	267,500	289,562	268,779	396,027	370,706
20	13,801	19,780	12,625	57,499	48,728

Note: η is the elasticity of the marginal social utility of private consumption.

In case A the social weights for private savings (ξ_k) are drawn from table 8-1 by use of the rule $\xi_k = [1 + \lambda_k(\eta)]/2$. In case B a uniform weight of 0.7 applies to all private savings.

the net present value is somewhat lower than both for $\eta = 1$. For $\eta = 2$, the higher social values placed on government outlays early in the project's life happen to be counterbalanced by the large net social benefits associated with the high incomes accruing to the project's comparatively poor beneficiaries when it attained maturity. When the social utility function is less concave, the fall in the mean social value of additional private income is apparently strong enough to cause a reduction in the project's net present social value. For similar reasons, the two methods of valuing private savings do not make much of a difference to the project's profitability.

In the case of fully employed nonfarm labor, the project's net present social value increases with η; that is, emphasis is placed on equity. The reason is that in the face of full employment of nonfarm labor the additional income accruing to nonfarm households is more modest, so that the ratio of additional farm to nonfarm income generated by the project is greater than in the previous case. In turn, this implies that the social value of additional household income is also greater, for the distribution of income generated by the project is more favorable to farm households when nonfarm labor is fully employed. Moreover, the higher is η, the stronger is this effect.

Conclusions

The analytical approach adopted here rests explicitly on the use of a simple general equilibrium model of the economy to determine a set of endogenous variables with the prices of all goods, domestic and foreign alike, parametrically given. Starting from an initial equilibrium for the economy, a "project" takes the form of a sequence of changes in the economy's exogenous variables and underlying technology. Within each unit of time, we assume that the system adjusts fully to the changes in the exogenous variables. The time path followed by the system's variables may be forged out of the chain of comparative static equilibria thus derived. This yields, among other results, the streams of consumption, savings, and government revenues, the changes in which are the arguments of the social welfare function. The criterion for accepting the project is that the net present social value of the change in national income, as given by equation (8.16), should be positive.

The implicit assumption of instantaneous and full adjustment to the exogenous disturbances of the period in question, and the neglect of all such disturbances in the future, which entails myopic expectations, are certainly very strong assumptions. Yet the "dual" route of using shadow prices is tarred with the same brush (Srinivasan, 1978). Perhaps nothing better can be done in practical analysis, but this does not exonerate the analyst from being clear about what he is doing. Even if a truly dynamic structure were available, it is far from obvious a priori that the costs of such intrinsically difficult refinements would be justified by the improvements in practical decisions they would bring about. Still, one is left feeling uncomfortable that the results should rest on such foundations—unless one seeks refuge in the convention that projects are small enough for these considerations not to matter.

A more particular point to emerge from the analysis is that the only interindustry linkages which have an induced—as opposed to a direct—effect on the endogenous variables are those expressed by the technology for producing nontradables. This follows immediately from the semi-input-output formulation, of course; but it does serve to highlight the fact that capturing multiplier effects in cost-benefit analysis requires, in principle, that the estimates of the relevant parameters pertain to the region in question. Since developing countries are usually regionally heterogeneous, this cannot be dismissed as a theoretical nicety. The analyst usually considers himself fortunate, however, if he has access to a fairly recent *national* input-output table. Again, the same applies in the case of households' expenditure patterns.

With regard to issues in social cost-benefit analysis, in this case study, variations in the elasticity of the marginal social utility of private consumption have relatively little effect on the project's present social value—at least within the range of values considered as "normal" in the literature. The project's main beneficiaries are poor, and hence socially deserving. The social rate of return exceeds 20 percent, however, even when all incomes are given the same social weight.

Despite the care lavished on incorporating all the project's effects in a static framework, our treatment of the intertemporal aspects of resource allocation has been rough and ready. Thus, in estimating the social value of private savings, we have resorted to one shortcut method recommended in the literature. As it turns out, the social profitability is robust to changes in this parameter. Were this not so, a detailed treatment of the dynamics of the economy would have been necessary.

Once the model has been built and estimated, it can be used to evaluate any project. Computationally, nothing more than matrix inversion is needed. Even so, this invites the retort that valuing the *direct* inputs and outputs of a project at their (appropriate) shadow prices is more straightforward still, if a set of such shadow prices is available. That, however, raises two questions.

First, what are the shadow prices that correspond to the primal model laid out here? Although this question goes far beyond the scope of this chapter, we illustrate the principles involved in the appendix which follows. To be sure, since the Muda project was large enough to affect the supply price of labor, the shadow prices for the economy at that point would not have been stationary. Hence, the use of shadow prices based on a constant supply price of labor, which is advocated (and usually valid) for small projects, would be incorrect in this case. Estimating the shadow prices appropriate to this case, however, would involve the same steps followed in the "primal" route taken here, so that there would be no saving in effort.

Second, how much would the above estimate of the project's social profitability differ from one obtained by wholesale use of the shortcut methods laid out in the various manuals on social cost-benefit analysis? We are now so well versed in the intricacies of a full analysis that any attempt on our part to arrive at an estimate by shortcut methods would be utterly compromised by the knowledge already gained. What is needed is an independent effort by a practitioner who is denied all information bar that required for a shortcut estimate; but no such estimate is available.

Appendix

In this appendix, we show how shadow prices may be derived for an economy such as the one just described. Following the terminology used in the beginning of the chapter, we now pursue the dual—as opposed to the primal—route to social cost-benefit calculations. Since the purpose of this exercise is purely expository, we employ a highly simplified version of the model in this chapter. The model presented here is identical to that in Bell and Devarajan (1980*a*). A more general version, as well as extensions to alternative specifications of the economy, may be found in Bell and Devarajan (1980*b*).

A Simplified Model

We make the following simplifications to the model in this chapter.

1. The economy consists of only three sectors, two (1, 2) producing tradables, the third a nontradable.
2. There is only one type of household in the economy.
3. All income is spent on consumption; there is neither saving nor investment.
4. The only source of government revenue is tariffs (there are no direct taxes). Changes in government income manifest themselves as changes in foreign borrowing.
5. Labor is available in perfectly elastic supply, at zero opportunity cost.

Consider a project that increases the output of sector 1 by X_1. By assumption 5, the output of sector 2, which produces a traded good, does not change. Letting X_i, C_i, and E_i stand for changes in the values of gross output, private consumption, and net exports, respectively, of good i, we obtain a set of material balances at the margin as a result of the project:

(8A.1) $$X_1 = a_{11}X_1 + a_{13}X_3 + C_1 + E_1$$

(8A.2) $$0 = a_{21}X_1 + a_{23}X_3 + C_2 + E_2$$

(8A.3) $$X_3 = a_{31}X_1 + a_{33}X_3 + C_3.$$

The project-induced change in income of our single household is:

(8A.4) $$Y = w(\ell_1 X_1 + \ell_3 X_3)$$

where w is the market wage rate, and ℓ_i the labor coefficient in sector i.

If domestic prices are normalized to unity, the no-savings assumption yields:

(8A.5) $$Y = C_1 + C_2 + C_3.$$

This normalization, along with the assumption of all value added being wage payments, constrains the input-output and labor coefficients to be in the following relationship:

$$\sum_{i=1}^{3} a_{ij} + w\ell_j = 1. \qquad j = 1, 2, 3$$

Finally, the assumption of linear expenditure curves (from chapter 8) requires

(8A.6) $$C_i = b_i Y. \qquad i = 1, 2, 3$$

Note that summing (8A.1)–(8A.6) tells us that a change in the value of net exports at domestic prices is zero:

(8A.7) $$E_1 + E_2 = 0.$$

Derivation of Shadow Prices

Shadow prices for the economy described by equations (8A.1)–(8A.7) are now derived in three steps. First, the project's social profit based on market prices is calculated (this corresponds to the primal route followed in chapter 8). Next, it is asserted that this social profit is equal to the value of the project's direct output less the value of its inputs, where "value" refers to shadow valuations. Finally, using the Little-Mirrlees prescription that the shadow price of nontradables should equal their unit shadow cost, we solve explicitly for the vector of shadow prices.

The social profit of a project is the change in government revenue plus the change in consumption, where the latter is weighted by λ, the social valuation of domestic consumption in terms of the numéraire (government income). According to assumption 4, the only source of government income is tariff revenue, which is defined as

$$(P_1 - 1)E_1 + (P_2 - 1)E_2$$

where P_1 and P_2 are the accounting ratios (ratios of border to domestic prices) of goods 1 and 2, respectively. Substituting equation (8A.7) and invoking assumption 3 (all income is consumed), we arrive at the following expression for social profit when taking the primal route:

(8A.8) $$U = P_1E_1 + P_2E_2 + \lambda Y.$$

In terms of the parameters of our model, equation (8A.8) may be written

(8A.8)′ $$U = [P_1(1 - \alpha_{11} - m\alpha_{13}\alpha_{31}) + P_2(-\alpha_{21} - m\alpha_{23}\alpha_{31}) + \lambda w(\ell_1 + m\ell_3\alpha_{31})]X_1$$

where $\alpha_{ij} = a_{ij} + b_i w\ell_j$ and $m = 1/(1 - a_{33} - b_3 w\ell_3)$. Now, U is also equal to the shadow value of the project's direct output less the shadow value of the direct inputs. Since the project is in sector 1,

(8A.9) $$U = [P_1(1 - a_{11}) - P_2 a_{21} - P_3^* a_{31} - w^* \ell_3]X_1$$

where P_3^* and w^* denote the shadow prices of the nontradable good and labor, respectively. Here, we assume that border prices are the appropriate shadow prices of tradable goods, which is the case for our model (see Blitzer, Dasgupta, and Stiglitz, 1981).

It should be noted that by equating (8A.8)′ with (8A.9), the X_1 cancels out, so that the shadow prices are independent of the scale of the project (as well they should be). Moreover, there are two unknowns in equation (8A.9), P_3^* and w^*. To solve explicitly for both requires another equation, which is provided by the Little-Mirrlees prescription equating the shadow price of nontradables with their unit shadow cost:

(8A.10) $$P_3^* = P_1 a_{13} + P_2 a_{23} + P_3^* a_{33} + w^* \ell_3 .$$

Solving for P_3^* and w^* from (8A.8)′, (8A.9), and (8A.10), we obtain the following expressions for shadow prices:

(8A.11) $$w^* = (1 + mb_3 w\ell_3)\left[P_1 b_1 + P_2 b_2 + \left(\frac{P_1 a_{13} + P_2 a_{23}}{1 - a_{33}}\right) b_3 - \lambda\right] w$$

(8A.12) $$P_3^* = m[(P_1 b_1 + P_1 b_2) w\ell_3 + P_1 a_{13} + P_2 a_{23} - \lambda w\ell_3].$$

These expressions show how interindustry and consumption effects influence the shadow prices of labor and nontradables, both through the additional domestic absorption of tradables and through the extra private income generated (the latter being scaled by λ). For example, consider the first term in equation (8A.11), $(1 + mb_3 w\ell_3)$. The m term can be thought of as an output multiplier for sector 3. The payment of a Malaysian dollar to a previously idle worker results in his spending b_3 of it on nontradables. But each unit increase in demand for sector 3's output leads to another round of demand of $(a_{33} + b_3 w\ell_3)$ owing to increased production and employment in that sector. When this process converges, it yields a multiplier for a *unit* increase in demand for sector 3 of

$$1 + a_{33} + b_3 w\ell_3 + (a_{33} + b_3 w\ell_3)^2 + \ldots = \frac{1}{1 - a_{33} - b_3 w\ell_3} = m .$$

However, the initial injection was not a dollar, but b_3, and each round of sector 3 output generates $w\ell_3$ worth of consumption expenditure. Hence, the total (direct and indirect) increase in consumption resulting from the hiring of one more worker and paying him a wage w is $(1 + mb_3w\ell_3)w$.

The shadow wage, though, depends on the social valuation of what workers do with this additional income. Of every dollar, they spend b_1 on good 1, which entails an opportunity cost to the government of P_1b_1. Similarly, their expenditure on good 2 leads to a drop in numéraire government income of P_2b_2. There is also the induced demand for tradables from workers' consumption of good 3, since the latter uses traded inputs in its production. The effect on government income of a dollar's increase in this induced demand is given by $(P_1a_{13} - P_2a_{23})/(1 - a_{33})$, which should be scaled by b_3, the fraction of income spent on nontradables. Thus, the total fall in government income from an increase of a dollar in private consumption is

$$P_1b_1 + P_2b_2 + \left(\frac{P_1a_{13} + P_2a_{23}}{1 - a_{33}}\right)b_3.$$

But the social benefit of a dollar's additional consumption, relative to a unit of government income, is λ. Therefore, since the shadow wage measures the net effect on social welfare of hiring an additional laborer, it is

$$w^* = (1 + mb_3w\ell_3)\left(P_1b_1 + P_2b_2 + \frac{P_1a_{13} - P_2a_{23}}{1 - a_{33}}b_3 - \lambda\right)w.$$

A similar analysis can be applied to interpret equation (8A.12).

It should be emphasized that the shadow prices derived here are appropriate only for economies like the one presented in this appendix. Specifically, the assumption that changes in government income take the form of changes in foreign exchange balances is crucial. This is equivalent to assuming that the government is unconstrained in its level of foreign borrowing. Alternatively, if the government *is* constrained in its foreign borrowing, so that it makes room for a project by altering its domestic expenditure, a different vector of shadow prices will result, although the principles used in deriving them are identical to those employed above. The latter and other adjustment mechanisms are treated in Bell and Devarajan (1980*b*).

9

Prospects and Public Policy

THE PREVIOUS CHAPTERS have dealt almost exclusively with the period from the start of large-scale construction work to the completion of the project. It is now natural to put our analytical apparatus to work again and look at subsequent developments—as far as the patchy evidence permits. Moreover, we will speculate on the prospects for the regional economy, paying particular attention to the economic and social organization of the paddy farming sector and the role of public policy in its various guises.

We begin with the paddy farming economy, which, along with public spending, is the engine of growth for the region as a whole. Next we discuss the growth of output and income in the region in response to changes in the paddy sector and any autonomous forces that may be at work, including public policy. Then we move on to the national context and discuss certain effects of public policy on the paddy sector and the influence of national economic growth on the Muda region.

The Paddy Farming Economy

There are numerous issues worthy of attention, and the scope of the discussion is kept deliberately broad, admittedly with the loss of a certain formal rigor. First, changes are likely in technology and factor supplies: mechanization, the introduction of new varieties of paddy, improvements in water control through a remodeled system of distribution, and shifts in the supply functions for family and day workers. Second is the future course of the price of rice, the effects of which ought to be a central concern in the formulation of public policy, alongside investment in irrigation and plant breeding. Third, there is the scope, if any, for private on-farm investments. Finally, some consideration should be given to the impact of these and other factors on the economic and social organization of paddy production.

Mechanization

Although four-wheel tractors were making slow but steady inroads into the market for land preparation long before the water began to flow through the irrigation system, the project gave a new impulse to that trajectory. The double-cropping regime eliminated the open-field grazing of draft animals in the off-season and drove up wage rates through its exacting demands for both labor and draft power.[1] This change in land-preparation technique reduced the wage bill for hired labor far below what it would otherwise have been, but it did not cause the wage bill to decline (see chapter 4). Nor did it result in the displacement of tenants; for the owners of the tractors are legally barred from purchasing land, and most tenancy contracts involve kin. Moreover, competition among these contractors was (and remains) so fierce that real rental rates fell in the late 1960s and early 1970s. Thus, the introduction of four-wheel tractors had little impact on the distribution of income among paddy farming households, and it did not prevent an increase, albeit a small one, in the share of value added paid to hired labor as a result of the project.

Toward the end of the 1970s the stock of four-wheel tractors began to decline, and the character of the machines changed markedly. After several continuous years of double-cropping, the land had become so soft that the heavy 60 and 70 horsepower machines, which accounted for most of the stock, were getting bogged down too often for profitable operation. These large machines were quickly sold off in favor of smaller ones rated at 20 to 30 horsepower. According to one local estimate, by 1977 only 40 of the larger sort remained in operation, while the stock of smaller ones had reached 400 or so.[2]

Furthermore, many farmers purchased a pedestrian tractor, the stock of which almost doubled, from just over 3,000 in 1974 to about 6,000 in 1978. Although pedestrian tractors do relatively better than four-wheel machines under heavily waterlogged conditions, the new and competitive stock of small four-wheel machines owned by non-Malays indicates that the shift to pedestrian tractors cannot be attributed to the change in soil conditions. It is much more plausible that the cause lay in certain imperfections in the capital market. After improving an existing house or constructing a new one, farmers have few other investment opportunities: only jewelry, a yard

1. Comparing the TIGER solutions for 1974 in the presence and absence of the project (table 4-3), we see that the model predicts a radical switch in land preparation technique as a result of the project, from predominantly animal draft power to exclusively tractors. At the same time, the predicted average wage rate for hired labor in the presence of the project is 37 percent higher than in its absence.

2. Jegatheesan, personal communication.

for drying paddy, a tractor, and a savings account. The surge in farm income owing to the project certainly left many households an ample surplus over their normal expenditure on consumption. With the passage of time, they could amass enough savings to make a down payment on, or an outright purchase of, a pedestrian tractor, even if better housing and gold jewelry for the lady of the household came first. Few of them, however, were likely to risk the far greater outlay on a four-wheel machine, and capital markets being what they are, no lender would consider extending a line of credit to them for such a purpose. Nevertheless, a pedestrian tractor does not cost much more than bringing in a contractor, and it probably conferred some prestige on its owner.[3] Certainly, it made more sense than leaving funds in a savings deposit to earn a negative real rate of interest.

Combine harvesters began to appear in 1974. By 1978 more than a hundred were in operation, and according to a World Bank estimate they harvested about 35 percent of the main-season crop in 1977–78. These machines represent very lumpy investments, of course, which are far beyond the means of any Malay farmers. Competition from traditional hand-harvesting methods, however, and the need to establish a clientele among farmers have kept contract rates competitive. It is thought that the natural and man-made features of the topography will restrict mechanical harvesting to at most 60 percent of the project's command area; fields that are very small, or prone to waterlogging, or sited within a close network of irrigation and drainage ditches will continue to be harvested by hand. Thus, while rising real wages have made the introduction of combines profitable, many of the project's farmers are unable to avoid higher labor costs by resorting to the contractors' combines. This will certainly alter the distribution of income among farm households. It will also produce capital gains by opening up a gap between the value of land which can be harvested by combine and that which cannot. There is, however, no reason for these gains to accrue especially to either land-abundant or labor-abundant farm households, since households in these categories are randomly spread over the project area. But the wage bill for hired labor will certainly be lower in the presence of a hundred or more combines than in their absence.

For farm mechanization as a whole, the long-term prospect offers both interesting contrasts to, and parallels with, the position in the early 1970s. In the early years of the project, most plowing was done by non-Malay

3. If pedestrian tractors had been the only source of draft power in the postproject steady state, aggregate net income from paddy production would have been about 2 percent lower (table 4-7). For the reasons advanced in chapter 4, this estimate of the efficiency loss is an outside one.

contractors using four-wheel tractors, while virtually all harvesting was done by Malays using sickles. In the next decade, Malay farmers will do most of the plowing with their own or rented pedestrian tractors, while harvesting operations will be shared about equally between laborers with sickles and contractors with combines. Land-preparation techniques will be less capital-intensive than in the early 1970s, while harvesting will be more so.

Labor Supplies

The analysis in chapter 3 reveals that the output of paddy is very insensitive to variations in labor supply conditions, as characterized by the reservation wage of family labor and the slope of the supply function for hired labor—at least over the range of alternatives set out in table 3-8. Some other key variables for the paddy farming economy are highly responsive to such variations, however, and the technique of land preparation is especially sensitive. The use of hired labor varies more gently but still appreciably, so that although the average wage rate is fairly stable in the face of changes in labor supply conditions, the total outlay on hired labor varies quite strongly.

Yet one must not jump to the conclusion that these qualitative findings for the early 1970s will hold good under the technological and labor supply conditions likely to rule in the 1980s. For TIGER is formulated to generate a competitive equilibrium, with the real supply price of family labor assumed to be constant over the entire range up to the limit of its availability. This assumption can be interpreted in two ways: either the marginal value of leisure is constant, or the households concerned have unlimited access to off-farm jobs at a parametrically given wage. If we leave aside, for the moment, the matter of interpretation, the level at which this reservation wage is pitched is of little consequence to the levels of output and net farm income (gross of the value of family labor inputs), provided it is not so high as to cause some land to be left uncultivated. If it is set high enough, then marginal land—in this case, plots that have acidic soil or that cannot be harvested by combines—will simply go out of cultivation; for the farmers concerned will do better by earning their "reservation wage" in the form that suits their fancy. Thus, there certainly exists some set of labor supply conditions, variations in which cause significant variations in paddy output. This set would undeniably change if we were to depart from our simplifying assumption of a constant reservation wage rate. In particular, if the reservation wage were to rise with family labor inputs while being everywhere less than the lowest wage at which any day laborer would work, the formulation in TIGER would understate the real supply price of labor at which marginal land would go out of cultivation. But be that as it may, there surely exists a regime in which paddy output and net farm income (gross of the value of family labor inputs) are sensitive to changes in the parameters that characterize labor supply conditions.

It is also likely that there is an intermediate regime, in which the real wage is high enough for both land preparation and harvesting to be fully mechanized, within the limits set by the topography, but all land is cultivated. Once this regime is fully established in place of that ruling in the early 1970s, the possibilities of substituting capital for labor are largely exhausted. Thus, further increases in the wage rate will have little effect on the demand for hired labor, which is used almost exclusively in these two stages of the crop cycle, until land begins to go out of cultivation. Hence, in looking at the position in the late 1970s and the prospects for the 1980s, it is very important, for certain purposes, to specify labor supplies to the paddy farming sector with some care.

Unfortunately, however, it is not possible to be precise about future labor supply conditions. In adjusting the relevant parameters for the various simulations in chapter 4, which dealt with steady-state configurations at the prices ruling in 1974, it was assumed that the real supply price of labor remained the same as in 1973. While this is hardly objectionable from one year to the next, it will not do for the longer term. The Malaysian economy has grown rapidly since 1974, as have real wages and employment, and the prospects for the next decade are for continued growth. Although labor is far from perfectly mobile across regions, these changes in the national economy will surely be reflected, if only weakly, in the supply price of labor to the Muda regional economy in general and its paddy farming sector in particular. Indeed, the World Bank has estimated that the nominal wage rate for paddy work in 1978 was about $7 a day, which in real terms is more than twice that in 1973. This was due to a favorable movement in the price of paddy (table 7-1) and the last stage of the extension of irrigation. It is certainly possible that this increase in farm income also raised the supply price of family labor for farming activities.

But it is one thing to establish the likely direction of changes in the parameters of the labor supply function, and quite another to quantify them. In the light of the foregoing discussion, we can hope to ascertain the conditions under which the two new regimes get established, even if we cannot say anything about the distribution of value added from paddy production without first specifying the critical parameters of the supply function for labor. The following experiments with TIGER are designed to shed some light on these matters.

The Paddy Economy in the Late 1970s: Some Experiments with TIGER

To create a picture of the paddy sector in steady-state equilibrium, given the conditions ruling in the late 1970s, it is necessary to make the following modifications to the specification of the model for 1974:

• Average yields during 1976–78 were about 5 percent greater than those in 1974 (Yamashita, Jegatheesan, and Wong, 1976; World Bank data), no doubt as a result of farmers' learning by doing as well as minor improvements in crop varieties. Accordingly, the yield response function for fertilizer inputs has also been shifted upward by 5 percent.

• Combine harvesting activities have been introduced, the coefficients for which have been taken from MADA (1970*b*) and World Bank data, suitably inflated or deflated to arrive at costs in 1974 prices. The cost of a new machine capable of harvesting five relongs a day is \$60,000. With an expected life of five years, a salvage value of 10 percent of the purchase price, an annual (real) interest charge of 8 percent on outstanding principal, and insurance and taxes, the annual fixed cost of such a machine worked out at \$14,000. The variable costs of harvesting a relong—fuel, oil, spare parts, and skilled labor—amount to \$14. Such a combine reduces the labor input needed to harvest and thresh a picul of paddy by 75 percent.

• Instead of the cropping intensity of 200 percent that was used in chapters 3 and 4, the long-run sustainable level of 185 percent has been used (World Bank estimate). If nothing else, the lower figure can be construed as a long-run average that reflects events such as the loss of the 1978 off-season crop because of the lack of water in the storage reservoirs. This is an upper bound on cropping intensity; land can be left idle in TIGER if that is a profitable action.

• Both parameters of the supply function for day laborers have been increased by 40 percent, as has the reservation price of family labor. This uniform shift in the real supply price of labor is admittedly no more than a rough guess. As argued above, its purpose is to provide clues about the switching of mechanization regimes rather than the distribution of value added in paddy production.

If we retain the set of input and output prices ruling in 1974, the structure of which is similar to that at the end of the decade, these changes yield the steady-state configuration for the late 1970s reported as solution AA of table 9-1. When this solution is compared with that for 1974 (table 4-3), output is greater by 3.5 percent, and aggregate net (nominal) income by 6.9 percent. The really striking difference, however, is that combines have displaced harvesting by hand on all the land that permits combine harvesting. As a result, total labor use is 23 percent lower, and the main burden of adjustment falls on hired labor, which is employed at only 15 percent of its 1974 level. Interestingly enough, draft animals have made a limited reappearance: just over one-quarter of all land is now prepared by buffaloes, largely because cropping intensity is at its upper bound of 185 percent, which allows a more relaxed scheduling of operations than the level of 200 percent permitted in chapter 4. The introduction of combines, coupled with a lower cropping intensity, makes for smaller peaks in the pattern of labor demand. Only in the six weeks of the off-season harvest does the wage rate exceed the reservation level, and the average wage rate is virtually the same as in 1974.

The significance of this finding is, of course, that the use of combines is limited solely by topographical constraints, even at the average wage rate predicted for that earlier year. This is given additional support by solution BB, in which the upper bound on cropping intensity is relaxed to, and binding at, 200 percent. Here, the tighter scheduling forces out virtually all buffaloes in favor of four-wheel tractors. The employment of both family and hired labor is higher, however, because there is more land to be profitably transplanted and harvested by hand over the whole annual cycle. The wage rate is above the reservation level in five periods, but the average is slightly lower than in AA. Because real wage rates were actually far higher at the end of the decade than those emerging from solution BB, we conclude that, with the exception of land posing insurmountable difficulties to combine harvesting, a fully mechanized regime was in the making.

Suppose, then, that combines were to be banned, but that all other conditions of the basic variant for the late 1970s were to hold. The prediction from solution CC, as one would expect, is that both the pattern of labor use and its total would be little different from that for 1974. All land preparation is mechanized, but the stock of tractors is lower, in keeping with the lower level of the upper bound on cropping intensity, which is binding. The interesting point about wage rates is that they exceed the reservation level in seven of the thirteen periods. The average wage rate is almost $4.50, the peak being nearly $5 during the off-season harvest. Land rents are slightly lower than those for 1974, but about 15 percent lower than those in solution AA.

A transitional but (it must be remembered) steady-state regime is represented by solution DD, in which the area that can be harvested by combines is limited to 25 percent of the whole. The employment of hired labor is smaller than that in solution CC by almost half, and the wage rate is somewhat less than $4 on average and exceeds the reservation level in only four periods. Land rents are about midway between those in solution A and those in solution CC.

All things considered, the labor supply conditions specified above for the late 1970s seem fairly plausible, if the early history of the project is any

Table 9-1. *Results from TIGER for the Late 1970s*
(in 1974 prices)

Item	1978 Model solutions			
	Base (AA)	Cropping intensity = 2.0 (BB)	No combines (CC)	Maximum 25 percent of area able to use combines (DD)
Aggregate net income ($ thousands)[a]	200,071	211,564	189,784	196,759
Total paddy production (thousands of piculs)	9,921	10,623	9,926	9,925
Total usage				
Nitrogen (thousands of pounds)	19,351	20,515	19,368	19,362
Family labor (thousands of days)	18,015	18,156	19,191	18,900
Day labor (thousands of days)	699	891	4,557	2,306
Total labor (thousands of days)	18,714	19,047	23,748	21,206
Stock of four-wheel tractors	871	1,387	1,248	1,243
Stock of two-wheel tractors	0	0	0	0
Stock of buffaloes	4,498	567	0	0
Stock of combines	394	460	—	164
Relongs harvested per combine	912	845	—	912
Days worked per four-wheel tractor	139	126	134	135
Days worked per two-wheel tractor	0	0	0	0
Land rents ($/relong/year)				
Nonacid soil	518	517	441	483
Acid soil	400	397	336	371
Additional value of combine land				
Off-season	31	3	0	52
Main season	3	26	0	4
Monthly wages ($/day)				
Nov. 26–Dec. 23	2.50	2.50	2.81	2.50
Dec. 24–Jan. 20	2.50	2.50	3.03	2.55
Jan. 21–Feb. 17	2.50	2.50	3.07	2.50
Feb. 18–Mar. 17	2.50	2.65	4.07	3.18
Mar. 18–Apr. 14	2.50	2.50	2.50	2.50
Apr. 15–May 12	2.50	2.50	2.50	2.50
May 13–Jun. 9	2.50	2.58	2.50	2.50
Jun. 10–Jul. 7	2.50	2.50	2.50	2.50
Jul. 8–Aug. 4	2.68	2.81	4.68	3.76
Aug. 5–Sep. 1	3.14	3.08	4.90	3.96
Sep. 2–Sep. 29	3.24	3.31	4.88	4.11
Sep. 30–Oct. 27	2.50	2.50	2.50	2.50
Oct. 28–Nov. 25	2.50	2.50	2.50	2.50
Average wage ($/day)	3.14	3.08	4.44	3.84
Annual wage payments ($ thousands)	2,191	2,743	20,240	8,859
Percentage of area using				
Combines	60.0	60.0	0	25.0
Buffaloes	27.3	3.0	0	0
Nitrogen (pounds/relong/season)	32	32	32	32

— Not applicable.

a. Defined as the net return to farm households, nonfarm landlords, and four-wheel tractor contractors. Wage payments to day laborers are not included.

1978 Model solutions				MUDA III	
No four-wheel tractors (EE)	Two-wheel tractors only (FF)	Paddy price 25 percent lower (GG)	Paddy price 25 percent higher (HH)	Base (JJ)	Labor supply parameters doubled (KK)
201,789	195,830	139,689	259,856	237,841	233,604
9,921	9,921	9,891	9,924	12,266	12,163
19,345	19,349	18,541	19,357	46,952	43,746
19,312	17,631	17,952	17,809	18,249	18,169
2,161	794	690	732	1,319	1,190
21,473	18,425	18,642	18,541	19,568	19,359
—	—	871	934	1,287	1,311
247	6,725	0	0	0	0
16,197	—	4,498	3,746	0	0
394	394	394	394	428	428
912	912	912	912	877	0
0	0	139	138	136	0
93	139	0	0	0	0
491	509	309	729	596	349
375	391	230	573	468	250
37	30	31	32	23	91
7	4	3	2	35	82
2.67	2.50	2.50	2.50	2.56	5.00
2.50	2.50	2.50	2.50	2.50	5.00
2.50	2.50	2.50	2.50	2.50	5.00
2.78	2.50	2.50	2.50	2.72	5.76
2.81	2.50	2.50	2.50	2.50	5.00
2.52	2.50	2.50	2.50	2.50	5.00
2.50	2.50	2.50	2.50	2.50	5.00
2.50	2.50	2.50	2.50	2.50	5.00
3.24	2.75	2.68	2.68	3.22	6.38
3.70	3.21	3.14	3.17	3.27	6.38
3.85	3.27	3.21	3.25	3.60	6.66
3.01	2.50	2.50	2.50	2.50	5.00
2.65	2.50	2.50	2.50	2.50	5.00
3.41	3.17	3.12	3.15	3.32	6.37
7,368	2,520	2,153	2,306	4,382	7,585
60.0	60.0	60.0	60.0	60.0	60.0
97.2	0	27.3	22.6	0	0
32	32	31	32	75	

guide, the adjustments in the system were by no means complete at that time. Be that as it may, the results presented in table 9-1 point to limited future employment opportunities for hired laborers. The formulation of the model assumes that they are able to find other remunerative work in the regional economy, and even with wage rates as low as $2.50 to $3.00 a day (in 1974 prices), that is what most of them will have to do. In this connection, it should be recalled that the sources of marginal labor supply are, in descending order of marginality, Thailand and Kelantan, southern Kedah, the fringe areas of the scheme, rural nonfarm households, and landless paddy farm worker households. The relevant solutions presented in table 9-1 suggest that all but the last category will be squeezed out by combines in the steady state, so that the assumption that alternative earning opportunities exist does not seem especially objectionable.

To round off this discussion, we shall look at the consequences of banning four-wheel tractors (solution EE) and allowing only pedestrian tractors (solution FF). Combines dominate in both cases. In the former, the buffaloes reign virtually supreme once more, and the employment of hired labor is about three times that in solution AA. The wage rate is above the reservation level in eight periods, but the average is only 10 percent above that in solution AA. Aggregate net income is slightly higher because households devote more of their time to work than to leisure in this labor-intensive regime. Solution EE is not much different from AA, except that pedestrian tractors entail higher costs than the unconstrained mix of four-wheel tractors and buffaloes. This is reflected in an aggregate net income that is 2.2 percent lower than in solution AA, which, as we noted above, is an outside estimate of the actual loss.

One final point: the dual variable for the topographical constraint on the use of combines is between 6 and 10 percent of the annual rental on nonacid plots in the off-season and a negligible proportion in the main season. (The position is reversed in solution BB, in which the cropping intensity is 200 percent.) Hence, increases in land values for land that can be harvested by combine do not threaten a major redistribution of wealth among farmers under the conditions ruling in the late 1970s.

Public Policy: Paddy Prices

The results reported in tables 3-8 and 4-3 point to the conclusion that, under the conditions ruling in the early 1970s, wide variations in the price of paddy would have had little effect on output. It is also clear, however, that paddy price policy is a powerful instrument for altering the incomes of paddy farming households, in both the short and the long run. Moreover, the results in chapter 7 suggest that changes in the price of paddy may bring about strong shifts in the distribution of income between farm and nonfarm households, although it is less clear that regional income as a whole is equally sensitive (table 7-10). Thus, it is important to look into this form of policy intervention in some detail.

From 1967 until 1973, the nominal protection afforded to the paddy sector varied between 10 and 30 percent and was especially large in the project's transitional period, 1971–73 (see the section "Estimation of National Parameters for Social Cost-Benefit Analysis" in chapter 8). A surge in the world price of rice in 1974 reversed the position temporarily; thereafter, the domestic price support level was doubled, and the sector has enjoyed only modest nominal protection since. The current World Bank projections for the 1980s are that the real world price of rice will continue to rise, exceeding the 1979 level by about 25 percent by the close of the decade. That being the case, the domestic price is certain to follow suit if the recent history of price policy is any guide.

The structure of relative prices did not change much between 1974 and 1978; perhaps it moved against paddy farmers a little, but not enough to matter for present purposes. Hence, solution AA can be retained as a basis for comparison when examining solutions GG and HH, in which the price of paddy is set 25 percent lower and higher, respectively, than its level in 1974. To take solution GG first, output is barely (0.3 percent) lower, while aggregate net income is fully 30 percent lower than in AA. Rents are lower still, with a decrease of more than 40 percent on both acid and nonacid plots. In all other respects, solutions AA and GG are virtually identical, inputs of labor and fertilizer being slightly lower in GG. Since the general characterization of labor supply conditions seems plausible, it is fairly safe to say that the burden of such a change in the price of paddy would fall solely on farmers and landlords. As the supply of land is inelastic, landlords would suffer larger proportional losses than tenants.

The converse case, a 25 percent rise in the price of paddy, is perhaps of greater topical interest. Here again, the main departure from solution AA is in aggregate net income, which is 30 percent higher, and its distribution, which has shifted in favor of landlords, rents being slightly more than 40 percent greater. Paddy output and fertilizer inputs are identical to their counterparts in AA, labor inputs are slightly lower, and almost 20 percent of the buffalo stock has been replaced by four-wheel tractors. To sum up, if the real price of paddy does increase by 25 percent over the next decade, it will have few effects on resource allocation within Muda's paddy sector. Farmers' income will rise (and landlords' somewhat more so), thereby stemming some of the permanent out-migration which would otherwise have occurred.

Public Policy: Improvements in the Irrigation System and Crop Varieties

Soon after the project had attained maturity, it became clear that the comparatively crude system for water distribution would impose limits on the yields farmers could attain; these limits were reached by the late 1970s. The present system (described in chapter 2) takes the form of field-to-field flooding over distances between outlet and drain of 1.2 to 2 kilometers, the average being about 1.6 kilometers. This makes water control precarious for individual farmers and creates uncertainty in the scheduling of their operations. The relatively low density of access roads also makes for higher transport costs.

One remedy for this situation is to construct a much finer network of tertiary canals, with additional drains, pumps, and access roads as necessary. This is the heart of the so-called Muda Phase II project, which is to cover about one-quarter of the entire project command area, the blocks concerned being those least well served by the existing system. In addition to this physical infrastructure, there will be a new system of agricultural extension, with frequent visits by field-workers, who are to monitor and coordinate the activities of the hundred or so farmers in each block of the new system. The physical structure should be complete in 1983, and paddy output in the blocks thus served should eventually be 25 percent higher than it would have been in the absence of the project.

Still further in the future, MADA intends to remodel the entire Muda system to conform to the same standards as Phase II. Thus, rather than simulate the effects of Muda Phase II in a literal way, which would involve some tedious and costly redesign of TIGER with little gain in insight, we have chosen to look at this very long-run alternative, hereinafter called Muda III, while keeping solution AA as our benchmark. The changes are:

• The long-run upper bound on cropping intensity is raised to 193 percent as a result of improved water control and use.

• The yield response functions are shifted upward to reflect improvements in varieties, extension, water control, and other supporting services. This shift is somewhat speculative in nature, and we have had to resort to the results of field trials conducted during the 1975 off-season for a variety called Sri Malaysia II. The response functions were:

$$\text{Acid soil} \quad Y = 313.8 + 2.855N - 0.0178N^2$$
$$\text{Nonacid soil} \quad Y = 323.8 + 4.038N - 0.0215N^2$$

where Y denotes yield in gantangs per relong and N nitrogen inputs in pounds per relong. These functions were linearized so that TIGER can accommodate them. Although it is unlikely that these functions will be an exact representation of the actual response of paddy output to fertilizer inputs in the late 1980s, they are certainly in the expected range. For example, the average yield over both soil types for the off-season crop would be 488 gantangs per relong if nitrogen were applied to the point at which its marginal product is zero. The average yield in the blocks covered by Muda Phase II is expected to be about 475 gantangs per relong.

The complete configuration is reported in table 9-1 as solution JJ. Paddy output and aggregate net income are 23 percent and 19 percent higher, respectively, than in solution AA. Higher cropping intensity is profitable as well as feasible and has caused buffaloes to be displaced entirely, while more labor, especially hired labor, is needed to transplant and harvest where machines cannot do the work. Perhaps the most marked difference is that fertilizer inputs are almost 2.5 times greater. The wage structure is essentially the same, the average for hired labor being only 6 percent larger than in solution AA. In light of the earlier comparison of solutions AA and FF, it is also clear that farmers' preferences for two-wheel tractors will have no effect on solution JJ, save a very small cut in net income.

We turn now to the issue of whether marginal land will go out of cultivation. Although all land is under cultivation in solution JJ, it is plain that the real supply price of labor to the paddy sector in the late 1970s will be far greater than $3.50 a day. To probe further, consider solution KK, in which both the parameters of the labor supply function have been doubled. Output is lower than in case JJ by almost 1 percent, and the employment of hired labor falls by 10 percent. Marginal land is just going out of cultivation, and any further increase in the real supply price of labor will result in a significant fall in output, the possibilities of substituting capital for labor being all but exhausted. The seasonal pattern of wages is much the same, but the average daily real wage paid to hired labor is $6.37 in 1974 prices, or about $8 a day in 1979 prices. Since the actual rate in 1979 was $7 a day, the possibility that some marginal land will be uncultivated in the decade to come cannot be lightly dismissed, especially if the price of paddy fails to rise relative to the prices of inputs going into paddy production.

The share of hired labor in value added in solution KK is slightly more than 3 percent, compared with less than 2 percent in solution JJ. Since KK is on the borderline between two regimes, however, it does not follow that this share will rise in line with the supply price of labor. For any increase in the latter will lead to a decrease in the employment of hired labor—a decrease proportionally greater than the associated drop in output as marginal land goes out of cultivation.

Since rents in solution JJ are about 16 percent higher than those in solution AA, there is a marginal shift in the distribution of income in favor of landlords and against farmers. With the labor supply conditions of solution KK, however, rents take a precipitate fall to two-thirds of their levels in solution AA and three-quarters of those in the postproject steady state (table 4-3). This is, of course, as it should be: with an inelastic supply of land and a constant rate of profit, an upward shift in the supply price of labor will lead to a fall in rents. Despite the intensification of agriculture, characterized here by Muda III, and the prospect of some increase in the price of paddy relative to the prices of production inputs, the rapid growth of jobs and real wage rates elsewhere in the economy will do much to keep rents within reasonable bounds.

In solution JJ the premium on land that can be physically harvested by combine is significant in both seasons. The combined premium is about 10 percent of the annual land rent, which is a windfall gain not to be sniffed at. In solution KK, however, the premium is about three times as large, and about half the annual land rent. If these estimates inspire some confidence, land that cannot be harvested by combine will become marginal indeed during the years to come.

The Economic and Social Organization of the Paddy Sector

The changes in resource allocation and technology in the late 1970s and those that can be expected with some confidence over the next decade will exert some influence over the economic and social organization of the paddy sector. Once we move away from the issue of farm mechanization, the basis for discussing the long-run evolution of the farm system becomes rather shaky. Such changes in organization are, however, of sufficient importance to warrant some discussion, even if speculative.

From the earlier chapters of this book, it is clear that the strategy of raising the incomes of paddy producers through the provision of irrigation, new varieties, agricultural extension, and price supports has been successful. Yet now that this intensive strategy of increasing paddy output has run most of its course, and the lives of most paddy farming households have changed for the better, the outcome must be placed in the wider setting of the Malaysian economy. Comparatively well-off as they are—taken as a group—Muda's paddy farming households are still much poorer than the national average: in 1975, the levels of per capita income defined on a comparable basis were US$330 and US$600 respectively (Bell, 1980, table 1), a difference we call the "income gap." Moreover, 80 percent of the farmers' income was derived from paddy cultivation. Thus, if paddy output will grow no faster than 3.5 percent annually, which is as much as Muda III and other sources of yield improvement can offer, while the rest of the economy grows at twice that rate or more, paddy farmers must acquire larger farms, diversify their sources of income, or press for better terms of trade, if their living standards are to lag no further behind the national average. The only question is: What scope is there for farms to increase in size and what form will diversification take?

Since the population is very young and the "momentum" effect is powerful, any decline in fertility will not make for lower natural rates of growth for at least fifteen or twenty years. Within that context the starting point is the growth of the farm population on a fixed land base. Moreover, the natural rate of increase of the paddy farming population is no lower than that of the population at large. But even if paddy holdings stayed at their present size, so that out-migration would have to run at 2.6 percent annually for at least another fifteen years, the income gap would tend to grow and thus further speed out-migration. As we saw in chapter 7, the population of paddy farming households grew at a mere 0.8 percent annually between 1957 and 1970, a period during which the incomes of paddy farming households lagged far behind those elsewhere in the economy. Even with heavy investments in water control and the diffusion of ever improving varieties, out-migration from Muda's paddy farms will continue apace.

It seems likely, then, that paddy holdings will not shrink much, and they will grow larger if the creation of jobs outside the paddy sector encourages a stream of migrants that outpaces the natural growth of the paddy farming population—although this is not likely until the late 1980s. But even if holdings and households stay at roughly their present size, and paddy producers' terms of trade remain constant, the already large relative income gap is likely to grow at 1.0 to 1.5 percent annually—unless incomes from other sources increase rapidly—since per capita GNP will probably grow at 4.0 to 4.5 percent annually.

The first possibility to be considered is that farmers will diversify into other crops and animal husbandry. Given their high income elasticity of demand, the growth of incomes in the towns, and a good transportation network, there should be no difficulty in finding markets to absorb these products at remunerative prices. As seen in chapter 2, however, the acute shortage of suitable land gives little scope for expanding supply. Perhaps the stall-feeding of selected stock has some promise, as does the systematic raising of poultry and vegetables, but it is highly improbable that the output of such products could be increased at a pace much faster than that of paddy over the long haul. Yet the income thus arising would have to grow at an initial rate of 12 to 15 percent a year just to keep the relative income gap where it is, even with the assistance of a 3.0 to 3.5 percent annual growth rate in paddy output.[4] Hence, the prospects for profitable on-farm diversification are certainly not sufficient to relieve the pressures—over and above those keeping operational paddy holdings at their present size—that make for out-migration.

As a second possibility, the government could reduce the income gap by

4. If per capita income elsewhere in the economy is growing, on average, at 4.5 percent annually, then the *ratio* of its level to that of paddy farming households will stay constant provided the nonfarm incomes of the latter grow at a rate y_o given by $4.5 = 3.0 \times 0.8 + y_o \times 0.2$, 0.8 being the weight attached to incomes from paddy cultivation, and the growth of paddy output being 3 percent annually.

propping up the farm-gate price of paddy above world levels and subsidizing farm inputs such as fertilizer. (Indeed, if the world price of rice grows more slowly than the prices of commodities the farmer purchases, keeping the paddy producers' terms of trade constant entails a subsidy anyway.) The long-run price elasticities of demand for, and supply of, paddy are rather low in Malaysia, and the sector accounts for only 2 percent of GNP (Bell, 1980). Hence, the long-run fiscal burden of raising the price of paddy relative to other commodities at a rate of 1 or 2 percent annually, which would keep the relative income gap constant, should not be intolerable. Domestic supply and consumption will alter little, so that the loss of tariff revenue on imports or the outlay on subsidizing exports to be dumped on world markets will also be modest. After some time, the domestic price of rice might get far out of line with world levels. But such happenings are not unheard of in northeast Asia, where small rice farms survive in the midst of relative or absolute affluence. Of course, the drawback of this form of intervention is that it bears heavily on all poorer households, since rice is a wage good. In view of the administrative difficulties entailed, however, there is no immediate prospect of subsidizing paddy farming households through lump-sum transfers, so the adverse effects of price support policies will have to be borne as necessary evils.

The remaining possibility is that part-time paddy farming will become the rule, as in contemporary Japan. With a good road network and the growth of off-farm jobs in Alor Setar and smaller towns, farmers might commute to work. Islamic rules and ethos make it less plausible that farmers' wives will commute to off-farm jobs in any number, although some of their unmarried daughters do so already. Moreover, recent and prospective changes in techniques of paddy cultivation make part-time farming, managed by the women of the household, a distinct possibility. The arduous task of land preparation is now fully mechanized, and combines will soon be harvesting rather more than half the land area. Women do the transplanting anyway and could probably cope with the heavier operations, given occasional help from their husbands, adolescent sons, and hired hands. Indeed, there appears to be no reason they should not develop and exercise the management and husbandry skills possessed by their husbands. After all, many widows do well enough as cultivators, and there is some recent fragmentary evidence that important farm decisions are usually made jointly by the married couple.[5] Thus, if part-time paddy farming were to develop, it seems plausible that the economic role of women would change. This does not necessarily pose any threat to the present system of patriarchy, even though some vestiges of the matriarchal, pre-Islamic order have survived to the present day.

5. Jegatheesan, personal communication.

The Regional Economy

As we have been at some pains to emphasize, the growth of the regional economy depends on the growth of the output of tradables (especially paddy), the growth of public expenditures, and favorable movements in the barter terms of trade. In this section we estimate the downstream effects of two sorts of public policy intervention—price support and agricultural intensification—using the models developed in chapter 6. We begin by constructing a picture of "1978"—admittedly a very sketchy one, but the data admit of nothing better. Using this as a base, we look at the effects of an increase of 25 percent in the price of paddy over its 1974 level. Then we turn to the impact of Muda III, the stylized Muda Phase II covering the entire command area. Putting these together, we obtain an estimate of the effects of the two principal policy instruments toward the paddy sector on the level and distribution of regional incomes over the next decade. Then we turn, once more, to the issue of whether the region will enjoy self-sustaining growth.

In "1978"

Tentative though our analysis for the paddy sector may have been (solution AA in table 9-1), the regional picture must be more tentative still. We start with the solution for 1974 as a base. The difficulty is that in looking at the effects of changes that involve prices or technologies, strictly speaking we cannot employ equation (6.14)″ in first-difference form because there are changes in the various matrices which appear therein (see the discussion accompanying equations (8.15)′ and (8.15)″ in "Applying the Model to the Muda Irrigation Project" in chapter 8). As up-to-date data are not available, however, what is done here is roughly hewn. We make guesses about the values of the exogenous variables; but since we are after *differences* from comparative static experiments, such roughness should not, in itself, be of great account. The base year 1978 is an updated version of solution H for 1974 reported in table 7-6, the structure of relative prices in the late 1970s being closer to that ruling in 1974 than in 1972.

The basis for amending it is most solid in the case of the paddy sector. The actual position seems to have been a combination of solutions AA and FF (table 9-1), weighted heavily toward the latter. Within the Muda project area, output was about 3.5 percent higher than in 1974 and the technology was much more mechanized. In the fringe areas, there are no grounds for supposing that either output or technology would have changed at all, since the factors imposing limits on improved yields within the project area itself during that period were all the more strongly at work outside it. Changes in the technology of paddy production are reflected in a sharp rise in the ratio of intermediate inputs to gross output in the aggregate, and in sectors 12 and 13 in particular, since they supply and maintain agricultural machinery. As for the distribution of value added, the employment opportunities for hired labor in solutions AA and FF are so limited that only landless worker households are likely to get employment in the sector; there is no place for workers from nonfarm households or the fringe areas, and certainly none for seasonal migrants. The wage rate ruling in 1979 has been used, after deflating it to 1974 prices. It is so far above that ruling in 1974 that, despite a fall in employment, the share of value added received by landless paddy worker households has risen. The shares of nonproject farm and nonfarm households have fallen with their loss of employment in paddy production, even though the absentee landlords among the latter enjoy higher rents. Table 9-2 provides a summary of the position.

A closely related change is the increase in throughput for the region's paddy mills. The milling capacity in place in the late 1970s is considered ample to cope with post–Muda II output, owing largely to the completion of new plants owned by the state corporation LPN. Hence, it seems plausible to assume that the whole of paddy output is milled locally in 1978.

In the case of the more speculative exogenous variables, public spending in Peninsular Malaysia by the federal and state governments combined rose by an estimated 37 percent in real terms between 1974 and 1978 (Treasury of Malaysia, various years). By simple, if not wholly plausible, extension, we assume that the same holds true for the Muda region.

The growth of population is beset by comparable uncertainties. In the absence of anything better, we have adhered to the assumption in chapter

Table 9-2. *Household Shares in Gross Output from the Paddy Sector* (Ω_5)

	1978		*Muda III*	
Household	*1974 prices*	*25 percent increase in paddy price*	*1974 prices*	*25 percent increase in paddy price*
Landless paddy workers	0.0212	0.0182	0.0227	0.0193
Labor-abundant paddy farms	0.1447	0.1449	0.1424	0.1430
Land-abundant paddy farms	0.4692	0.4699	0.4616	0.4637
Nonproject farms	0.0883	0.0883	0.0728	0.0728
Nonfarm	0.0520	0.0541	0.0481	0.0488
Total	0.7754	0.7476	0.7476	0.7476

7 that the farm and nonfarm populations were growing respectively at 1 and 2 percent annually over the period in question.

That leaves us with the output levels of the remaining sectors producing tradables and the magnitudes of investment deliveries, lump-sum taxes, and private transfers. Whatever is done to the first two items must seem highly arbitrary. Casual observation in 1977 and inspection of the changes in these variables over the 1967–74 period suggest that their growth in the late 1970s was modest. In the interest of simple arithmetic, therefore, all the exogenous values of output, exports, and investment deliveries have been increased by a round 10 percent, save investment deliveries in the form of livestock from sector 10, which, in keeping with the demise of the draft animal, have been set at zero. Lump-sum taxes and private transfers are unlikely to have changed much in real terms over this period, so they have been kept at their values in 1974.

The picture for 1978 that emerges from this set of assumptions is set out in tables 9-3 and 9-4, which also provide the flows for 1974 for comparison. For what it is worth, our rough and ready figuring suggests that real gross output was about 20 percent higher in the late 1970s, largely because of the rapid growth in real public expenditures and the increase in output of the rice milling sectors. In aggregate, per capita household income is a little less than 10 percent higher, with a strong shift in favor of nonfarm households. Since the farming population rises slightly faster than farm output and farmers face higher rents, the per capita incomes of paddy farming households are slightly lower in 1978 than in 1974. In reality, therefore, it seems safe to say that if the per capita incomes of paddy farmers did grow in the late 1970s, they did so at a meager pace. Landless paddy worker households are somewhat better off, thanks to the sharp increase in real wage rates.

Public Policy: Paddy Prices

With "1978" as base, we now examine the effects of increasing the prices of both paddy and rice by 25 percent. Table 9-3 shows that, in aggregate, real gross output is almost 3 percent higher. Thus, the associated shift in exogenous prices, which raises the terms of trade index from 120 to 143 (1972 = 100), has a slightly expansionary aggregate effect. The sectors which respond most are other agriculture, transport, services, distribution trades, and rural housing, especially those catering to farm households. The effects of this shift on household incomes are also strong (table 9-4) and slightly different from those produced by changes in the region's terms of trade during 1967–74, as analyzed in chapter 7. In the aggregate, real household incomes are almost 10 percent higher. Paddy farming households enjoy the largest proportional boost, as one would expect, with the

Table 9-3. *Real Gross Output: 1978 and Muda III*
(millions of Malaysian dollars in 1972 prices)

	1974	1978	
Sector	*1974 terms of trade* (1)	*1974 terms of trade* (2)	*Paddy price 25 percent higher* (3)
1. Commercial rice mills	107.00*	142.45*	142.45*
2. Small rice mills	68.00*	72.36*	72.36*
3. Food processing	8.20*	9.02*	9.02*
4. Fish processing	7.33*	8.06*	8.06*
5. Paddy production	170.56*	172.81*	172.81*
6. Fishing	34.00*	37.40*	37.40*
7. Estates rubber	3.90*	4.29*	4.29*
8. Smallholder rubber	22.30*	24.53*	24.53*
9. Rubber processing	2.32*	2.55*	2.55*
10. Other agriculture	24.72	28.59	32.37
11. Sawmilling	9.76	10.85	11.18
12. Agricultural machinery services	3.12	13.78	13.78
13. Machinery repair	6.90	11.35	11.82
14. Manufacturing n.e.c.	21.56	24.75	25.86
15. Road transport	22.71	26.59	29.64
16. Rail transport	1.45	1.43	1.52
17. Hotels and restaurants	17.13	21.79	23.96
18. Entertainment	6.50	8.59	9.47
19. Services n.e.c.	2.07	2.18	2.67
20. Private health	4.92	5.46	6.11
21. Distributive trades	51.50	58.15	62.43
22. Petty trading	5.49	5.75	6.75
23. Electricity	6.13	7.36	7.75
24. Water	3.37	3.98	4.35
25. Posts and telecommunications	3.17	4.57	5.00
26. Residential construction	18.56	20.50	20.83
27. Other construction	21.71	26.82	26.95
28. Government, irrigation	8.37*	8.37*	8.37*
29. Government, education	53.38*	73.40*	73.40*
30. Government, health	12.82*	17.58*	17.58*
31. Government, other	49.91*	68.46*	68.52*
32. Traditional financial services	8.67	8.88	10.85
33. Modern financial services	2.95	4.86	5.05
34. Urban housing	20.90	26.70	28.96
35. Rural housing	33.70	36.05	41.15
Total	845.06	1,000.17	1,029.78

Note: An asterisk denotes exogenously fixed levels of gross output.

Muda III				
1974 terms of trade (4)	*Paddy price 25 percent higher* (5)	$\frac{3}{2} \times 100$ (6)	$\frac{4}{2} \times 100$ (7)	$\frac{5}{2} \times 100$ (8)
195.30*	195.30*	100.0	137.1	137.1
85.93*	85.93*	100.0	119.0	119.0
9.02*	9.02*	100.0	100.0	100.0
8.06*	8.06*	100.0	100.0	100.0
209.49*	209.49*	100.0	121.2	121.2
37.40*	37.40*	100.0	100.0	100.0
4.29*	4.29*	100.0	100.0	100.0
24.53*	24.53*	100.0	100.0	100.0
2.55*	2.55*	100.0	100.0	100.0
31.31	35.94	113.2	120.0	126.0
11.04	11.44	103.0	105.0	125.4
14.37	14.37	100.0	104.2	104.2
12.09	12.67	104.1	107.0	112.0
25.95	27.32	104.4	105.0	110.3
30.04	33.80	111.4	113.0	127.1
1.52	1.62	106.2	106.2	113.2
23.78	26.48	110.0	109.1	122.0
9.48	10.70	111.4	112.0	126.0
2.49	3.08	122.4	114.2	141.2
5.94	6.74	112.0	109.0	123.4
65.94	71.20	107.3	113.4	122.4
6.41	7.63	117.3	111.4	133.0
7.83	8.31	105.3	106.3	113.0
4.29	4.74	109.3	108.0	119.1
5.15	5.69	109.4	113.0	125.0
20.82	21.24	102.0	102.0	104.0
25.91	26.07	100.4	97.0	97.2
11.33*	11.33*	100.0	135.3	135.3
73.40*	73.40*	100.0	100.0	100.0
17.62*	17.62*	100.0	100.0	100.0
68.54*	68.61*	100.0	100.1	100.2
10.31	12.74	122.1	116.1	143.4
5.80	6.03	104.0	119.3	124.0
29.27	32.11	108.4	110.0	120.2
39.79	46.07	114.1	110.3	128.0
1,136.98	1,173.54	103.0	114.0	117.3

Table 9-4. *The Level and Distribution of per Capita Household Incomes*
(Malaysian dollars in 1972 prices)

	1974	1978		Muda III	
Household	*1974 terms of trade*	*1974 terms of trade*	*Paddy price 25 percent higher*	*1974 terms of trade*	*Paddy price 25 percent higher*
Landless paddy workers	266	293	311	360	382
Labor-abundant paddy farms	398	387	459	450	546
Land-abundant paddy farms	690	667	810	780	958
Nonproject farms	469	471	506	483	522
Nonfarm	970	1,103	1,172	1,188	1,275
All	692	737	819	812	914

distribution of income among them moving strongly against landless paddy worker households, which gain little extra wage income and make heavy outlays on the consumption of rice. Households in the fringe areas, which specialize more in the production of rubber, have a proportional gain of only about 7 percent. The surprise, however, is that nonfarm households also gain—by 6 percent. Higher rents have only a small part in this. The important impulse is from the increase in real spending by farm households, coupled with the fact that the increase in rice prices has only mild effects on costs in the nonfarm sectors of the economy, either for industries or households. Outlays on rice are a fairly small proportion of nonfarm households' budgets at the income levels ruling in the late 1970s, and their cost of living index rises by only 2 percent compared with about 5 percent for farm households.

We conclude that a 25 percent rise in the prices of paddy and rice, whether it stems from a comparable increase in the world price of rice or the government's drive to raise farm incomes, would give quite a strong impulse to regional incomes, but with a rather ambiguous effect on income distribution. Of course, if a rise in the world price of rice provides the occasion for the government to raise the domestic support price, there can be no claim that additional (partial equilibrium) subsidies are involved. In these circumstances, both policymakers and the people of the Muda region can have their cake and eat it.

Public Policy: Improvements in the Irrigation System and Crop Varieties

Building on the analysis above (see the corresponding subsection under "The Paddy Farming Economy," p. 231) and retaining the assumption that the structure of relative prices in 1974 continues to hold, we now consider the effects, relative to 1978, of remodeling the entire Muda system to the standards of Phase II. This involves an increase in paddy output, with accompanying changes in the structure of costs and the distribution of value added, which are laid out in table 9-2. All the extra paddy output is assumed to be milled in the region. There are no changes in any of the remaining exogenous variables for 1978, save the *decrease* in the lump-sum taxes on paddy farming households resulting from additional irrigation subsidies.

In the aggregate, real gross output rises by 14 percent, led by rice milling, other agriculture, irrigation, and financial services. Household income increases by a little more than 10 percent, virtually the same increase achieved by raising paddy prices by 25 percent. From table 9-4, it appears that paddy farmers and nonproject farm households do slightly less well from our stylized Muda III than the increase in the price of paddy, and conversely for paddy worker and nonfarm households. This prompts the intriguing question: Would it not be better to subsidize farmers through paddy price support policy than to continue with agricultural intensification? An answer requires a full-blown cost-benefit analysis along the lines of chapter 8, but we will not attempt that here.

If the two policies were combined, tables 9-3 and 9-4 suggest, the outcome would be approximately the (logarithmic) sum of the outcomes of each pursued separately. Compared with 1978, the combined policy would raise the region's average per capita household income by 24 percent. Paddy farming households would enjoy an increase of 43 percent and landless paddy farm worker households one of a little more than 30 percent. Farming households in the fringe areas would gain least, both absolutely and relatively (barely 10 percent). Nonfarm households, riding on a flow of favorable downstream effects, would enjoy an increase of 16 percent.

Although these calculations are merely tentative, the results derived above are probably good guides to the order of magnitude of the effects of these two most important policy interventions in the paddy sector on the variables of interest. Aside from increases in real public spending, the prospects for self-sustaining growth in the regional economy remains to be discussed.

Self-sustaining Regional Growth

Self-sustaining regional growth, then, must stem from a continuous stream of private investment in the sectors of the economy producing tradables (or "manna from heaven" technical progress). Yet in dealing with the prospects for the next decade, we have concentrated, for the most part, on public investments that have already raised or promise to raise farm output and incomes.

The discussion of farm mechanization has naturally dealt with private investments in tractors and combines, but a more systematic scrutiny of private on-farm investment opportunities is now called for. We have already pointed to certain imperfections in the capital market that restrict the investment opportunities open to Muda's farmers. Moreover, the paddy farming system is such that there are few ways of putting savings to profitable use on the farm itself. The limits to the growth of the farm sector through private investment have already been closely approached at the time of writing. Furthermore, since the growth of the sectors producing paddy and other tradables drives the growth of the regional economy, there can be no simple presumption that farm households will find ready investment opportunities in the nonfarm sectors of the regional economy, which, for the most part, produce nontradables.

As for the other agricultural sectors, output in the smallholder rubber sector has continued to grow under the impetus of replanting with high-yielding clones. But it is difficult to argue that these investments were financed by Muda's paddy farming households out of extra savings generated by the project. Part of these savings may have been invested in livestock, fruit trees, and vegetable production to satisfy local and extraregional demand; but it cannot have been a very large part, given the scarcity of suitable land and the low incremental capital-output ratios for these activities. If the additional savings generated (directly and indirectly) by the project were to lead to self-sustaining regional growth, they would have to finance investments that would not otherwise have been undertaken in the nonfarm sectors producing tradables. For, as we have been at pains to point out, investments in the sectors producing nontradables are induced, in the first instance, by increases in the demand for such goods arising out of increases in output and incomes in the tradables sectors of the regional economy. They will not be undertaken autonomously, except by the government, whose outlays in the region do not depend on the tax revenues it collects there. As shown in chapter 7, the project did bring about a large increase in both output and investment in the nontradables sectors of the regional economy. But what was measured there took the form of a once-and-for-all adjustment in the level of output rather than a steady growth of output once the project's immediate benefits had been attained.

There are only three ways the project might produce downstream effects on the local economy which could be construed as a process of self-sustaining growth. First, having accumulated some savings, farm households and perhaps some nonfarm households may find themselves able to invest in some nonfarm production activities shunned by other investors—activities they could not enter before owing to imperfections in the capital market. Second, the expansion of local demand may create a market sufficiently large to warrant local production of items that were previously imported from other regions. This possibility requires increasing returns over some range of output (so that the relevant cost curves are U-shaped). Third, the initial surge in output and incomes generated directly and indirectly by the project may produce a heady business climate in the local economy. Their "animal spirits" thus aroused, investors may then invest locally as opposed to elsewhere in the national economy.

None of these possibilities strikes us as very plausible for the Muda region, which lacks the skilled work force and much of the basic infrastructure for industrial development and is poorly located in the economy's transport system. The conclusion that the project's effects were of a once-and-for-all kind seems to us virtually inescapable, with the implication that the task of sustaining the region's growth will fall to those who direct public policy. As argued in chapter 5, additional private savings are destined largely for investment in other regions of Malaysia, either by direct placement or through the operations of financial intermediaries. Indeed, as shown in chapter 7, this capital outflow was significant even in the preproject period, when it amounted to about $17 million annually. Under the influence of the project and a sharp rise in the price of paddy, it reached nearly $74 million a year in 1974. With due allowance for error in our estimation procedures and adjustment for price changes, it is hard to believe that the capital outflow from the region did not double in real terms between the pre- and postproject periods.

The National Economy

There is no need to labor further the magnitude of the induced capital outflow from the Muda region. Of interest is the impact of public policy toward the paddy sector on "uncommitted government income," since the aim is to displace imports of rice with domestic production, which happens to enjoy a protective tariff, as well as to increase farm incomes.

To arrive at an answer, however rough, recall the analysis of the original project (Muda I) in chapter 8. If the nominal tariff on rice imports stays at about 10 percent over the project's lifetime, the total change in annual uncommitted government income is an *increase* of about $10 million (in 1972 prices) in the steady state, or about 8 percent of the total change in steady-state value added stemming from the project (table 8-2). That there should be a net gain in uncommitted government income despite the tariff is due in large part to an elastic structure of direct taxes and stiff indirect taxes on those commodities for which income elasticities of demand are high. Thus, in the new equilibrium established after the project was completed, there is an additional tax burden on households and firms. Naturally, only an analysis based on a general equilibrium model, of which that in chapter 8 is a simple example, permits such an assertion to be made with some confidence—although we hasten to add that we are stating only a necessary condition for such assertions.

Since the investment cost of Muda III is relatively modest, benefiting as it does from the facilities provided by Muda I, there are grounds for believing that this follow-on project will not be a heavy charge against uncommitted government income. Only if the domestic price of rice is kept well above the corresponding world price, presumably because the government wishes to subsidize paddy farmers, does there seem to be a real threat that undertaking the project would mean lower uncommitted government income in the postproject steady state. The partial equilibrium argument suggests that the cost of the associated subsidy will be relatively modest, however, even if the nominal tariff is quite large. Moreover, the results reported in table 8-2 show a small increase in uncommitted government income because of Muda I in 1973, when the nominal tariff on rice was almost 30 percent. We conclude, therefore, that Muda III is unlikely to be a heavy drain on the public purse, even if it is combined with a fairly high protective tariff.

One further aspect of the national setting concerns us here: the growth of output and jobs over the next decade. So long as GNP continues to grow at about 7 percent a year and the creation of jobs keeps pace with the growth of the labor force, as has been the case in the past, the pressures on the regional economy caused by population growth can be relieved by out-migration and there is scant chance that farm holdings will shrink much, if at all. What is more, factors outside the region will govern the supply price of labor to the region's production activities to a greater extent. The mechanized techniques of paddy cultivation now widely employed in the field (and obdurately reproduced by TIGER) have surely taken hold for good. All this suggests that in pursuing the issues set out in this chapter, we have pressed our analytical models to the limits of their usefulness.

Concluding Remarks

Despite the caveats voiced above concerning the limitations of our comparative static analysis and the speculative way in which key parameters have been estimated, we advance two qualitative conclusions without much hesitation. First, because conditions in the rest of the economy are bound to raise the real supply price of labor to the paddy sector, the course of mechanization charted above is inexorable. The main losers will be the seasonal migrants from Thailand, Kelantan, and the region's peripheral areas, as well as those farmers who are stuck with harvesting by hand.

Second, the combination of Muda III and a 25 percent improvement in paddy producers' terms of trade will just about maintain the relative income gap at its level in the mid-1970s. This combination represents something close to the practical limits to public policy intervention in favor of paddy farmers over the 1980s and beyond. Once its full effects have been felt, the income gap will begin to widen. Since the prospects for sustained and independent growth in the region's nonfarm sectors are not good, it follows that heavy, permanent out-migration from Muda's farming community cannot be delayed much beyond 1990.

10

Epilogue

IN THIS STUDY we set out to quantify the indirect, or downstream, effects of a particular agricultural project and to ascertain how they are distributed among the people of the surrounding region. Our primary finding is that when the project was complete, the indirect effects were large: about 80 cents of additional value added was generated in the region for each dollar of value added generated directly. Furthermore, the incomes generated indirectly accrued largely to the nonagricultural households of the region, which were generally better placed to participate in nonfarm entrepreneurial activities. Each dollar of downstream value added was probably supported by a little more than a dollar of additional investment in plant and equipment, spread appropriately over the sectors that expanded as a result of the project. These findings are, of course, subject to the frailties of the data on which our quantitative analysis is based, as well as the limitations of the analysis itself, but nevertheless the qualitative findings seem robust.

Since the details of our findings have been adequately summarized in chapter 1 and in the conclusions to individual chapters, we will focus here on some of the more important issues that arise from our work: policy issues, which have some bearing on the selection and design of agricultural projects, and issues pertaining to methods of project appraisal, monitoring, and evaluation. Any lessons that may emerge, we hope, will be of some value to those who design, implement, and evaluate agricultural projects. Indeed, they may also be of value to those wishing to embark on similar studies of their own.

Some Policy Issues for Project Design

The extent to which development projects are able to bring about improvements in the distribution of incomes and set in train a process of self-sustaining growth is now widely recognized as an important aspect of project design. Governments and financing agencies seek policies that encourage development, often with the specific intent of recovering the costs of projects. We begin with income distribution.

Income Distribution

To review briefly the evidence, primarily from chapters 4 and 7, there is no doubt that the project's direct effects led to a very substantial increase in the incomes of both landless workers and paddy farmers living within the command area of the Muda irrigation scheme. In 1974 their real earnings from paddy production were more than two-thirds larger than they would have been had the project not been undertaken (table 7-7). Moreover, the distribution of these direct benefits between landless workers, labor-intensive and land-intensive farms, and absentee landlords was equitable, despite the widespread adoption of tractor mechanization between 1967 and 1974. This happy outcome is mainly attributable to the existence of fairly competitive markets for labor and draft power and, with the advent of the project, an increase rather than a fall in the wage/land rental ratio. This finding certainly runs counter to the observation, perhaps especially common in South Asia, that irrigation projects often worsen the distribution of farm incomes.

Yet, on reflection, there are scant grounds for being startled. In Muda there is no class of big farmers and landlords to bully tenants into vacating their leases, to wield market power when hiring labor, or to corner the credit and new inputs rationed out by a creaking administrative system. On the contrary, the ownership of land is spread widely, tenancy relationships usually involve kith and kin, the markets for variable inputs are fairly free of imperfections, and the administration is relatively open and efficient in its dealings.

The introduction of tractors did reduce the wage bill for hired labor below what it would otherwise have been (table 4-7), but it did not cause such payments to be lower than they would have been in the absence of the project. Nor did it result in the displacement of tenants by the owners of four-wheel tractors, because the latter were legally barred from cultivating land. Again, contrary to observations elsewhere, the introduction of tractors had little impact on the distribution of income among paddy farming households and failed to prevent an increase in the share of output paid to hired labor; it must be attributed, at least in part, to the social and inheritance customs of the Muda area.

Chapter 7 revealed that indirect effects were very much skewed toward nonagricultural households. In fact, they largely offset the beneficial distributional consequences of the direct effects, so that although all households were considerably better off as a result of the project, the distribution of income within the region barely changed (table 7-7). This finding is clearly of some importance to project analysts intent on improving the distribution as well as the absolute levels of incomes in project areas, for it implies that the indirect effects of a project cannot easily be ignored. Accordingly, much more attention than is typical of project design and appraisal work needs to be given both to who are the likely beneficiaries of a project's indirect effects and to those complementary policies that might help ensure a more equitable distribution of these secondary benefits.

Of necessity, our analysis of the distributional consequences of the project has been based on a relatively short period. Nevertheless, we conclude this section with some thoughts about the longer term impact of the project on household incomes. In so doing, we risk venturing beyond the reasonable limits of our analytical framework and our data base.

That the project had an initially equitable impact on small and large farms and on landless workers stems in large part from the favorable movement in the wage/land rental ratio. But as indicated in chapter 4, actual land rents in the immediate postproject period were only a quarter of their long-run equilibrium values, and there is evidence that they have risen substantially since then. Whether incomes will continue to be equitably distributed will depend on whether wages rise proportionally with land rents, an outcome that will probably depend more on the continued buoyancy of the demand for labor in the rest of Malaysia than on the demand for labor within the Muda region itself. In chapter 9 the evidence for 1978 suggests that real wages were rising faster than land rents, despite the introduction of combine harvesters.

The widespread introduction of mechanized harvesting will have income distributional consequences of its own. But since land-abundant and labor-abundant paddy farms are distributed evenly over the project's command area, there are no solid grounds for supposing that the reductions in cost offered by combines will benefit the one proportionally more than the other. The losers will be households whose plots cannot be harvested by combines because of the topography of the land. The owners of combines, being Chinese, cannot purchase land, so that tenants and marginal farmers will be displaced only if combines make managing cultivation so much easier that owners are encouraged to resume management of land formerly leased out and to purchase additional land when it comes on the market. Without denying that combines may save on management time and skills, we believe that the nature of the market for land leases and the comparatively even distribution of wealth in the paddy farming community make such a process of concentration unlikely, at least in the near future. It is quite certain, however, that combine harvesters will displace hired labor on an enormous scale. Although there will remain a niche for workers from landless paddy farming households and perhaps some of those from nearby nonfarm households, the seasonal migrants who come for harvesting, whether from near or far, will probably lose this source of earnings entirely by the early 1980s.

Once we move away from farm mechanization, the basis for discussing the evolution of the farming system over the long run becomes rather shaky, especially in regard to the distribution of income and wealth. In chapter 9 we argued that to keep the income gap within bounds over the next decade, either farmers will have to diversify their sources of income into activities that are growing rapidly, or there will have to be heavy policy intervention in the form of new investment in the irrigation scheme coupled with higher price supports. The former is an implausible prospect, and the latter can stave off the inevitable until 1990 at best. In any event, once the income gap begins to grow, out-migration will accelerate to the point at which farm holdings begin to increase in size, and so provide higher incomes.

Self-sustaining Growth

In chapter 5 we calculated that in 1972 there was a net capital outflow from the region of about $35 million, and in chapter 7 we arrived at a figure of about $18 million for 1967. We went on to estimate that this outflow increased fourfold to $74 million in 1974. The latter figure was equivalent to 72 percent of regional household savings in that year, a substantial leakage of investible resources which, had things been otherwise, might have been used to foster the region's further development.

Part of the reason for this large capital outflow must lie in the attractive investment opportunities available elsewhere in Malaysia, for the national economy was growing very rapidly from 1967 to 1974. But the limited opportunities for investment within the region itself were also a major cause. Similarly, the relative isolation of the region and the attendant high transport costs limit the potential growth of a significant industrial base. Nonfarm households are likely to continue to find greater scope for entrepreneurial investments elsewhere in Malaysia, especially now that the most lucrative local investments have already been made in providing the extra goods and services demanded by farm households as a result of the project.

For the private sector, such capital outflows represent the efficient use of a scarce resource, but they do not bode well for the ability of a relatively backward rural area to retain the resources necessary for self-sustaining development. In the absence of substantial investment of domestically generated capital, there can be no self-sustaining growth, and the region will continue to seek the benevolence of the federal government to crank it from one income plateau to another, either through additional projects or price support policies. Thus, the initial irrigation project is being followed by Muda II, and presumably this will be followed by Muda III. On each occasion, the investments may be justified on the grounds that Muda remains a relatively backward area by Malaysian standards.

Such scenarios are, of course, common for rural areas throughout the developing world, but it is particularly discouraging that the Muda region should be so entrapped. Unlike many regions, it has a well-developed infrastructure and its entrepreneurs have demonstrated their ability to respond to new investment opportunities. Our study bodes ill for the notion held by many rural development experts—for example, Lele (1975)—that project investments can, by themselves, initiate a process of self-sustaining growth.

Policies to Increase Downstream Benefits

We set out to analyze the relative strength of the production (or interindustry) and household demand linkages in generating the downstream benefits of the project. As it turned out (chapter 7), the effects of increased incomes on household demand accounted for about two-thirds of the total downstream benefits, despite the increased demand for inputs associated with the new technology in paddy production. This finding supports Mellor's (1976) position on the importance of rural household demand linkages in the development process, and may well apply, at least in a qualitative way, to other projects and other rural regions. Moreover, this conclusion is instructive in relation to the choice of complementary policies that might increase the downstream benefits from investments in rural areas.

The income (and expenditure) elasticities of demand for locally produced goods and services by different classes of household vary appreciably. Specifically, low-income households tend to have the highest income elasticities of demand for locally produced goods and services,[1] and hence

1. Using the FAO/World Bank survey data, we obtain the following estimates of household expenditure elasticities (quintiles are defined over per capita household expenditures):

	Nonfood consumer goods	
	Locally produced	*Imports*
Landless paddy workers	3.70	1.93
Bottom quintile	4.10	2.22
Second quintile	4.07	2.01
Third quintile	2.95	1.77
Fourth quintile	2.26	1.59
Top quintile	1.54	1.25

income increases in their hands generate larger downstream benefits in the region than income increases received by wealthy households. In Muda, the distribution of incomes among agricultural households is relatively equal, and alternative distributions of the value added generated directly by the project would have had little impact on the size of the project's downstream benefits. Indeed, in experiments conducted with the model, but not reported in the study, we found that lump-sum taxes that transferred income from land-abundant households to landless workers had a negligible effect on the region's total income. However, this finding stems largely from the unusual degree of equality among Muda's farmers, particularly the relative uniformity of farm size, and the efficient markets for labor and draft power. These distributionally benign results are unlikely in situations in which the markets are less perfect and the ownership of productive assets less equal. In such cases, a keen focus on who receives the direct benefits of a project may yield handsome dividends by increasing the project's downstream benefits.

Given the importance of household demand linkages in generating downstream benefits, the sectors producing the nontradable goods and services to meet such demands must have elastic outputs if the downstream benefits are to materialize. An imperfectly elastic supply structure will lead to higher prices and producer rents rather than increases in real downstream value added. In Muda, the important demand-driven sectors that expanded to meet increased household demands were trade, entertainment, travel, residential construction, and financial services. We assumed that these sectors were readily able to expand output to meet new demands at constant average costs—not an unrealistic assumption for Muda, which has a well-developed infrastructure and an abundance of capital and entrepreneurial skill. But Muda is hardly typical in these respects; elsewhere these sectors might require special assistance in obtaining credit, skilled personnel, and access to markets if they are to respond fully to increases in local demand.

Recovery of Project Costs

Despite the government's maintenance of price support policies that afforded paddy production modest nominal protection, there was a definite net surplus in the government's revenue from the region when the project reached maturity. Indeed, from tables 8-2 and 8-3, we see that incremental income accruing to the government in the 1975 steady state lies between $2 million and $11 million a year in 1972 prices, depending on whether there was full employment of nonfarm labor. This annual, steady-state surplus in government revenue arose from a progressive tax structure—primarily sales and commodity taxes and income tax—which fell heavily on the region's nonagricultural households, the principal recipients of the downstream benefits of the project. The government's total net outlay on the project was about $220 million in 1972 prices. Thus, quite apart from the increases in private incomes generated by the project, on the more favorable assumption of unemployed nonfarm labor the government will be fully recompensed for this outlay if the discount rate is less than 5 percent, even though the water charges levied on Muda's farmers covered less than half the scheme's operating costs. Even on the less favorable (full employment) assumption, the project more than pays for its current operating costs when everything has been taken into account. Seen in this light, our findings cast some doubt on the wisdom of those who doggedly demand sufficiently stiff direct taxation of a project's direct beneficiaries (through, for example, water charges) to ensure that the government "recovers" its costs in full.

Some Methodological Issues

In conducting this study we have wrestled with a number of methodological problems and have adopted procedures that, to some readers, may seem unusual or perhaps unnecessary. Consequently, we will dwell for a moment on some of these matters and draw what lessons we can for future studies of this kind. Moreover, we suspect that some of these lessons are also relevant to the appraisal, monitoring, and evaluation of projects in a more operational setting.

Defining the Region's Boundary

In studies of this kind, or indeed in seeking a regional perspective of a project's full impact at the design stage, there arises the difficult problem of defining the region that surrounds the project. The regional boundaries adopted in chapter 2 owe much to natural geography, and indeed the region is geographically well defined. However, the semi-input-output method, which underpins the measurement of the project's downstream effects, rests critically on the distinction between activities producing tradable and nontradable goods. Across national frontiers, this is not a difficult distinction to make; but in a regional context, it demands that the boundaries be drawn with some care.

In the present case, the only activities producing nontraded goods within the region that present serious difficulties are road transport, electricity, and certain kinds of financial services. If the region's demands for transport services should rise, trucks and buses could be brought in from other regions to meet it, perhaps at the cost of extra congestion on the roads. Similarly, any increase in the demand for electricity must be met elsewhere, either by generating extra output or switching supplies, since the Muda region has only a transmission system. And although the firms providing certain specialized business and financial services have offices in Alor Setar, most of the work is done at headquarters in Kuala Lumpur or Pinang. In chapters 6 and 7 these interregional trades were ignored in the case of transport and treated as noncompetitive import flows in the other two cases.

The point to be made here, however, is that changes in the region's demands for such goods are felt partly by other regions, and national production must adjust because they cannot be traded internationally. The same applies, although with less force, to the two "ragbag" sectors: livestock, fruits, and vegetables; and manufacturing not elsewhere classified. High transport costs and product quality limit the extent to which these goods are actually tradable with other regions, let alone internationally. Similarly, noncompetitive imports may include some manufactures produced elsewhere in Malaysia but not readily tradable internationally. The treatment of this issue in chapter 7 is an undeniably rough and ready compromise between the fully traded and nontraded cases.

The upshot of all this is that a balance must be struck between extending the region's boundaries to include virtually all activities producing nontradables and keeping them sufficiently tight to preserve the integrity of the region's labor markets. As it turns out, the definition of the region adopted early in this study has done rather well. The steady-state downstream effects felt by other regions owing to changes in the region's demand for nontraded goods are quite modest. At the same time, seasonal movements of migrant labor are also relatively small. Of course, the fact that the provincial capital, Alor Setar, lies in the middle of the region's natural boundaries is largely responsible for this happy result. The lesson for other such studies is clear: a project's surrounding region must include a town (or towns) that are the focus of transport, services, and distributive trades, since the project's downstream effects will, for the most part, change the outputs of these activities.

Need for and Limitations of the Models

Although we chose the Muda irrigation project for study in part because of its relatively rich data base, we had nevertheless to resort to models to picture the pre- and postproject situations in both the paddy farming sector and the regional economy. Some of this modeling work could have been avoided had more complete and suitable data been available for the pre- and postproject situations. In addition, however, our models were required to construct pictures of the economy as it would have been in 1974 without the project and to conduct experiments to separate the effects of the project itself from changes in prices, other autonomous forces, and the interactions between them. It is probable that other studies of this kind would also require a hefty modeling effort, particularly if the project is sufficiently large to have major, and therefore less predictable, ramifications for many aspects of the local economy.

Accordingly, we review below some key features of the models we have used and discuss both their limitations and suitability for future applications.

The three microstudies of the farm economy on which the analysis in chapters 2, 3, and 4 is based span a comparatively brief interval in the system's evolution. Hence, the only processes we can observe with any clarity are short term. TIGER, the linear programming model that constitutes part of our analytical apparatus, yields accounts of changes over time in the form of comparisons of steady-state equilibria not too far removed from the conditions ruling in 1973, the year to which it has been calibrated. In the absence of suitable time series data and a satisfactory quantitative theory of dynamics and the processes of accumulation applicable to a peasant economy, there is no need to be apologetic for having espoused simple comparative statics. Yet in using TIGER in chapter 9 to assess the impact of the two most important public policy interventions in Muda's paddy farming economy in the 1980s, we surely stretched the beast to the very limit of its ability (if not beyond). Moreover, the aggregative nature of the model precludes, a priori, detailed predictions about the distribution of income among paddy farmers. Similarly, tentative and crude adjustments to the supply function for labor are rather a shaky basis for assertions about the future share of hired labor in the value added from paddy production. Nevertheless, as argued in chapter 3, models such as TIGER have inherent advantages in specification over the more traditional econometric formulations and are computationally manageable. Although we cannot claim originality for the form of TIGER, we have put the beast to work in a somewhat unusual way in disentangling the several elements contributing to the stream of direct benefits flowing from the project.

In attempting to measure the downstream effects of the project, we were after something rather new, for such effects are more often discussed than quantified. Although the basic framework employed for this purpose, semi-input-output analysis, is not novel, the uses to which it has been put are rather unconventional. In particular, the extensions developed in chapters 6 and 8 to analyze the effects of changes in the region's barter terms of trade and the effects of the project on the entire economy, respectively, have a modest claim to some originality. As was the case with TIGER, however, none of these formulations strayed outside the familiar territory of comparative statics, and for precisely the same reasons.

Although the strengths and weaknesses of the semi-input-output method are sufficiently well known to require no lengthy discussion here, two specific issues should be addressed. First, if the assumption of constant average costs of production in the sectors producing nontradables seems rather farfetched in the face of rapidly rising real wages, especially in the context of the experiments in chapter 9, it should not be forgotten that we have ignored technical progress altogether. Insofar as gross investment saves on labor inputs, costs per unit of output will rise, if at all, more slowly than real wage rates.

Second, the treatment of the relationship between private savings and investment varied according to the circumstances, and this is as it should be, since capital (as finance) is the most mobile of all factors in the regional context. The savings of the region's households and firms will be invested in the region itself only if profitable investment opportunities are to be found locally; otherwise, they will finance investment (and perhaps consumption) elsewhere in the national economy, either through direct transfers or through the banking system. For this reason, we did not require regional savings to equal regional investment in chapters 6 and 7, preferring instead to estimate investment separately and then set its level exogenously in those two models. Only when we turned, in chapter 8, to a social cost-benefit analysis of the project, for which the context is the whole of the national economy, was it natural to set savings equal to investment in the hypothetical "extended region" defined for the purpose.

The models we have formulated to analyze the project in its regional and national setting are linear, and though their demands for data are heavy, we were confronted with comparatively few difficulties in estimating their parameters from the available empirical evidence. In the light of the frailties of some of our data, however, a move to more complex nonlinear models would necessarily have involved the wholesale guesstimation of key parameters. We see no profit in such speculation, especially since the extra insight afforded by nonlinear specifications is, in the present case at least, open to question and computational tractability is certainly sacrificed.

TIGER and its close cousins are now among the stock-in-trade of agricultural economists when tackling diverse issues at the farm and sectoral levels. Advances in the techniques of formulation and computation have made such models far easier to construct and use, and their strengths and defects are relatively well understood. If the evaluator is charged with examining ex post the size and composition of the direct benefits accruing from a project, a methodology of this kind has much to recommend it. But where the estimation of a SAM is concerned, which provides the basis for our regional model and its variants, far more caution is advisable. Only in the case of extremely large public projects will it be worth going into such regional details. Perhaps the most natural and promising circumstances for such an effort are those in which public investment is allocated by systematic regional planning. There the estimation of a regional SAM is an invaluable, if not essential, first step in the process of evaluating ex ante not only the projects that are contenders for the investment program but also other public policies. Conversely, in an ex post framework it provides the basis for estimating the size of the actual multiplier effects.

Implications for Project Monitoring

Although we may have established that our choice of models was appropriate for the purposes at hand, we have also frequently alluded to the frailties and shortcomings of our data base. Indeed, we have set out in appendix A some of our more important criticisms of the primary empirical material available to us, and in chapter 5 and appendix C we have tried to give some flavor of the difficulties of constructing and estimating our regional models and of the effort required. Nevertheless, the data base we have used was unique in that it was collected with the clear intention of measuring both the direct and indirect effects of the project, albeit without a sufficiently developed analytical framework and over too short a time. These considerations prompt at least two important questions. First, if efforts to evaluate projects in this way are worthwhile, what are the implications for data collection? Second, if such massive efforts are not practical, what can be done to improve the monitoring and evaluation exercises that are now a common feature of many development projects?

The first question relates to measuring the progress of projects through survey work and other forms of data collection. Suitable studies should begin long enough before the start of construction to collect sufficient material for a clear picture of the preproject situation. Such work is not incompatible with normal prefeasibility or preparation studies and indeed could be of material assistance. That such studies are being planned or executed (for example, in Nigeria) suggests that this lesson is already being heeded in some places. Data collection should continue at intervals until most adjustments resulting from the project intervention are complete. In many situations this may be long after the end of construction or direct implementation. Without time series information spanning a sufficient period, it is difficult, if at all possible, to obtain a real understanding and a measure of the magnitude of the structural changes that have taken place, particularly those resulting from new technology or having to do with the processes of accumulation. This applies with particular force to agricultural projects in places where the annual environmental variation is large. It would seem, then, that if there is a serious intention to monitor and ultimately evaluate a project, the allocation of time and resources must be fully commensurate.

These general precepts contain other implications. Those who undertake these evaluation studies must approach their work with a clear idea of what they are trying to do. It matters little whether the aim is to investigate the so-called direct changes or to measure the project's full impact, including the contributions of different components. In either case, there must be an explicitly stated set of questions to be answered and a carefully selected analytical framework before the data collection begins. The asking of questions and the collection of data merely because they appear interesting will not advance our understanding of the processes of change unless they are part of a systematic and consistent analytical structure. Although projects commonly have monitoring and evaluation components, it seems to us that little thought has been given to precisely what it is that should be monitored or evaluated or the likely role that properly conducted studies could play in the formulation of future policy.

In this regard, and from the results presented in chapter 7, it seems clear that in rural areas such as Muda there is greater profit in concentrating on the determinants and measurement of household incomes and expenditures and the composition of final demand than in studying the niceties of interindustry transactions. Of concern are not only the households that are the direct "target" of the project, but also those deriving their livelihood from sectors not directly affected or living on the geographical fringe of the project area. The latter, it seems, may easily be relative losers in the sharing of the incremental benefits brought about by the project.

Our answer to the second question must, perhaps, be less satisfactory. Although the largest projects appear worthy of the extensive studies and analysis we have outlined, few projects are undertaken on such a scale. Nevertheless, the capital cost of the Muda project, even in current terms, is no longer considered particularly large. Furthermore, the allocation of resources to monitoring and evaluation is often 1 or 2 percent of total project costs and may be larger. Much can be done with such resources if they are wisely used. Indeed, if funds were collected from two or three projects and then spent on just one, more ambitious studies would be possible. Surely it is better to understand a few carefully chosen projects—which are in some sense representative of a wider group, are technically novel, or are expected to yield unpredictable results—than to have a far from perfect understanding of the effects of many.

Indirect Benefits and Project Appraisal

In view of the well-defined boundaries of our region, most of the indirect costs and benefits of the project—with the exception of changes in government income—were captured within the regional economy itself. The project generated substantial savings, however, much of which flowed out of the region to the benefit of the rest of the Malaysian economy. Some of the increased "imports" into the region as a result of the project certainly generated additional income in Malaysia, while the additional capital and labor absorbed within the region relinquished opportunities for productive employment elsewhere. In chapter 8 we attempted to measure these wider repercussions of the project on the rest of Malaysia through a full social cost-benefit analysis. Our finding was that the project returned a very handsome social return.

In conducting this social cost-benefit analysis, we employed an extension of the semi-input-output model developed in chapter 6, and the indirect or downstream benefits of the project were explicitly incorporated into the calculations of the project's benefit streams. Although Bell and Devarajan (1980*b*) have shown the equivalence of this "primal" method and the more standard shadow pricing ("dual") approach to be found in the literature, it is worth reemphasizing the conclusions in chapter 8 concerning the way in which downstream benefits should be handled in project appraisal.

Both in principle and when applied to the stream of direct inputs and outputs associated with a project, the shadow prices used in a full cost-benefit analysis account fully for all downstream effects because they contain, among other things, all the information obtainable from a semi-input-output analysis. In practice, however, shadow prices are not usually calculated on the basis of such complete information, but are derived using the shortcut methods proposed in the standard literature. Whether shadow prices derived in this way adequately capture all the downstream benefits (costs) is a moot point, which we have not been able to test in this study for the reasons given at the end of chapter 8. Furthermore, shadow prices are often computed on a more ad hoc basis in the operational work of international lending institutions and government departments, and it may well be that the full downstream benefits (costs) of a project are not adequately captured by such operational rules-of-thumb. We hasten to add, however, that our work in no way justifies the reckless application of a "fudge factor" to the estimated direct benefits of a project. Rather, we recommend the more careful derivation of shadow prices from national data or, if resources permit, the estimation of a regional SAM and the application of the kind of semi-input-output approach to social cost-benefit analysis that we have attempted in chapter 8. Certainly, we see this latter approach as being desirable for the appraisal of very large-scale public projects.

Appendix A
Primary Data Sources

THIS VOLUME has drawn heavily on diverse sources of data, both primary and secondary. The secondary material includes books, manuscripts, and published reports by other scholars working on Malaysia and the varied output of the many arms and agencies of the Malaysian government, notably the Department of Statistics, the Ministry of Agriculture, and the Muda Agricultural Development Authority. These secondary sources are readily available and are listed in the references at the back of the book. In this appendix we provide a brief description of the original empirical material collected during the FAO/World Bank field studies, which we have used in constructing our models. We shall comment on some of the strengths and weaknesses of this material and point out certain improvements that would be desirable if another study of this kind were to be initiated. We deal first with the FAO/World Bank household survey and then with the smaller studies. Finally, we make brief mention of other data that were available to us through reports that are not generally available.

The FAO/World Bank Farm Household Sample Survey

With the exception of a few small pilot areas, the Muda scheme was implemented in annual phases starting in 1970, and each phase was related to a distinct geographical area. When the FAO/World Bank farm household survey was undertaken (November 1972 to November 1973) it was therefore possible to distinguish five main groups of farmers, namely, those who double-cropped for the first time in 1970 (or earlier), 1971, 1972, 1973, and 1974, respectively. This phasing of the project permits what is essentially a cross-sectional survey to provide a crude proxy for the trajectory of the project's impact. The farmers who were scheduled to come into the scheme in 1974 (that is, those who had not begun to double-crop at the time of the survey) represent, with certain qualifications, the "without project" situation, while those from earlier phases provide a basis for inferring the train of adjustments to the new economic circumstances created by the project. This basic stratification was augmented by taking into account the results of experimental data, which showed that the two basic soil types in the project area (acid and nonacid) significantly influenced crop yields.[1] Thus, the conceptual frame was a matrix with ten cells—five phases, each with two soil types. For the purposes of field investigation, however, the design was reduced to eight cells, owing to the impossibility of distinguishing ex ante between the 1973 and 1974 phases. In other words, when the survey was designed early in 1972, it was possible to identify only those farmers who were still single-cropping. Fortunately, it was unnecessary to subdivide this group into those who would double-crop for the first time in 1973 and those who would do so in 1974 or later.

Since there was no adequate sampling frame, a two-stage procedure was adopted. In the first stage, 700 rural population census enumeration blocks within the project area were randomly sampled by using a sampling fraction of about 23.2 percent.[2] All families in the 162 blocks in the sample were canvassed, and preliminary details of occupation, type of farm soil, and total annual income were collected. These families (14,788 in all) were then sorted into paddy farming and non–paddy farming categories. The former were arranged into a stratified sampling frame based on the two criteria of project phase and soil type. Finally, the sample was selected in such a way that the sample size in each stratum was proportionate to the standard deviation of mean annual cash income in that stratum. The steps in arriving at the final sample size are set out in table A-1.

The survey was of the "cost route" type, based on a weekly interview of each household in the sample together with opening and closing inventory (stock-taking) surveys. Thus, in principle, each household in the survey was interviewed a total of fifty-four times. The final sample size of 839 households therefore implied a total of 45,306 potential interviews, of which more than 90 percent were actually undertaken. The fieldwork was executed with great care and under close supervision to safeguard the quality of the data. The data were also extensively checked and validated by computer both during and after the survey.

The survey was originally intended to span two complete crop seasons. The agricultural year in the Muda project area begins in September, when the land is prepared for the main season crop, and ends in August, when

1. FAO (1975) provides an analysis of the relevant data, which were supplied by the government agricultural research station at Alor Setar, as well as details of the soil classification in the project area. A more extensive account of the soil classification may be found in Soo (1972).

2. This sampling fraction of 23.2 percent is based on the proportion of enumeration blocks. Because the blocks are not of uniform area or population, the 162 blocks used contained only 21.4 percent of the population.

the off-season crop is harvested. Ideally, the household survey should have begun in late August or early September and continued for twelve months, thereby spanning two complete crop cycles. It was not possible to start the survey until November, however. Thus, neither the first part of the 1972–73 main season nor the latter part of the 1973–74 main season was covered. Although this weakness would be important for some types of analysis, it does not place undue restrictions on the use of the data in a programming framework, provided the conditions of production did not vary greatly between the two partial main seasons embraced by the data. Fortunately, conditions did not change appreciably; in particular, a sharp rise in the price of paddy in late 1973 took place after farmers had made their key planting decisions for the 1973–74 main season.

Fortune, however, was not consistently sweet. Although the sampling design allowed the single-cropping strata to be divided into two groups at the end of the survey—those who remained as single-croppers and those who double-cropped for the first time during the course of the survey—the sample actually drawn was distorted by the spontaneous uptake of double-cropping by many who would have remained single-croppers except that unusually heavy rainfall in some parts of the scheme area during April and May of 1973 permitted an unirrigated second crop to be grown. Although this did not affect the quality of the data, it did weaken the statistical basis for treating the single-cropping farmers as a control group.

The questionnaires covered a comprehensive range of variables (in excess of 700) concerning households' socioeconomic characteristics, their assets, and flow variables, such as farm inputs and outputs, income from all sources, and itemized expenditures. In this sense, it was truly comprehensive and one of the first surveys of its kind that was statistically representative of an entire project. Nevertheless, from our specific analytical viewpoint there were certain weaknesses, which must be addressed. Of course, the comments below have the benefit of hindsight and do not detract from the pioneering nature of this survey.

First, the survey omitted to measure actual farm size, relying instead on farmers' estimates of field and farm size. Such a procedure inevitably permits bias and inaccuracy, and given the importance of factor endowments in determining income, it is a serious flaw. More specifically, in the light of the importance of new varieties of paddy as revealed in chapter 4, the survey should have observed the varieties grown by farmers much more closely than it did. Indeed, the entire set of activities in the matrix of agricultural technology (such as planting densities and quantities of inputs) should have been more precisely monitored. Observations on wage rates and the costs of various forms of draft power also needed greater attention. As we have pointed out in various chapters (notably chapter 3), these weaknesses have sometimes forced us to give preference, largely on

Table A-1. *Household Survey Sample Size (Paddy Farmers Only)*

Stratum	*Soil type*	*Double-cropping phase*	*Population of stratum from first stage*	*Mean annual cash income (dollars)*	*Standard deviation*	*Coefficient of variation*	*Sample size*[a]	*Adjusted sample size*[b]	*Sample size first used*[c]	*Sample size after O.I.*[d]	*Losses from sample fatigue*	*Final sample size at C.I.*[e]
Is.	Acid	Single	—	—	—	—	—	—	—	—	—	13
It.	Acid	1973	—	—	—	—	—	—	—	—	—	19
I.	Acid	Single	465	646	614	95.05	18	36	43	35	3	32
IIs.	Nonacid	Single	—	—	—	—	—	—	—	—	—	26
IIt.	Nonacid	1973	—	—	—	—	—	—	—	—	—	134
II.	Nonacid	Single	1,694	739	710	96.08	70	140	168	173	13	160
III.	Acid	1972	367	1,201	1,247	103.83	27	27	32	29	1	28
IV.	Nonacid	1972	1,779	1,463	1,748	119.48	182	182	218	171	23	148
V.	Acid	1971	516	966	822	85.09	25	25	30	29	2	27
VI.	Nonacid	1971	2,288	1,300	1,300	100.00	174	174	209	191	26	165
VII.	Acid	1970	454	1,007	1,011	100.40	27	27	32	53	8	45
VIII.	Nonacid	1970	1,837	1,403	1,860	132.57	200	200	240	250	52	198
Total			9,400	1,184	1,406	118.75	723	811	972	931	128	803[f]

— Not applicable.

Note: The *s* and *t* in the stratum numbers represent single-croppers and those in transition from single- to double-cropping, respectively. It was impossible to distinguish those categories ex ante, and thus the sample design was based on the parameters for strata I and II, the breakdown into the subcategories shown in the final sample size column being made only at the end of the survey (see FAO, 1975).

a. Sample size calculated by the formula:

$$\left(\frac{1}{F} - n_1\right) + \left(\frac{S_2}{S_1F} - n_2\right) + \left(\frac{S_3}{S_1F} - n_3\right) + \ldots \left(\frac{S_8}{S_1F} - n_8\right) = X.$$

where X = maximum sample size (given), S = standard deviation, F = sampling fraction, and n = stratum population.

b. Strata I and II were doubled because at the beginning of the survey they consisted of only single-croppers but at the end of the survey would consist of both single-croppers and the group of farmers who started double-cropping in 1973.

c. Sample was increased by 20 percent in each of the strata to allow for nonresponse and sample fatigue.

d. The stratum sample size after the opening inventory (O.I.) is in some cases greater than the sample size first used owing to changes made following differences in response between the first stage and the O.I.

e. C.I. = closing inventory.

f. As calculated for paddy farmers only; the total sample also included 36 paddy farm workers.

grounds of consistency, to results from the model over those from the surveys. We would have greatly preferred to avoid such choices.

The focus of our study is regional, including the changes in the regional distribution of income and the extent to which consumer spending underlies the downstream effects of the project. Consequently, it would have been desirable to have more than one class of nonfarm household in order to explore more fully the changes in income distribution brought about by the project. This was not possible because data were lacking. If the household sample had been representative of the region rather than the project this serious restriction would have been lifted. To the extent that a regional basis is appropriate for evaluating the effects of projects, it is clear that those households not *directly* affected by the project should be studied on an equal footing with those that are.

The timing of the FAO/World Bank survey was unfortunate in that it covered the project during a year of transition. As any sensible assessment of the effects of a project necessarily involves comparisons of what went before with what came after, the analysis we have attempted would have been better served by two surveys, one in the period before the project began and the other at some stage after it was completed. Naturally, some monitoring of the course of technological and exogenous changes during the intervening period would also have been desirable.[3]

The clear lesson to be drawn from our labors is that project evaluation of the type we have attempted requires both a long and broad view. Long, in that empirical measurement should begin before the project has commenced and should not end until the project is completed. Broad, in that the relevant impact of a project extends beyond its direct beneficiaries to include all people living in the project's command area or those on its periphery. Given the expense of such surveys, this may be a counsel of perfection; in practice, some compromise may have to be struck in the light of the particular circumstances.

Other FAO/World Bank Surveys

At the outset of the FAO/World Bank investigations, the overall objective was to generate a quantitative description of the overall impact of the Muda project. It was originally thought that this might be achieved by constructing a set of suitably disaggregated sector accounts for the decade ending in 1974. In practice, this proved too ambitious, and instead a partial approach was followed, which pieced together, largely through specially

3. During 1971, when the survey design was under discussion, the desirability of doing both an early and a late survey was appreciated and proposed, but at the time there was inadequate support for this more ambitious approach.

designed surveys, evidence of the course of economic events in key regional economic sectors from the early 1960s to the early 1970s.

This work suffered from unfortunate timing in the same way as the household survey. Furthermore, the absence of an explicit analytical framework caused more emphasis to be given to the measurement of ends than to the investigation of means. In particular, the linkages between different economic sectors in the regional economy were relatively neglected. Once again, however, the attempt was both along the right lines and somewhat novel; for the frameworks we employed to analyze the changes in the region's economy were themselves largely undeveloped at the time the FAO/World Bank studies were conceived and implemented. The following treatment of these studies is not exhaustive, and we dwell only on those sectors for which the data exhibit either notable weaknesses or strengths. As it turned out, the FAO/World Bank studies addressed sectors that accounted for about 62 percent of the gross regional product.

Regional Imports and Exports

Inasmuch as the region was essentially a producer and exporter of primary produce, with relatively limited processing, and most other tradable goods were imported, the FAO/World Bank studies attempted to measure the relevant flows. Had these efforts been more successful, the estimation of these flows as residuals in the SAM (chapter 5) would have been unnecessary. Nevertheless, these studies were not slight. There was a painstaking examination of the data in the detailed station reports of Malaysian Railways on the flow of goods by rail. Historical data on road tonnage were provided by earlier official road transport surveys. Current data on road traffic and tonnage flow were obtained by conducting two seven-day traffic surveys on the sole significant route out of the region, a road on its southern boundary.

Although the resulting data on the movement of goods by road gave details on commodities, the frequency with which it had been collected (only twice in twelve months) and the highly seasonal nature of the flows rendered the annual estimates tenuous indeed. Even with the addition of other historical data, we could get only a qualitative impression of the commodity composition of trade. Furthermore, although the data obtained from the railway authorities was satisfyingly complete, it suffered from the twin weaknesses of prior aggregation by Malaysian Railways and the relative insignificance of rail as a bulk carrier.

Financial Institutions

In view of the importance of financial institutions as mobilizers of savings, data were collected on changes in the level of savings deposited with these institutions and on the composition of their loan portfolios broken down by four major loan (account) groups: paddy farmers and closely allied activities; those engaged in other agriculture; manufacturing and services; and all others, which included state enterprises of all types. This information was collected with the assistance of Bank Negara, the Malaysian reserve bank, and provided a unique but partial basis for some of our key prior estimates in the SAM. The nature of this work is more fully described in FAO (1975), where its primary weaknesses are outlined. First, savings accounts with commercial banks are operated by their holders much as current accounts (because the former are not interest bearing), so that the distinction between consumption and savings is blurred. Second, informal financial intermediaries (moneylenders, pawnbrokers, and the like) and the quantitatively more important finance houses, which are sources of capital for more risky undertakings, were not included.

Rice Milling

An obvious candidate for study was the rice milling sector, which with good reason was expected to be among the first to feel the impact of the increased production generated by the project. Accordingly, each of the thirty-three commercial mills in the region was canvassed, and supplementary material was obtained from the Kedah Rice Millers Association, the Lembaga Padi dan Beras Negara (State Paddy and Rice Marketing Board), and the Employees' Provident Fund. There were also 303 small mills in the region, which are legally restricted to milling solely for domestic consumption. These were canvassed by a postal questionnaire, to which the response was a very satisfactory 66 percent. The few government-owned mills were treated like the more numerous commercial mills. The data on paddy throughput, employment, and investment for 1962–72 are largely complete, but no information on the structure of costs and the composition of assets was gathered. Here, then, we have a clear case of the failure to examine interindustry dependence. This forced us to seek our technology coefficients from doubtless reliable but not always regionally specific sources. Lacunae of this kind have frequently troubled us, and we highlight them here to forewarn others.

Retail Trade

As FAO (1975) makes clear, it was realized early on that the retail sector would be important to study, partly because of the dominance of services in the region's nonfarm economy and the weakness of other economic sectors, particularly manufacturing, and partly because strong growth in consumer demand would bear directly on retail activity. Data on the retail sector were not readily available, however, and a special investigation was undertaken.

The sampling frame for this study consisted of three main strata: urban, semiurban, and rural businesses.[4] For the urban stratum, the frame obtained from the Department of Statistics listed all retail and wholesale establishments in Alor Setar in 1970.[5] Since the region was growing rapidly, this was likely to be out-of-date, but it was beyond the time and resources of the study to create a new and updated frame. After the retail survey had been concluded, however, the Department of Statistics made available data from their Retail and Wholesale Censuses for 1970 and 1972. This material was used to calculate the urban shop population in 1972, estimate certain details for 1971, and corroborate the study's results for 1970. For the semiurban businesses, no sampling frame existed, and one had to be compiled from scratch in the early part of 1973: a listing of all retail establishments (from which businesses established in 1973 were excluded) in every town in the region with a population exceeding 1,000 and designated as a town council or town board area. For the rural area, the sampling frame was derived from the first stage of the main household survey. Thus, both the semiurban and rural sampling frames were up-to-date at the time of the survey in the second half of 1973, and estimates based on the results are reasonably accurate. Both the urban and semiurban strata were studied by means of postal questionnaires, and details of the responses are:

	Urban	*Semiurban*
Number of retail businesses in 1972	467	555
Usable responses	131	150
Response rate (percent)	28	27

The rural area sample contained 173 businesses, that is, 9.1 percent of the estimated total population. Successful interviews were, however, conducted with only 113 businesses.

The resulting data enabled us to arrive at independent estimates of the gross value of output for the retailing sector and, indirectly, wholesale services. In addition, the FAO/World Bank survey included a limited but nevertheless valuable substudy of the most important classes of retail

4. Owing to practical considerations of time and resources, definitional problems, and the likelihood of obtaining usable material, several types of business were excluded: timber dealers, restaurants and coffee shops, tailors, barbers and hairdressers, petrol stations, laundries, photographers, and tire shops. No effort was made to measure the substantial turnover of itinerant hawkers or stall holders. Markets are mostly populated by fruit, vegetable, fish, and meat stalls, which were not surveyed.

5. Because the Department of Statistics defines as "urban" those places with populations exceeding 75,000, the urban stratum was limited to Alor Setar.

traders, which yielded evidence on some key aspects of incomes, consumption, and savings of the region's nonfarm households. In particular, the data gave us some confidence in our estimates of high rates of saving by this class of household from the SAM. Valuable as they were, however, these data again point to the need for collecting region-specific data on the effects of large-scale projects.

Small-scale Manufacturing

Early in 1975, after the initial FAO/World Bank studies were completed, it became clear that it was necessary to carry out a survey of small-scale manufacturing establishments. With limited time and money, it was impossible to proceed in a rigorous way, and there was no pretence at aiming for representativeness. Instead, the objective was to canvass firms making a wide variety of products, especially those using locally produced inputs or those selling mainly to households or other firms in the region. The enumeration was done by a small group of students from the University Sains, who were instructed each day where to go and what kind of firm to look for, and the authors checked the returns each day. At the end of three weeks, about fifty usable questionnaires had been completed, and the work finished there.

The questionnaire itself covered the history of the business, employment, wages, inputs, outputs, destination of sales, and investment outlays. As we had hoped, the information thus collected proved valuable in estimating certain cells of the SAM.

Other Important Data Sources

We briefly set out here (without prejudice to the numerous secondary sources we have drawn upon) the debt we owe to three additional studies, the results (rather than the data) of which were available to us in a very detailed form.

In chronological order, the first is the *Farm Economic Survey of the Muda River Project, 1966/67* (Malaysia, 1967). This study told us much of what we know about the preproject situation. Using a two-stage sample design similar to that used in the Malaysian Census of Agriculture in 1960, this sample survey covered 2,476 farming households, largely in the area that subsequently came under the command of the project. It collected extensive information on the demographic, tenurial, and technical features of paddy production, but regrettably omitted to quantify the levels of the main inputs such as labor and fertilizer.

The second is a report on the data from the 1970 Muda Agro-Economic Survey (Doering, 1972) of 386 households, which was implemented under the auspices of MADA. Though not statistically representative, the survey paid careful attention to the quantification of inputs, the use of new varieties, and the level of prices. It was designed to study the management practices of farmers who were introduced to double-cropping in the small pilot schemes established by MADA before the launching of the full-scale project.

Third, there are the detailed statistical tables resulting from a joint farm management research project by MADA and the Tropical Agriculture Research Center of Japan. This study is fully reported in Yamashita, Jegatheesan, and Wong (1976), and the detailed statistics made available to us provided a data base from which we derived vital technical coefficients for TIGER. The survey covered 101 households and was carefully executed between February 1974 and March 1975, with interviews conducted weekly. In many ways it was closely modeled on the FAO/World Bank household survey, but differed in two important respects. First, the sample was purposive and cannot, therefore, be used to yield statistically reliable estimates of the project area. Second, great attention was paid to the agronomic aspects of the farming system. Although the use of these data for econometric purposes (see appendix B) revealed some data deficiencies, it was a valuable and rich source of vital material.

Conclusions

We have benefited from rich primary data sources, even though our work has, perhaps not surprisingly, exposed weaknesses in them. We regard these shortcomings as valuable pointers to the way in which future data collection, perhaps under the auspices of monitoring and evaluation units, might profitably be changed.

Appendix B
In Support of TIGER

IN CHAPTER 3 we laid out our reasons for selecting a mathematical programming approach to the paddy farm sector, discussed the theoretical basis of TIGER, and obtained an initial solution for 1973. In the interests of continuity, however, our presentation omitted detailed reference to both the physical environment and our derivation of the key parameters of the model. Consequently, for the benefit of the interested reader and in support of our arguments in chapter 3, we provide below some supplementary data, describe the way we estimated the response of paddy to nitrogen, and recount the manner in which the labor supply function was estimated.

Some Supplementary Data

Annual rainfall in the project area has a slightly bimodal distribution, with the lower peak in May and the higher in September. Annual total rainfall in the project area is variable, but rarely less than 75 inches or more than 100 inches. The average monthly distribution of rainfall (in inches) derived from recordings at fifteen rainfall stations in the project area from 1960 to 1972 was: January 1.7, February 2.2, March 3.5, April 6.0, May 9.8, June 8.6, July 8.4, August 9.1, September 13.4, October 13.3, November 8.9, December 4.2, and annual 89.1 (FAO, 1975, vol. 2, table 1). Accumulated moisture is critical to the maturation of paddy, and the time required for the critical levels to be reached limits the choice of transplanting dates under rain-fed conditions. For a rice variety with a 140-day growing period, we derived (from FAO, 1975, vol. 2, table 1) the following average cumulative rainfall totals (in inches, 1960–72) from transplanting to flowering and from transplanting to harvesting for selected transplanting dates in the main season:

	Transplanting date					
	July 1	*Aug. 1*	*Sep. 1*	*Oct. 1*	*Nov. 1*	*Dec. 1*
Transplanting to flowering	24.6	29.0	30.9	24.4	13.8	6.9
Transplanting to harvesting	37.9	40.2	37.4	27.3	25.8	9.8

The scheduling of planting was discussed at some length in chapter 3, but the full sequence of cultural operations for paddy production as specified in TIGER were not spelled out. Accordingly, and still assuming a 140-day variety, we set out in table B-1 the complete sequence of cultural operations as specified in TIGER.

Chapter 3 made it clear that the utilization rates for most inputs into paddy production are determined by the model. A few such inputs are not, however, subject to farmers' decisions in TIGER; these inputs, their fixed values, and their costs are set out in table B-2.

Paddy Response to Nitrogen

Fertilizer trials in the Muda area have generally shown good response to nitrogen for high-yielding varieties. Experimental yield data for the three years 1971 to 1973 averaged over the important soil classes in the project are set out in table B-3.

In a quadratic approximation, the data summarized in table B-3 indicate

Table B-1. *Timing of Cultural Operations for Paddy in TIGER*

Operation	*Fortnight performed*[a]
Nursery seeding	t
Land preparation with tractors	$t + 2$
Land preparation with buffaloes	$t + 1$ and $t + 2$
Transplanting	$t + 2$
First pesticide application	$t + 4$
Second pesticide application	$t + 6$
First fertilizer application	$t + 2$
Second fertilizer application (topdressing)[b]	$t + 4$
First weeding	$t + 4$
Second weeding	$t + 6$
Harvesting	$t + 10$

a. See chapter 3, the subsection "Nursery Scheduling Decisions," for an explanation of these periods.

b. Applicable only when 15 or more pounds of nitrogen are used.

Table B-2. *Fixed Input Requirements in TIGER per Relong of Paddy*

Input	*Quantity*	*1973 cost (dollars)*
Nursery fertilizer (pounds)	4.0	1.10
Seed (gantangs)	3.8	—[b]
Pesticides, BHC[a] 6 percent equivalent (pounds)	10.0	3.90
Labor for transplanting[c] (man-days)	7.94	—[b]
Labor for pesticide application and weeding[c] (man-days)	3.54	—[b]

a. Benzene hexachloride.
b. Prices determined endogenously in TIGER.
c. Divided equally between periods $t + 4$ and $t + 6$.
Source: FAO/World Bank survey.

optimal levels of nitrogen use, at 1974 price ratios, of 43 pounds per relong in the off-season and 59 pounds per relong in the main season. Similar estimates for the newer variety C4-63 indicate even higher optimal levels of nitrogen use, namely 60 pounds per relong in the off-season and 91 pounds per relong in the main season. MADA's recommendations to farmers in recent years have been based on these kinds of results, and yet farmers persist in using substantially less nitrogen (about 35 pounds per relong) as confirmed in both the FAO/World Bank survey and the 1974–75 agroeconomic survey (Yamashita, Jegatheesan, and Wong, 1976).

On the one hand, it is not unusual for experimental results to overstate the case for nitrogen response, because experimental stations maintain higher standards of husbandry than those encountered under farm conditions. On the other hand, attempts to measure nitrogen response under farm conditions are usually confounded by a lack of accurate and suitably controlled farm data. For example, in a regression analysis of cross-sectional farm survey data from the Muda area, Doering (1973) found that fertilizer use was both statistically insignificant and perverse in explaining

Table B-3. *Nitrogen Response for the Variety Bahagia*

Nitrogen (pounds per relong)	*Off-season yield (gantangs per relong)*	*Main-season yield (gantangs per relong)*
0	341	266
20	367	305
30	376	321
40	382	332
50	386	341
60	388	347
70	387	349
80	383	348

yields. We obtained similar results by using the FAO/World Bank survey data. The 1974–75 agroeconomic survey, however, provides a much improved data set, which is well suited to a fresh analysis of the yield response to nitrogen under farm conditions.

Specification of the Model

The data available from the 1974–75 survey include measurements of paddy yield and nitrogen usage, soil pH, soil class, paddy variety, and the dates of nursery seeding, transplanting, and harvesting. These data are available for both the 1974 off-season and the 1974–75 main season crops.

The data on paddy variety are of limited value because many farmers reported the local names for varieties, which are not uniform throughout the project area. Consequently, we use the number of days between transplanting and harvesting as a proxy variable. Moreover, since the soil classes are very broadly defined and largely reflect soil acidity, we use the pH value as the relevant variable. Our sample for analysis consisted of thirty-six and thirty-three complete sets of field observations for the off-season and main season crops, respectively. The sample means for each variable are set out in table B-4.

It is well established in the Muda area that paddy grown on acid soils yields significantly less than that grown on nonacid soils (see, for example, Samy and others 1970). It is also well known that high-yielding varieties (characterized by their short growing period from transplanting to harvest) are more responsive to nitrogen than the slower maturing local varieties. Thus, when we analyze nitrogen response from cross-sectional farm data, these variables must enter the analysis together with the amount of nitrogen used. We hypothesize that both the soil acidity and variety variables will have important shift effects on the intercept of the nitrogen response function—positive for soil pH and negative for growing period. Further-

Table B-4. *Sample Means of Model Variables for Main-season and Off-season Crops*

Item	*1974–75 main season*	*1974 off-season*
Yield (gantangs per relong)	346	334
Soil pH	4.63	4.63
Growing period (days)	110	111
Nitrogen use (pounds per relong)	33.8	37.0
Transplanting date (day of year)	272	—

Source: Data provided by MADA.

more, we expect these variables to have interaction effects with, and thereby affect the rate of response to, nitrogen. In particular, we expect soil pH and nitrogen to interact positively on yields, while growing period and nitrogen are expected to interact negatively.

A suitable model to capture these effects may be written as:

$$Y = \beta_0 + \beta_1 A + \beta_2 G + \beta_3 N + \beta_4 N^2 + \beta_5 AN + \beta_6 GN \qquad \text{(B.1)}$$

where Y = paddy yield (gantangs per relong)
A = soil pH
G = growing period from transplanting to harvest (days)
N = nitrogen use (pounds per relong)

Our hypotheses are that β_0, β_1, β_3, and β_5 are positive, and β_2, β_4, and β_6 are negative. The form of equation (B.1) is the usual quadratic nitrogen response function, but in addition β_1 and β_2 are the intercept shift parameters for A and G, respectively, and β_5 and β_6 are the corresponding interaction parameters between A and G and N.

This model is well suited to the off-season crop, where the date of transplanting is narrowly determined by the arrival of irrigation water. In the main season, however, farmers have a greater range of transplanting dates from which to choose, though the later they plant the greater is the likelihood of encountering water deficiencies as the dry season advances. In the first section of this appendix, we have already set out the relationship between accumulated rainfall from transplanting to flowering and harvesting and the date of transplanting. But our knowledge of the water requirements of paddy in the Muda region is too limited to formulate detailed hypotheses about the effect of transplanting dates on yields. Consequently, and in order to retain flexibility in the main season analysis, we simply expand the model by adding the following terms to equation (B.1):

$$Y = \text{(B.1)} + \beta_7 T + \beta_8 T^2 + \beta_9 TN \qquad \text{(B.2)}$$

where T is the transplanting date (day of year). This formulation permits a possible nonlinear intercept shift effect of T on the nitrogen response function, and also captures any interaction effects between T and N. We might expect β_9 to be negative, since the utilization of nitrogen by rice depends on an adequate water supply during the tillering stage.

The agronomic models described above were fitted to the farm data by the use of ordinary least squares regression. Problems of multicollinearity arose because of high correlations between the interaction terms and the amount of nitrogen. The problem was most severe for the off-season crop data, but even for the main season crop it led to large standard errors for the estimated parameters. Multicollinearity does not bias the results, however, so that if the estimated parameters are a priori plausible, we may still place some confidence in the results.

1974–75 Main Season Results

The appropriate model for the main season crop was defined in equation (B.2). The estimated parameters and their *t*-statistics (in parentheses) are:

$$\begin{aligned} Y = 6{,}150 &+ \underset{(1.92)}{3.606A} - \underset{(0.03)}{2.09G} + \underset{(0.52)}{6.408N} - \underset{(0.19)}{0.08285N^2} + \underset{(1.62)}{2.301AN} \\ &- \underset{(0.74)}{0.07256GN} - \underset{(0.67)}{41.088T} + \underset{(1.97)}{0.07422T^2} - \underset{(2.03)}{0.01411TN}\,. \quad (0.18) \end{aligned}$$

$$R^2 = 0.61$$

The signs of all the parameters are in accord with our expectations, except those for T and T^2, for which we offered no hypothesis. When N is set at its sample average value, a little arithmetic shows that yields are affected by T as shown in table B-5, regardless of the level of G or A. Apparently yields do decrease as the transplanting date advances, but only up until day 280, after which they begin to pick up again. These figures may, of course, simply reflect the peculiarities of the 1974–75 rainfall pattern, but the estimated parameters on T and T^2 are significant at the 5 percent confidence level.

To evaluate the reasonableness of the parameter values, it is helpful to tabulate yield predictions for different values of the explanatory variables. Such values are presented in table B-6 for acid and nonacid soils (A equals 4.0 and 4.9, respectively), for modern and traditional varieties (G equals 105 and 125 days, respectively), as well as for the sample average values of A and G. For each situation, the yield response to nitrogen is shown on the assumption that the transplanting date T is fixed at its sample mean value of 272.

Table B-5. *Changes in Yield with Changes in the Date of Transplanting for the 1974–75 Main Season*

Changes (in days) from mean transplant date[a]	*Change in yield (gantangs per relong)*
− 21	+ 58
− 14	+ 31
− 7	+ 12
+ 7	− 5
+ 14	− 2
+ 21	+ 8
+ 28	+ 25

Note: Mean transplanting date was day 272.

a. A minus sign indicates "early"; a plus sign indicates "late."

The results are very plausible in terms of the magnitudes of the soil acidity and varietal effects. Furthermore, the last column of table B-6 shows that the average nitrogen level used in the sample (33.8 pounds per relong) was very close to the economic optimum. At this level of nitrogen use, the predicted average yield is about 334 gantangs per relong. When adjusted for the 17 percent deduction for chaff and other waste that was removed in the yield measurements, this corresponds to a farm yield of 390 gantangs per relong. This figure is entirely consistent with other data on average yields, such as the FAO/World Bank survey. Furthermore, although the response to nitrogen peaks at a level consistent with farmers' behavior, there is nevertheless a convincing correspondence between the last column of table B-6 and the experimental data reported for the main season crop in table B-3.

The multicollinearity problem prompted us to run an additional regression in which the interaction terms AN and GN were omitted. The estimated equation (t-statistics in parentheses) is:

$$\begin{aligned} Y = \underset{(1.73)}{4{,}895} &+ \underset{(2.03)}{71.95A} - \underset{(3.28)}{4.49G} + \underset{(0.86)}{16.09N} \\ &- \underset{(1.48)}{0.06555N^2} - \underset{(1.73)}{32.83T} + \underset{(1.81)}{0.060054T^2} - \underset{(0.69)}{0.04481TN}. \end{aligned}$$

$$R^2 = 0.53$$

This regression has better t-statistics, but the response relationships between yield, soil, pH, variety, and transplanting date are not much different from before. Given this insensitivity and our confidence in the signs of the parameters, we conclude that our estimates in table B-6 are credible. A larger and more diversified sample might well reduce the standard errors of the parameter estimates, but lead to little change in their values. Thus the results reported in table B-6 are those used in our linear programming model of the Muda region.

1974 Off-season Results

Multicollinearity problems were much more severe in the 1974 off-season crop data. The best results were obtained when the GN interaction term was omitted from equation (B.1), but even then the results were statistically entirely nonsignificant.

$$\begin{aligned} Y = \underset{(0.10)}{44.77} &+ \underset{(0.85)}{58.93A} + \underset{(0.12)}{0.22G} \\ &+ \underset{(0.02)}{0.2296N} - \underset{(0.43)}{0.0131048N^2} + \underset{(0.07)}{0.1458AN}. \end{aligned}$$

$$R^2 = 0.08$$

Table B-6. *Estimated Yield Relationships for the 1974–75 Main Season*

Nitrogen level (pounds per relong)	Acid soils: Modern variety	Acid soils: Traditional variety	Nonacid soils: Modern variety	Nonacid soils: Traditional variety	Sample average
	Gantangs per relong				
0	261	219	264	222	252
10	294	238	318	262	296
15	305	241	339	275	312
20	311	240	355	284	323
25	313	235	368	290	331
30	311	226	376	291	334
35	305	212	381	288	333
40	295	195	380	281	328
	Pounds of nitrogen (N) *per relong*				
N for which yield is maximum	25	16	38	29	31
Optimal *N* at 1974 prices[a]	22	14	35	26	29

a. The relevant 1974 prices are $10 per 44 pound bag of 46 percent *N*, which equals $0.5 per pound of *N*. Paddy is priced, before deductions, at $27 per picul, which equals $1.13 per gantang. The corresponding price ratio *PN/PY* is 0.44.

We report this result with some trepidation, and only because the magnitudes of the yield response data are not wholly implausible (table B-7). The sign for the *G* parameter is perverse, but the yields in the last column of table B-7 do bear some resemblance to the experimental results obtained for the off-season crop in table B-3. Our results, however, suggest a much lower response to nitrogen, with an optimal input level, at 1974 prices, of only 18 pounds per relong, compared with 59 pounds from the experimental data. This is also inconsistent with farmers' behavior. As table B-4 shows, farmers applied an average of 37 pounds of nitrogen per relong in the 1974 off-season, higher even than the 1974–75 main season average application of 33.8 pounds per relong. Clearly, a much better sample of data is required before any firm conclusions can be drawn about nitrogen response in the off-season crop. Consequently, and as we have made clear in chapter 3, for the purposes of TIGER we have simply scaled the 1974–75 main season results by the average yield differential between the two seasons observed in the FAO/World Bank survey.

Table B-7. *Estimated Yield Relationships for the 1974 Off-season*

Nitrogen level (pounds per relong)	*Acid soils*		*Nonacid soils*		
	Modern variety	*Traditional variety*	*Modern variety*	*Traditional variety*	*Sample average*
		Gantangs per relong			
0	304	308	357	361	342
10	311	315	365	369	350
15	313	317	368	372	353
20	315	319	371	375	355
25	316	320	372	376	356
30	317	323	373	377	357
35	316	320	374	378	358
40	315	319	374	378	357
		Pounds of nitrogen (N) per relong			
N for which yield is maximum	31	31	36	36	35
Optimal *N* at 1974 prices	14	14	19	19	18

Estimation of the Four-week Day Labor Supply Function for TIGER

Day labor comprises landless paddy workers and rubber smallholders from regional sources; seasonal workers from other parts of Malaysia, such as Kelantan, Baling, and Sik; and migrant workers from Thailand. We treat this labor as a single pool and postulate the following structural model:

$$S_t = \alpha + \beta W_t \tag{B.3}$$

$$W_t = a + bS_t + cD_t \tag{B.4}$$

where S_t = supply of day labor in four-week period t
W_t = wage of day labor in four-week period t
D_t = total demand for labor in four-week period t (day and family labor).

Equation (B.3) is the labor supply function, and it is hypothesized that supply is affected only by the current wage.

In the wage determination equation (B.4), the supply of day labor and the total demand for labor differ by the amount of farm family labor used. We assume, however, that the total demand for labor in period t is predetermined by planting and mechanization decisions made at the beginning of the crop season, so that the amount of day labor used in each period is determined by the family's work-leisure tradeoff and the cost of day labor.

Solving (B.3) and (B.4) for their reduced forms gives:

(B.5) $$S_t = \frac{\alpha + a\beta}{1 - b\beta} + \frac{c\beta}{1 - b\beta} D_t$$

(B.6) $$W_t = \frac{a + \alpha b}{1 - b\beta} + \frac{c}{1 - b\beta} D_t.$$

The system has only one exogenous variable, and while the supply function is exactly identified, the wage equation is underidentified. Fortunately, we are only interested in estimating the structural parameters of the supply function, so that given estimated reduced forms $S_t = \hat{A} + \hat{B} D_t$ and $W_t = \hat{E} + \hat{F} D_t$, the parameters of the supply function (B.3) are obtained as $\alpha = \hat{A} - \beta\hat{E}$ and $\beta = \hat{B}/\hat{F}$.

Data for the average farm were extracted from the FAO/World Bank farm survey and multiplied by the number of farms in the Muda project. The observations for S_t were derived from the average farm as the difference between the amount of labor hired in, regardless of source, and the total amount of family labor employed on other paddy farms. The wage W_t was calculated by dividing total payments for hired labor by the amount of labor hired for each period. The resulting data are set out in table B-8.

There are clearly problems with these data for periods 3 and 10: the wage rates are excessively low compared with the peak quantities of labor supplied. Both periods correspond to times of peak harvesting, and we

Table B-8. *Labor Market Data, 1972–73*

Four-week period	*Wage (W_t) ($ per day)*	*Supply (S_t) (thousands of days)*	*Demand (D_t) (thousands of days)*
1[a]	1.88	61.8	850.0
2	1.50	212.1	842.1
3	1.30	1,330.2	2,827.9
4	2.19	504.7	1,121.6
5	1.20	106.2	612.3
6	1.54	453.9	1,585.7
7	1.80	208.1	1,061.6
8	1.60	13.4	580.9
9	1.45	384.8	1,168.6
10	1.85	1,116.7	2,799.6
11	2.52	653.7	2,165.9
12	2.33	497.7	1,822.7
13[b]	2.88	134.6	989.2

a. Starting November 26, 1972.
b. Ending November 25, 1973.
Source: FAO/World Bank survey.

surmise that there may have been some confusion in the survey over the treatment of exchange labor contracts, which were still common in 1972–73. Consequently, we decided to drop these two observations from the sample and to fit the model to the remaining eleven periods.

The reduced form estimates were:

$$S_t = -150.679 + 0.38189D_t$$
$$(1.88) \quad (6.01)$$
$$R^2 = 0.89$$

$$W_t = 1.3076 + 0.0005083D_t$$
$$(3.45) \quad (1.69)$$
$$R^2 = 0.49$$

where S_t and D_t are measured in thousands of days. From these are obtained the structural estimates of the labor supply function: $S_t = -1133.021 + 751.2787W_t$ or $W_t = 1.51 + 0.001331S_t$. The supply elasticity is 4.86 when evaluated at the mean for the eleven periods, and a million man-days of labor gives a wage of \$2.84 a day, which is also of a sensible magnitude.

Appendix C
Notes on Estimating the Social Accounting Matrix

THIS APPENDIX describes in some detail how we set about estimating the transactions in the SAM and the sources we used in arriving at those estimates. A sense of decency and restraint enjoins us to avoid going deeply into the origins of certain of them, a delicacy most applied economists would share. Yet we have not fought shy of setting out the general principles that have shaped our estimates, even if the practical procedures cause us to blush a little. Those who feel inspired to follow our example may therefore find some profitable lessons in what follows.

The structure of these notes is given by the columns of the SAM, that is, the outlays of each category of account in table 5-2. Thus, the following categories will be taken seriatim: households, firms, federal government, state government, combined capital account, production activities, and rest of the world. The subsections within each of the above follow the same order, and production activities (sectors) are identified here by the number of their account in the full SAM (table 5-4).

Households

Interhousehold Transfers

There are two items in this submatrix: compensation for domestic services and the retirement of historic debt. The former involves solely a payment by land-abundant to landless households, which we have guesstimated on that intangible basis, "local knowledge." The latter involves payments by agricultural to nonfarm households, the estimates for which come directly from the FAO/World Bank survey data.

Federal Taxation

Agricultural households pay *zakat* taxes and license fees for vehicles and radios. The estimates of both are drawn directly from the FAO/World Bank household survey data, except that the distribution of the latter among the four classes of households is in proportion to their aggregate incomes. Since the latter are rather small, we dispensed with the bother and expense of creating specific estimates directly from the field data.

Nonagricultural households pay a variety of taxes, some as households and some as unincorporated enterprises. In total, noncorporate incomes in the region assessed for tax were $62.345 million in 1972 (FAO, 1975, table 55). For the nation as a whole, the average tax rate on such incomes in that year was 10 percent (Treasury of Malaysia, *Economic Report, 1974–75*, table 5.3), which yields a total tax payment of $6.235 million. License fees for vehicles have been estimated by taking the population of registered vehicles (FAO, 1975, table 68) and the national average license-cum-tax fee of $220 in 1972 (Department of Statistics, September 1974, pp. 129, 179). Radio license fees have been interpolated from the growth in the number of licenses nationally (Treasury of Malaysia, *Economic Report, 1975–76*, table 6.2) and license payments (Department of Statistics, September 1974, p. 179), using a rough-and-ready adjustment for the region's population and living standards. Business registration fees also bear heavily on unincorporated businesses, which we treat as households. The estimate for the region has been derived from the national figure by scaling the relevant levels of population and income. Employees' contributions to the Employees' Provident Fund (EPF) have been estimated from sundry interviews with businessmen in the region and the authors' limited sample survey of firms (appendix A). In the absence of information on certain sectors, the minimum contribution of 5 percent has been assumed. To arrive at net contributions, the gross contributions thus derived are then scaled down in the proportion holding nationally (Department of Statistics, September 1974, p. 181).

State Taxation

Agricultural households pay land taxes and quit rents; these estimates have been drawn directly from the FAO/World Bank household survey data.

Nonfarm households pay various fees and duties, the estimate of which is taken from FAO (1975, table 54) by using relative levels of population to scale the figure for the whole of Kedah into that for the part of Kedah falling within the region's boundaries.

Household Savings

These have been estimated as residuals, in each case by taking total incomes and subtracting consumption expenditures and direct taxes.

Household Final Demands: Agricultural Households (H_1–H_4)

Since the final demands of agricultural and nonagricultural households have been estimated from different data sources, they are treated separately. The four types of agricultural households in the SAM are: landless households (hereafter referred to as H_1); labor-abundant farm households, or H_2; land-abundant farm households, or H_3; and farm households outside the Muda project's command area, or H_4.

The main source for estimating the final demands of these four groups is the FAO/World Bank household survey, which reports mean expenditures on various commodities for the paddy farming population of the Muda scheme area. However, some adjustments have to be made to the data before we derive our estimates:

- The survey was undertaken in 1973, whereas the SAM is based on 1972. We have adjusted the 1973 expenditures by a scale factor that reflects the change in both prices and incomes between these years.
- The survey classified its households according to three categories: single-croppers, double-croppers, and landless farm workers. Moreover, they referred to farms within the Muda scheme area only. We have fitted these data into our H_1–H_4 framework by a weighting method.
- Expenditures in the survey are in market prices, whereas the SAM coefficients are estimated in producer prices. We have calculated appropriate trade and transport margins with which to deflate the survey figures to obtain final demands at producer prices.
- Finally, the 9-commodity breakdown in the household survey has been disaggregated to fit our 35-commodity schema.

Each of these adjustments is now described in detail.

DEFLATING FROM 1973 TO 1972. To transform 1973 survey data to 1972 SAM expenditure figures, we have assumed that real expenditure on each commodity remained a constant fraction of real income. In symbols, if Q_i denotes purchases (in quantity terms) of commodity i, Y nominal income, and P the aggregate price level, all in 1973, then

$$Q_i = b_i \frac{Y}{P} . \tag{C.1}$$

With primes denoting 1972 values, our assumption implies

$$Q_i' = b_i \frac{Y'}{P'} . \tag{C.2}$$

Table C-1. *Expenditures and Price and Cost of Living Indexes*
(1973 Malaysian dollars)

Survey commodity	*Single-croppers*			*Double-croppers*		*Landless workers*	*1972–73 price index*
	Small	*Large*	*Small and large*	*Small*	*Large*		
Home-produced foods	373	400	386	475	521	229	
Nonprocessed foods	269	336	300	273	357	284	1.163
Processed foods	249	283	265	260	313	213	1.085
Alcohol and tobacco	57	71	63	67	87	46	1.019
Clothing	72	94	82	77	115	63	1.217
Durables	26	90	55	37	66	19	1.083
Fuel and electricity	24	31	27	27	31	20	
Services	90	153	119	86	158	47	1.043
Other	76	128	100	87	152	35	
Total in 1973 prices	1,236	1,586	1,397	1,389	1,800	956	
Total in 1972 prices[a]	1,105	1,424	1,251	1,240	1,612	851	
Change in cost of living index, 1972–73 (percent)	11.9	11.4	11.7	12.0	11.7	12.3	

a. The total in 1972 prices is obtained by deflating expenditures on each commodity by the 1972–73 price index. The 1972–73 cost of living index is the percentage increase in total mean expenditure when measured in 1973 instead of 1972 prices.

However, the available data (and the SAM's requirements) are in terms of expenditures rather than quantities purchased. If P_i and P_i' are the prices of good i in 1973 and 1972, respectively, combining (C.1) and (C.2) gives

$$P_i'Q_i' = (P_i'/P_i)\frac{Y'/P'}{Y/P} P_iQ_i \,. \tag{C.3}$$

Thus, for each commodity, to obtain 1972 expenditure figures from their 1973 counterparts, we scale the latter by the ratio of real incomes and the ratio of prices between the two years.

The ratio of real incomes in 1972 to 1973 has been derived by first determining the change in the aggregate price level and multiplying its reciprocal by the change in nominal incomes. In particular, a cost of living index for each of the survey's household types has been computed by using the survey's mean expenditure figures as weights. The individual price changes have been derived from the consumer price index (CPI). Each of the first three survey "commodities," home-produced foods, nonprocessed foods, and processed foods, represents a mix of different products. Their prices are determined by taking a weighted average of the changes in price of each of their components, the weights being the mean expenditure levels, as reported by the survey. The mean expenditure levels and the resulting cost of living indexes for the survey's households and commodities are given in table C-1.

Because we are dealing with agricultural households here, the changes in nominal income can be calculated by looking at farm revenues and costs, as reported by the FAO/World Bank survey. It has been assumed that revenues from nonpaddy sources were the same in 1972 and 1973, and that farm costs rose at the rate determined in the notes on paddy sector technology (see below). The ratio of real incomes between 1972 and 1973 is determined by dividing the ratio of nominal incomes in the ratio of price levels. The results of these calculations are summarized below (all figures are sample means):

Single-croppers			*Double-croppers*		
Small	*Large*	*Small and large*	*Small*	*Large*	*Landless workers*
1.05	1.00	1.02	0.94	0.86	1.09

We can now adjust the expenditure figures given by the survey to 1972 consumption levels (at 1972 prices). Each commodity's 1973 expenditure is deflated by its own inflation rate (from table C-1) and then scaled by the 1972–73 real incomes ratio. The results are given in table C-2.

Table C-2. *1972 Expenditures per Household at 1972 Market Prices*

Survey commodity	*Single-croppers*			*Double-croppers*		
	Small	*Large*	*Small and large*	*Small*	*Large*	*Landless workers*
Home-produced foods	340	348	342	388	390	217
Nonprocessed foods	243	289	263	221	264	266
Processed foods	241	261	249	225	248	214
Alcohol and tobacco	59	70	63	62	73	49
Clothing	62	77	69	59	81	56
Durables[a]	26	95	61	43	66	21
Fuel and electricity	25	31	27	25	26	22
Services	91	147	116	78	130	49
Other	73	119	94	76	121	35
Total	1,160	1,425	1,284	1,166	1,385	927

a. Includes jewelry.

MAPPING INTO H_1–H_4 FRAMEWORK. The survey data give the sample proportions of the survey's household types in H_1, H_2, and H_3. The percentages are:

	Landless	*Single-croppers*	*Double-croppers*
H_1	100		
H_2		54	34
H_3		46	66

No H_4 households were in the sample. Since these households did not enjoy the direct effects of the Muda irrigation scheme, we have assumed they have expenditure patterns similar to those of single-croppers in the scheme area.

All that remains, then, is to estimate the 1972 populations of landless farm workers, single-croppers, and double-croppers. These have been derived as follows:

FAO (1975, p. 13) reports farm populations of 51,571 in the scheme area and 23,357 outside.

Of the scheme area farmers in the sample, 3.95 percent were single-croppers and 95.05 percent double-croppers. Applying this breakdown to the global population, we get:

$$
\begin{aligned}
51{,}571 \times 0.0495 &= 2{,}553 \text{ single-croppers} \\
51{,}571 \times 0.9505 &= 49{,}018 \text{ double-croppers} \\
& \ 51{,}571 \text{ total.}
\end{aligned}
$$

To obtain the number of landless households, we note that the survey strata population represented 21.4 percent of the scheme area population and included 796 landless workers. If we "blow up" this sample, we get 3,431 workers in the region, but we must first correct for underreporting. The strata population included 9,399 scheme farmers. If this represented 23.2 percent of the population, there would be 9,399/0.232 = 40,513 farmers in the scheme area. However, we know independently that the area contained 51,571 farmers, so that the underreporting adjustment factor is 51,571/40,513 = 1.273. Hence, the estimate of the number of landless farm workers is 3,431 × 1.273 = 4,367.

Finally, we know that there are 23,357 farming households outside the scheme area, all of which we have assumed to behave like single-croppers. To obtain the number of landless farm workers outside the scheme, we have assumed the same ratio of farm workers to farmers as inside the scheme. This ratio is:

$$4{,}367/51{,}571 = 0.085.$$

Therefore, the population of nonscheme landless farm workers is:

$$23{,}357 \times 0.085 = 1{,}976.$$

The total nonscheme population is:

$$23{,}357 + 1{,}976 = 25{,}333.$$

The population breakdown connecting our two household classifications is shown in table C-3. "Small" and "large" refer to the labor-abundancy and land-abundancy criteria that distinguish H_2 from H_3. That is, we have classed all single-croppers in H_2 as "small" and all single-croppers in H_3 as "large." The reason for this is that we can generate separate sample expenditure means for small and large farmers, thereby making our total expenditure estimates more accurate.

ESTIMATING FINAL DEMANDS AT PRODUCER PRICES. Given the mean expenditure levels and the population sizes, total expenditures on commodities by the four household types in 1972 *market* prices follow at once. To convert these expenditures to producer prices, retail, wholesale, and transportation margins must be subtracted. These margins have been determined as follows:

The retail margins on various commodities in Western Malaysia have been computed from Department of Statistics (1972*c*), table 32. The assumption that those margins apply to the region as well as justified by the fact that the retail margin (for all commodities) in Alor Setar is the same as that in all of Western Malaysia (ibid., table 31). When, as in most cases, the commodity group consists of several commodities, a weighted average has been taken, the weights being mean expenditures for the whole sample in the FAO/World Bank survey.

The wholesale margins are specific to the region, being drawn from Department of Statistics (1972*b*). We have used the same weights as for retail margins.

The transport margins have been derived mainly from the notes to the transactions matrix entries, except for alcohol and tobacco, clothing, and durables, for which we use the FAO/World Bank survey of retail business.

Table C-3. *Population of Households by Socioeconomic Type*

	Single-croppers			*Double-croppers*		
	Small	*Large*	*Small and large*	*Small*	*Large*	*Landless workers*
H_1						4,367
H_2	1,379			16,666		
H_3		1,174			32,252	
H_4			23,357			1,976

Final demands in producer prices have been calculated with the use of these margins (see table C-4). The margins on home-produced items have not been added to final demand for trade and transport services because there is no outlay for these services.

MAPPING TO 35 COMMODITIES. We have disaggregated the nine commodity groups to fit the SAM's 35-sector framework by taking the proportions of mean expenditure in each commodity group allocated to the various sectors.

The estimates of final demands by rural households follow at once, with the following amendments. Demand for rural housing services has been assumed to be imputed rent, which is determined from the FAO/World Bank survey (see notes below on sector 35, residential construction). Demand for transport and distribution services has been derived from the margins calculated in the preceding section. The breakdowns between road and rail transport and between organized and petty trading have been guesstimated. The latter division, however, is not altogether arbitrary, since the entire gross value of output of the petty trading sector is delivered to final demand.

Household Final Demand: Nonagricultural Households (H_5)

The estimation of nonagricultural households' (H_5) final demand was undertaken in three steps. First, we derived a figure for the total expenditure of these households by estimating their incomes net of savings and taxes. Second, we applied a set of Engel (expenditure) coefficients to this figure, where the coefficients have been derived from the 1973 *Household*

Table C-4. *Final Demand at Producer Prices*
(thousands of Malaysian dollars in 1972 prices)

	Household class			
Survey commodity	*Landless* (H_1)	*Labor-abundant* (H_2)	*Land-abundant* (H_3)	*Nonproject* (H_4)
Home-produced food	804	5,875	11,048	7,139
Nonprocessed food	925	3,232	7,068	5,308
Processed food	817	3,559	7,276	5,450
Alcohol and tobacco	183	930	2,041	1,310
Clothing	215	935	2,376	1,509
Durables	71	440	1,377	1,135
Fuel and electricity	96	451	878	674
Services	214	1,423	4,378	2,806
Other	153	1,366	4,054	2,265

Expenditure Survey (HES) of Western Malaysia. We used data relating to urban households in the nationwide HES, implicitly assuming that these households have similar tastes to the region's nonagricultural households (which are, in fact, concentrated in urban areas). Third, the 31-commodity HES framework was mapped into our 36-commodity scheme, along with the appropriate transformation from market to producer prices.

ESTIMATING TOTAL EXPENDITURE. From the 1970 *Post-Enumeration Survey*, mean monthly incomes for nonagricultural households in the region were: urban, \$354.13 and rural, \$214.25. Since the populations of these two subgroups were 16,065 and 27,982 households, respectively (FAO, 1975; Anand, 1977), total annual nonagricultural income in 1970 was:

$$12 \times [(354.13 \times 16{,}065) - (214.25 \times 27{,}982)] = \$140.21 \text{ million.}$$

We know that taxable income grew 25.02 percent in the next two years (FAO, 1975, table 55), so that nonagricultural income in the region in 1972 has been estimated to be

$$\$140.21 \times 1.25 = \$175.29 \text{ million.}$$

Unfortunately, this figure does not square as closely as we would wish with our own independent estimates of nonagricultural incomes from production activities (the value added row in the SAM) less savings and taxes. Therefore, we have "compromised" and set total expenditure (incomes, net of taxes and savings) by nonagricultural households at \$180 million.

APPLYING ENGEL COEFFICIENTS. To estimate the Engel coefficients, we have used the expenditure data (for 32 commodities) of the 2,000 or so *urban* households in the HES. An ordinary least squares regression of expenditure on commodity i, C^i, on total expenditure, E, yields a coefficient β^i (and intercept γ^i), which we have then applied to our total expenditure estimate for the region to get final demand for commodity i.

MAPPING TO THE SAM FRAMEWORK. In mapping HES expenditures to our 36-sector framework, we have made the following adjustments:

Expenditure on domestic services has been excluded because we treat this as an interhousehold transfer in our model.

Per capita expenditure on hotel and restaurant services was thought to be higher for the nation than for the region. Consequently, this figure (HES commodity 31) has been scaled down to about one-third of its original size.

In cases where the mapping is not one-to-one (for example, when one HES commodity encompasses two SAM sectors), independent evidence was used to make allocations among the sectors. For example, final demand for regional meat and poultry (HES commodities 5 and 6) was calculated by subtracting *agricultural* households' final demand from the value of meat and poultry from regional slaughterings. The latter figure was obtained from work on the sector "other agriculture." The residual expenditure from the total on commodities 5 and 6 has, therefore, been allocated to imports.

In moving from market to producer prices, the procedures are similar to those described above. For imported commodities, we assumed that only the retail margin accrues to the region. Therefore, the producer price is actually the wholesale price. We have estimated the weighted retail margin to be 12.1 percent. Of the total demand for retail trade services, we have assigned an arbitrary $1 million to "informal distribution."

Firms

Profits Distributed to Households

After payment of corporate taxes and Employees' Provident Fund (EPF) contributions, it has been assumed that one-third of the remainder is paid as dividends, and solely to nonfarm households.

Corporate Taxes

In 1972 the average rate of corporate taxation was 41.2 percent (World Bank estimates). Applying this rate to gross profits and adding employers' contributions to EPF yield total taxes paid to the federal government.

Corporate Savings

Corporate savings were determined as a residual (see above, "Profits Distributed to Households").

Repatriated Profits

Sectors 16, 22, 26, 36, and 42 "repatriate" all their posttax profits, since the incorporated firms among them are owned by agents residing outside the region. Sectors 27 and 35 were reckoned to be two-thirds extraregionally owned, and the associated outflow of profits was estimated accordingly.

Federal Government

Transfers to Households

Transfers to households take the form overwhelmingly of the imputed value of the goods and services provided to households through current expenditures. In the case of irrigation, farmers do not pay the full marginal costs of operating the scheme, but the paddy production activity is charged in full. There must therefore be an offsetting transfer to preserve farmers' incomes at their actual level. According to the FAO/World Bank household survey, actual payments were $1.427 million. Hence, the subsidy is the gross value of output in irrigation less actual payments, which has then been distributed among household classes 1–3 in proportion to their output of paddy. For education, health, and other government services, the gross value of output has been allocated among household classes in proportion to their respective populations, except that nonfarm households receive half of all outlays on education (46 percent of all primary school places and 56 percent of all secondary are in Alor Setar and Kangar [FAO, 1975, table 4]).

Federal Government Savings

These are capital outlays of various kinds, including those on the Muda scheme (FAO, 1975, tables 7 and 52). It has been assumed that the federal government did not sell any assets in the region. Certainly, there were no sales of physical assets, and sales and redemptions of bonds can be safely ignored.

State Government

Transfers to Households

Transfers to households have been estimated and allocated in the same fashion as in the case of the federal government.

State Government Savings

The estimate of capital outlays has been taken directly from FAO (1975, table 53).

Repayments of Debt

Repayments of principal and interest on loans taken from the federal government have been guesstimated from the sketchy information in Malaysia, *Economic Report, 1974–75*.

Combined Capital Account

Outlays on Domestically Produced Investment Goods

The sectors involved are the livestock component of other agriculture (19), manufacturing not elsewhere classified (23), residential construction (35), and other construction (36).

Since we are concerned with gross investment, deliveries from sector 19 are simply the value of the increase in stocks. The estimate for the region has been constructed from data in various issues of the Ministry of Agriculture and Fisheries, *Agricultural Statistical Digest*, by interpolating from the data for Kedah and Perlis on the basis of the area of paddy land for buffaloes; stocking densities in Perlis for cattle, goats, and sheep; and the Chinese population for pigs.

Three subsectors of sector 23 concern us: truck bodies, industrial machinery, and fabricated structures. The estimates have been taken directly from the Department of Statistics, *Survey of Manufacturing Industries* (1972).

Deliveries from the construction sectors have been pieced together from FAO (1975, table 63), which gives private construction (but excludes rural construction, which was estimated from the FAO/World Bank household survey data); and the Department of Statistics, *Survey of Construction Industries*, which reports outlays by government and quasi-government agencies.

Imports of Investment Goods

The investment goods take the form of vehicles, plant, and machinery. Gross investments by farmers have been estimated from the FAO/World Bank household survey data. Taken from FAO (1975) are investments in rice milling (tables 7 and 65); in bus transport (table 71); and four-wheel tractors (table 67). Deliveries to other sectors have been pieced together from the *Survey of Manufacturing Industries, Survey of Construction Industries, Annual Fisheries Statistics*, and special unpublished tabulations of the financial accounts of the federal and state governments prepared for FAO (1975).

Production Activities

The estimation of the full interindustry transactions matrix of thirty-five sectors was a major task involving extensive use of the primary data sources described in appendix A, the statistical tabulations in FAO (1975), a wide range of other secondary data sources, and frequent guesstimation based on our own local knowledge. We do not endeavor here to lay out the full set of estimates, but provide instead a decently detailed explanation of our methods and sources for eight selected sectors, which exemplify procedures, problems, or proclivities that have influenced our treatment of the available material.

The sectors we have chosen are:

	Column number in table 5-4
Commercial rice mills	10
Paddy production	14
Other agriculture	19
Road transport	24
Distributive trades	30
Residential construction	35
Government, education	38
Rural housing	44

Within each sector we deal first with the estimation of gross output, then intermediate inputs, and finally value added and its allocation across institutions. Thus a whole sector (column) is treated completely before proceeding to the next. To save space, we shall use the symbol M (for mega) to denote million.

Commercial Rice Mills (10)

The importance of this sector to the regional economy is sufficient justification for its inclusion here.

GROSS VALUE OF OUTPUT. FAO (1975, table 65) gives the number (thirty-six) of such mills in the region in 1972 together with their throughput of paddy. The *Survey of Manufacturing Industries* (SMI) for 1972 provides estimates of the gross value of output (GVO) and paddy throughput for all forty-one mills in Kedah and the five in Perlis. Thus, scaling the Kedah figures from the latter source and adding those for Perlis provides an estimate of gross output for the region of $87.906M.

GROSS VALUE OF INPUTS. First, the value of paddy throughput in Kedah and Perlis reported by the SMI (1972) has been rendered consistent with the regional estimate from FAO (1975) by a scaling procedure similar to that used for gross output. The resulting value of throughput has then been multiplied by a volume-weighted average of the paddy price reported by the SMI for Kedah and Perlis mills to give, at market prices, a regional gross value of paddy throughput of $74.20M. Trade and transport margins have been deducted with the use of farm-gate prices estimated for 1972 by MADA and local information, respectively. Other input values have been computed either directly from the SMI or from data obtained from the Kedah Rice Millers Association and the small survey of industrial establishments undertaken by the authors (appendix A). In sum, these calculations reveal the cost structure set out in table C-5, in which the allocation between imports and domestic production sectors is based on the authors' knowledge of the region and the commodity composition of imports in FAO (1975, table 50).

ALLOCATION OF VALUE ADDED. It is then a simple step to calculate regional value added: 87.906M − 78.922M = $8.984M. Using evidence from the authors' survey and more detailed information from the accounts of one or two rice mills in the region, we have allocated this value added to institutions as follows: nonfarm households (H_5), $2,952M; firms, $6.032M. The figure attributed to nonfarm households comprises wage payments to unskilled and semiskilled labor, while the remainder is profits that go to firms.

Paddy Production (14)

Because this sector is of central importance, we have set out our procedure at some length.

Table C-5. *Cost Structure for Commercial Rice Milling*
(millions of Malaysian dollars)

Sector of origin	*Producer prices*	*Trade margins*	*Transport margins*[a]	*Market prices*
Domestic				
Paddy production	68.445	4.857	0.899	74.201
Manufacturing n.e.c.	0.059	0.005	—	0.064
Road transport	1.038	—	—	—
Distributive trades	4.862	—	—	—
Water	0.014	—	—	0.014
Posts and telecommunications	0.073	—	—	0.073
Other construction	0.131	—	—	0.131
Modern financial services	0.292	—	—	0.292
Imports				
Packing materials	2.356	—	0.130	2.486
Spares	0.944	—	0.004	0.948
Fuels and lubricants	0.441	—	—	0.441
Other materials	0.196	—	0.005	0.201
Financial services	0.071	—	—	0.071
Total	78.922	4.862	1.038	78.922

— Not applicable.

a. Partially based on the National Input-Output Table for West Malaysia, 1970 (hereafter referred to as NIOT).

GROSS VALUE OF OUTPUT. The gross value of output has been directly determined by taking regional paddy production from table 2-5, adjusting for chaff, waste, and moisture, and multiplying by the appropriate farm-gate price. To the resulting total, an estimate of the trade margins on paddy delivered to small-scale mills has been added, to give an estimate of gross output of $143.156M.

PADDY PRODUCTION. Purchases by the paddy production sector from itself are made up exclusively of seed. Drawing on table 2-4 for the relevant areas under paddy, using MADA prices and FAO/World Bank survey prices for purchased seed (allowing a small storage loss between seasons for single-cropping farmers), and taking the seeding rates from World Bank data and the *Farm Economic Survey* (Malaysia, 1967), yields a total cost for seed, at market prices, of $1.487M. After adjustments for trade and transport margins on purchased seed only, this gives $1.202M at producer prices.

OTHER AGRICULTURE. Paddy production purchases all services of draft animals from "other agriculture." For farming households within the project's command area, the relevant figures have been drawn directly from the FAO/World Bank household survey. The volume of these services for farming households outside the project has been derived from data on single-cropping households in Doering (1972), after revaluation to 1972 prices. This yields total regional expenditures on animal draft power of $1.249M.

AGRICULTURAL MACHINERY SERVICES. The FAO/World Bank household survey reveals that the average paddy farming household used 6.199 hours of four-wheel tractor time in 1973. From an examination of changes in the regional stock of such machines, in the price of such services, and in the gross cropped area during 1972–74 we concluded that the average level of utilization per household in 1972 was little different from that in 1973. Taking 6.199 hours and multiplying by the farm population in 1972 and by the price per hour (given by an interpolation between field observations in 1970 and 1974 from MADA (1970*a*) and the authors' fieldwork, respectively), we have total regional expenditures on tractor services of $2.561M.

MACHINERY REPAIR. Purchases of machinery repair services by paddy farming households represent outlays on the maintenance of two-wheeled tractors and transport vehicles, which we have assumed to be purchased exclusively for business purposes. Expenditures on two-wheel tractor maintenance by project paddy farmers have been drawn directly from the FAO/World Bank household survey. Deflating the result by the change in the import price index for transport equipment between 1972 and 1973 (Treasury of Malaysia, *Economic Report, 1974–75*) and by the change in regional two-wheel stocks (FAO, 1975, table 67) and then deducting the associated transport costs gives an estimate of total expenditures on maintenance of \$0.235M. Using a similar procedure, but omitting any deduction for transport costs, we have assessed vehicle maintenance costs at \$0.521M.

For nonproject farm households (which engage in both rubber and paddy production), the gross output per holding was very close to the average income of single-cropping households within the project. The relevant expenditure figures were taken from the FAO/World Bank household survey, deflated for price and stock changes, and then adjusted for the share attributable to sector 17, to estimate expenditures on vehicle maintenance by nonscheme households of \$0.064M.

Hence, aggregate purchases of machinery repair services have been estimated at 0.235M + 0.521M + 0.064M = \$0.820M at producer prices.

MANUFACTURING N.E.C. All input-output tables have ragbag (politely, "residual") sectors, and that misfortune falls upon manufacturing n.e.c., which includes, among other activities, the digging of bat guano. Using data from the FAO/World Bank household survey and procedures appropriate to the subject, but similar to those used for sector 22, we have estimated total deliveries of bat guano to paddy production at \$0.278M gross of transport costs. At an assumed cost of \$0.20 per tonne-mile and an average haul of fifteen miles, transport costs are \$0.018M. There are, of course, no trade margins.

OTHER CONSTRUCTION. From the FAO/World Bank household survey, we have a direct estimate of the capital value of farm buildings for the project's paddy farmers. On the assumption that this is the same for all classes of paddy producing households in the region, and that annual maintenance expenditure is 1 percent of the capital value, aggregate regional outlays are \$0.127M.

IRRIGATION. Although irrigation charges are collected together with land taxes, irrigation is a productive input. Hence, we treat it as a separate production activity, the entire output of which is purchased by paddy production. The relevant figure of \$8.466M has been calculated from FAO (1975, tables 8 and 56).

TRADITIONAL AND MODERN FINANCIAL SERVICES. Using FAO/World Bank household survey values, we have estimated aggregate interest payments by project paddy farming households to be \$0.620M after deflating for increases in area and the price of farm inputs between 1972 and 1973. As before, figures for nonproject paddy producing households have been estimated by using FAO/World Bank household survey values for single-cropping households, which yield a total outlay of \$0.049M. Hence, the regional aggregate is \$0.669M. Assuming that all borrowing by nonproject households was from traditional sources and using the relevant proportions from the FAO/World Bank household survey for project households, we have allocated \$0.458M and \$0.211M to sectors 41 and 42, respectively.

IMPORTS. From FAO (1975, table 66) we obtain detailed quantity estimates of regional imports of agrochemicals. These, together with prices derived from the FAO/World Bank household survey, MADA (1974), and Treasury of Malaysia, *Economic Report, 1974–75*, and with a suitable estimate of transport costs and trade margins taken from the national input-output table for 1970 (NIOT) give regional imports of agrochemicals of \$5.052M in producer prices.

Outlays on fuel and lubricants of \$1.003M have been estimated in a similar fashion. These outlays, together with those for all other intermediate inputs, are summarized in table C-6.

VALUE ADDED AND FACTOR COSTS. Value added is, residually, \$120.171M. Our SAM in table 5-4 has five classes of households, to which we allocate value added in the following manner. First, we estimate the incomes from paddy production for nonproject farm households (H_4) and

Table C-6. *Cost Structure for Paddy Production*
(millions of Malaysian dollars)

Sector of origin	*Producer prices*	*Trade margins*	*Transport margins*	*Market prices*
Domestic				
Paddy production	1.202	0.280	0.005	1.487
Other agriculture	1.249	—	—	1.249
Agricultural machinery services	2.561	—	—	2.561
Machinery repair	0.820	—	0.016	0.836
Manufacturing n.e.c.	0.269	—	0.018	0.287
Road transport	0.227	—	—	—
Rail transport	0.002	—	—	—
Distributive trades	1.338	—	—	—
Other construction	0.127	—	—	0.127
Irrigation	8.466	—	—	8.466
Traditional financial services	0.458	—	—	0.458
Modern financial services	0.211	—	—	0.211
Imports				
Agrochemicals	5.052	0.707	0.190	5.949
Fuel and lubricants	1.003	0.351	—	1.354
Total	22.985	1.338	0.229	22.985

— Not applicable.

nonfarm households (H_5) independently. We then take the remaining value added and distribute it among landless, labor-abundant, and land-abundant paddy farming households (H_1, H_2, and H_3), respectively, in proportion to their incomes from paddy production. Although we have retained the proportional distribution of income from the FAO/World Bank survey, the latter source provides direct estimates of income that appear to be too low.

We have computed the income of H_4 from paddy production by taking the output of paddy from table 2-5 and multiplying by the farm-gate price of paddy. From this we have deducted input costs estimated partly as above and partly from single-cropping values given by the FAO/World Bank household survey. The resulting estimate of net income from paddy production accruing to H_4 is \$15.143M.

The income from paddy production accruing to H_5 consists in large part of land rents. These have been estimated directly from the FAO/World Bank household survey at \$5.064M.

Deducting the incomes calculated above for H_4 and H_5 from total value added would leave a residual sum to be allocated to the remaining classes of household. This sum, however, is too large because it contains wage payments to hired labor supplied by households within the region and migrants from Thailand and elsewhere in Malaysia. We have estimated these payments as follows. From evidence derived from the FAO/World Bank household survey, we have calculated the total wage bill and then the share accruing to regional landless households (H_1). Deducting the former from the latter leaves a wage bill of \$9.447M attributable to H_4, H_5, and migrants (whom we treat as an imported intermediate good; see chapter 5). From data on the flow of migrants from Thailand and Kelantan (collected by MADA), we have estimated their share to be \$2.401M. On the basis of some rough and ready figuring, we have allocated that part of the total wage bill then remaining among H_4, H_5, and other migrants drawn into the region from its more immediate peripheries. The resulting incomes are summarized in table C-7.

After deducting from value added the factor payments described above, the remaining \$90.517M has been allocated to H_1, H_2, and H_3 in proportion to their share of income calculated directly from the FAO/World Bank household survey data. The shares are 2.499, 19.636, and 63.330 percent, in ascending order.

Other Agriculture (19)

The "other agriculture" sector is something of an agricultural ragbag, comprising production activities ranging from services of draft animals to the production of specialist fruit and vegetable crops. There is little profit

Table C-7. *Allocation of Value Added and Factor Costs to Households*
(millions of Malaysian dollars)

Source	Household: Landless (H_1)	Labor-abundant (H_2)	Land-abundant (H_3)	Non-project (H_4)	Non-farm (H_5)	Migrants
Wage payments	—	—	—	2.000	2.702	4.745[a]
Rents	—	—	—	—	5.064	—
Paddy production	2.647[b]	20.797[c]	67.073[c]	15.143	—	—
Total	2.647	20.797	67.073	17.143	7.766	4.745
Number of households	4,367	18,045	33,426	25,333	44,047	—

— Not applicable.
a. Added to imports.
b. Includes $1.682M of wage payments.
c. Net of wage and land rental payments between H_2 and H_3.

to be derived from laying out the full details of our estimation procedures, which were necessarily tailored to the particular circumstances of Muda and the peculiarities and shortcomings of the available data. We thus proceed at a brisk pace, providing detail only where understanding would otherwise be severely impaired.

GROSS VALUE OF OUTPUT. Table C-8 lists the estimates of the composition of gross output, which we subsequently treat seriatim.

The value of draft animal services has been taken directly from our work on sector 14. The gross output of meat other than poultry has been

Table C-8. *Gross Value of Output for Other Agriculture by Subsector*
(millions of Malaysian dollars)

Subsector	Gross output
Draft animal services	1.249
Meat other than poultry	8.514
Poultry meat	0.384
Milk and eggs	0.513
Livestock exports	2.912
Changes in animal stocks	−3.406
Kampong production	3.093
Specialist crops	4.022
Total	17.281

determined from data on livestock slaughterings and prices in the *Agricultural Statistical Digest* (1972), appropriately adjusted to represent the region, together with the value of on-farm slaughterings taken directly from the FAO/World Bank household survey. The values for poultry meat and milk and eggs have been derived by taking the per capita consumption of home production figures from the FAO/World Bank household survey, assuming them to be constant across the entire population of the region, and deflating by the corresponding elements of the consumer price index (see above, "Household Final Demands").

Livestock exports have been estimated by using the volume flows by road and rail out of the region (FAO, 1975, tables 49 and 50), weighted by the type and composition of the regions' animal population. Prices have been obtained from a price survey undertaken as part of the entire set of FAO/World Bank field studies, while trade and transport margins have been taken from the Department of Statistics, *Survey of Distributive Trades* (1968) and our own local knowledge, respectively.

Changes in animal stocks have been derived from various issues of the *Agricultural Statistical Digest* and some rough-and-ready figuring about productive life and birth rates. Prices have been taken from the FAO/World Bank price survey mentioned above.

Kampong (family fruit and vegetable) production has been treated in a manner similar to poultry, milk, and eggs, except that production outside the project has been estimated on an area basis using unpublished work sheets prepared for Wong (1966).

Drawing on this latter source, the *Agricultural Statistical Digest*, the FAO/World Bank price survey, and data on rubber intercropping from RISDA, we have estimated the outputs of some ten specialist crops. Trade and distribution margins were derived from the *Survey of Distributive Trades* (1968) and the NIOT.

GROSS VALUE OF INPUTS. Estimates have been derived subsector by subsector in much the same way as gross output. Where the standard references mentioned above were wanting, we resorted to the NIOT and, *in extremis*, to guesstimation based on our own fieldwork. The results are summarized in table C-9 and the associated footnotes.

ALLOCATION OF VALUE ADDED. Total value added in this sector is $14.556M, which we have distributed across the five classes of households by examining each component separately. For each component we applied a set of weighted proportions (derived from the FAO/World Bank studies) which reflect the ownership of the underlying factors of production. The results are set out in table C-10.

Table C-9. *Cost Structure for Other Agriculture*
(millions of Malaysian dollars)

Sector of origin	*Producer prices*	*Trade margins*	*Transport margins*	*Market prices*
Domestic				
Small rice mills[a]	0.389	—	0.005	0.394
Fish processing[b]	0.025	0.001	0.001	0.027
Other agriculture[c]	0.130	—	—	0.130
Machinery repair[d]	0.003	—	—	0.003
Manufacturing n.e.c.[e]	0.124	0.068	—	0.192
Road transport	0.038	—	—	—
Rail transport	0.016	—	—	—
Distributive trades	0.222	—	—	—
Other construction[f]	0.174	—	—	0.174
Other government services[g]	0.148	—	—	0.148
Traditional financial services[h]	0.150	—	—	0.150
Modern financial services[i]	0.067	—	—	0.067
Imports				
Agrochemicals[j]	1.047	0.125	0.039	1.211
Animal feed[k]	0.181	0.011	0.002	0.194
Veterinary medicines[l]	0.008	—	—	0.008
Fuel and lubricants[m]	0.003	—	—	0.004
Total	2.725	0.205	0.047	2.725

— Not applicable.

a. Bran only. Derived from NIOT. Transport margin based on \$0.2 per tonne-mile, an average distance of 10 miles, and an estimated price of \$168 per ton in 1972.

b. Fishmeal only; based on work for sector 13. Trade margin of 6 percent (*Survey of Distributive Trades*, 1968). Transport margin estimated using \$0.2 per tonne-mile and average distance of 40 miles and total tonnage of 60.

c. Seeds and seedlings only.

d. Transport margin negligible.

e. Trade margin of 35 percent (*Survey of Manufacturing Industry*, 1971).

f. Assumed to be 1.0 percent of capital value of farm buildings used in other agriculture. Capital stock derived from FAO/World Bank household survey and *Agricultural Statistical Digest* (1973).

g. Based on FAO/World Bank household survey, which yields a total of \$0.156M for veterinary services and drugs. The latter is treated as imports.

h. Derived from FAO/World Bank household survey.

i. FAO (1975), table 58; a 10 percent rate of interest is assumed.

j. FAO (1975), table 66; trade and transport margins are calculated as for sector 14.

k. Concentrates of \$0.221M less \$0.027M from sector 13. Trade margin of 6 percent (*Survey of Distributive Trades*, 1968). Transport margin based on \$0.2 per tonne-mile, price per ton \$337, and average distance of 15 miles.

l. See note g above.

m. Trade margin from NIOT. Other data from FAO/World Bank household survey and Rim (1969).

Road Transport (24)

Our estimates for this sector exemplify the way in which, in the absence of complete regional data, we have developed regional estimates from the NIOT.

GROSS VALUE OF OUTPUT. The vehicles contributing to the output stream from this sector are trucks, buses, and taxis. Cars and motorcycles are treated separately in the estimation of final demand. From the *Yearbook of Transport Statistics* (Department of Statistics, 1972) and FAO (1975, table 68) and after weighting vehicle populations by class and engine capacity, we have estimated that the region accounted for 4.42 percent of national gross output in 1972. Multiplying this percentage by national gross output in 1970 and inflating for 20.8 percent nominal growth between 1970 and 1972 (Treasury of Malaysia, *Economic Report, 1974–75*, table 1.2) yields a regional gross output of $19.606M in 1972. This squares quite well with the row sum of regionally estimated demands, namely, $17.552M. For the sake of consistency, we have adopted the latter figure.

GROSS VALUE OF INPUTS. Again from NIOT, we have derived a set of regional coefficients, which yield total inputs of $5.543M at producer prices. Of this, machinery repair (sector 22) and manufacturing n.e.c. (sector 23) account for $1.478M and $1.200M, respectively. FAO (1975, tables 68 and 71) provides independent regional estimates for sector 22 of $1.662M and for sector 23 of $1.346M.

VALUE ADDED AND FACTOR COSTS. Value added is $12.009M. From FAO (1975, table 71) and the associated unpublished working notes on bus companies operating in the region, profits have been estimated at $0.554M. Many of the region's trucks are owned by haulage companies and large firms in various sectors. Moreover, their haulage rates are largely determined by market forces, and their operating procedures are not subject to the stringent government controls applied to bus companies. It is therefore reasonable to assume that haulage is considerably more profitable than bus operation. From FAO (1975, table 68) trucks outnumbered buses in the region by almost two to one in 1972. Hence, profits from haulage have been set at $1.500M. Thus, the income accruing to nonfarm households (H_5) is 12.009M − 0.554M − 1.500M = $9.995M.

Distributive Trades (30)

We have chosen this sector to lay out our procedures in considerable detail, partly because it contributes more to regional gross output than any

Table C-10. *Disaggregated Allocation of Value Added in Other Agriculture*
(millions of Malaysian dollars)

Household	*Draft animal services*	*Animal stock changes*	*Pigs*	*Other livestock*[a]	*Kampong production*	*Specialist crops*	*Total*
H_1	0.012	−0.023		0.077	0.117		0.183
H_2	0.237	−0.436		1.381	0.485		1.657
H_3	0.766	−1.308		4.145	0.898		4.501
H_4	0.192	−0.654		2.072	0.680	3.754	6.044
H_5		−0.454	2.623				2.169
Total							14.554

a. Includes all livestock (except pigs), poultry meat, milk, eggs, and livestock exports.

other sector except paddy production and commercial rice milling, and partly because our data base is relatively strong.

GROSS VALUE OF OUTPUT. For urban and semiurban establishments, we have used the FAO/World Bank survey of retail trade (see appendix A) and estimated the urban gross value of output for each type and location of shop. Using the FAO/World Bank survey terminology, we have defined the gross output for a retail establishment as: Profit + Commission + Workers' wages + Workers' allowances + Shop rent + Business insurance + Fuel and power + Transport + Other business expenditures.

We have obtained regional gross output by multiplying the average for each urban shop type by the number of such shops in the region, which is given in FAO (1975, table 72), or in the *Survey of Retail Establishments in Western Malaysia*. The resulting estimates are set out in table C-11. Excluded are urban market stalls, which sell predominantly meat, fish, fruits, and vegetables. The appropriate population, together with total turnover, is given in FAO (1975). When 12.3 percent is taken as the average retail margin on the commodities mentioned above (Department of Statistics, 1972*c*), the resulting estimate of gross output for urban market stalls is \$0.794M. We have increased this estimate to \$0.992M to account for the fact that urban stalls comprise only 80 percent of all such stalls in the region (FAO, 1975).

We have estimated gross output for semiurban shops by procedures and sources identical to those above (table C-11). For semiurban shops selling clothing and footwear and medical supplies no sample data were available, although the *Survey of Retail Establishments* provides estimates of annual turnover and population. To calculate gross output for these two shop types

Table C-11. *Regional Gross Value of Output for Shops by Type and Location*
(millions of Malaysian dollars)

Shop type	*Urban*	*Semiurban*
Clothing and footwear	0.911	0.095
Medical supplies and liquor	0.219	0.119
Electrical	0.576	0.203
Furniture	0.175	0.162
Sundries (general stores)	0.963	0.158
Firewood and charcoal	0.017	—
Stationers	1.107	0.059
Chemists	0.175	—
Building materials	0.182	—
Hardware	—	.487
Total	4.325	1.282

— Not applicable.

we have thus assumed a retail margin equal to that of semiurban sundry shops.

The FAO/World Bank survey of retail trade omitted two significant types of urban retail enterprises: petrol stations and motorcycle retailers. The gross outputs for these have been calculated as follows:

FAO (1975) reports 4.049 million gallons of petroleum sold in the region in 1972. At $2.00 per gallon and a 23 percent retail margin, total gross output is $1.863M. On the further assumption that half of this represents household consumption, the resulting gross output of gasoline stations is $0.932M.

From nonsurvey evidence, FAO (1975) estimates the turnover of motorcycle shops to be $7.675M, and the *Survey of Retail Establishments* reports a retail margin of 10.6 percent, which yields a gross output of $0.818M.

Rural retail shops numbered 1,843 in 1972 and had a total turnover of $25.987M (FAO, 1975). Assuming the same retail margin for all rural retailers as for sundry retail shops (the latter is given by FAO, 1975) yields a gross output of $0.738M for rural retail shops.

There are no regional data on slaughtering and meat processing. From NIOT, the ratio of gross output to input for the slaughtering industry in Western Malaysia was 185.762M/178.300M = $1.075M. Since, as part of our work for sector 19, the regional input of livestock has been independently estimated at $10.967M, the gross output for this component is: 0.075M × 10,967M = $0.823M.

Total gross output in retail trades is simply the sum of the seven elements detailed above: 4.325M + 0.992M + 1.282M + 0.932M + 0.818M + 4.738M + 0.823M = $13.911M.

We have estimated gross output for wholesale services by examining the demand for these services. All intermediate demands for sector 30 have been assumed to be intermediate wholesale demand, which totals $15.516M. Urban final demand has been calculated by imputing a wholesale margin on goods bought for final consumption by urban households. The margins have been drawn from *Principal Statistics of Wholesale Establishments in Alor Setar* (Department of Statistics, 1972*e*). These margins yield $2.443M on domestic goods and $1.304M on imports, or a total urban final demand for wholesale services of $3.747M. Using the same procedure, we have estimated that rural households spent $2.737M on wholesale services. But estimates in FAO (1975) suggest that only 62.5 percent was purchased from domestic wholesalers, which yields $1.539M. In addition, demand by exporters for wholesale services has been calculated to be $6.162M. Thus, total gross output in wholesale services is given by summing the four components above: 15.516M + 3.747M + 1.539M + 6.162M = $26.964M.

For the sector as a whole, we have added the gross outputs of retail and wholesale services in the region: 13.911M + 26.964M = $40.875M.

GROSS VALUE OF INPUTS. We have obtained the input structure for urban and semiurban retailers from the FAO/World Bank retail survey, while the costs for petrol stations and motorcycle shops have been derived from the authors' survey (see appendix A). Regarding market stalls, we concluded that their only significant cost was rent. The *Financial Statements* of the state governments for 1970 (Malaysia, 1971*a* and *b*) reveals that state governments in the region received revenues from market stalls in that year of $0.141M, which is equivalent to $480 per year per stall. Assuming that rents did not rise and knowing that there were 371 registered stalls in the region in 1972, we have an input from sector 43 of $0.178M. Again, we have derived the cost structure for rural retail shops directly from the FAO/World Bank survey. Lacking specific data on wholesalers, we have assumed them to have the same input coefficients as retailers selling the same product. Finally, we have used NIOT to calculate the input structure for slaughtering. Our results are summarized in table C-12.

Table C-12. *Cost Structure for Distributive Trades*
(millions of Malaysian dollars)

Sector of origin	*Producer prices*	*Trade margin*	*Transport margin*	*Market prices*
Domestic				
Manufacturing n.e.c.	0.010	—	—	0.010
Road transportation	0.352	—	—	0.351
Distributive trades	0.006	—	—	—
Electricity	0.698	—	—	0.698
Water	0.012	—	—	0.012
Posts and telecommunications	0.058	—	—	0.058
Other construction	0.050	—	—	0.050
Other government services	0.007	—	—	0.007
Traditional financial services	0.565	—	—	0.565
Modern financial services	0.230	—	—	0.230
Urban housing	1.873	—	—	1.873
Rural housing	0.619	—	—	0.619
Imports				
Chemicals	0.028	—	—	0.028
Food and beverages	0.067	0.004[a]	0.001[a]	0.072
Metal products	0.022	0.002	—	0.024
Total	4.597	0.006	0.001	4.597

— Not applicable.
a. One percent transport margin, 5.6 percent trade margin, based on NIOT.

ALLOCATION OF VALUE ADDED. Value added is 40.875M − 4.597M = \$36.278M. Using the *Census of Wholesale and Retail Establishments 1970* (Malaysia, 1972), we have calculated the regional ratio of profit to gross output for both wholesaling and retailing and then applied this ratio as appropriate to derive wholesale and retail profits of \$16.664M and \$3.909M, respectively. Hence, total profits have been estimated at \$20.573M, and wages and salaries at \$15.705M. However, firms are represented only in wholesaling, and they account for about half of gross output in that activity. Thus, we have estimated the distribution of value added to be \$27.946M to H_5 and \$8.332M to firms.

Residential Construction (35)

We have included residential construction not only because it lays claim to a significant share of regional gross output but also because we were able to use our regional data to improve significantly on the available official statistics in estimating regional gross output.

GROSS VALUE OF OUTPUT. Utilizing the *Survey of Construction Industries* (Malaysia, 1972, table 2.1) and scaling for the region, we obtain a gross output of \$12.282M. To this figure must be added the output of rural residential construction (building and improvement of rural houses) which is not covered, even implicitly, by the *Survey of Construction Industries*.

From the FAO/World Bank household survey, we calculate that double-cropping households made average outlays in 1973 of \$81.12 on new construction, improvements, and repairs, while single-cropping households spent \$35.09. Using the latter figure to represent all farm worker and nonproject rural households and the former figure for project farm and nonfarm households alike, we have estimated the additional output at \$7.363M. Further utilizing the FAO/World Bank household survey and knowledge from our own fieldwork, we have determined that this \$7.363M includes some \$0.700M for hired labor, but excludes about \$1.400M for the imputed value of family labor. Accordingly, the regional value of gross output of rural residential construction has been estimated to be 7.363M + 1.400M = \$8.763M in 1973. Adjusting for strong positive changes in real incomes and the prices of inputs, we calculate output to have been \$5.774M in 1972. Thus, regional gross output for the sector becomes 12.282M + 5.774M = \$18.056M.

GROSS VALUE OF INPUTS. We have derived the sector's cost structure mainly from the *Survey of Construction Industries* and supplementary data from FAO (1975, tables 58 and 61) concerning financial services. Total costs of intermediate inputs have been estimated at \$11.476M.

VALUE ADDED AND FACTOR COSTS. Value added is 18.056M − 11.476M = \$6.580M. From the *Survey of Construction Industries* the ratio of wages and salaries to gross output is 0.1921. Hence, the total outlay on wages and salaries in the urban sector is \$2.409M, which accrues wholly to H_5. For the rural sector, we have taken the proportion of households in each rural group multiplied by the value of family labor used, and then distributed the value of hired labor across households in proportion to their earnings from skilled labor activities as given by the FAO/World Bank household survey. This yields the following allocation of factor costs in the rural sector:

	\$ millions
Landless farm workers (H_1)	0.103
Land-abundant paddy farms (H_2)	0.479
Labor-abundant paddy farms (H_3)	0.623
Nonfarm households (H_4)	0.457
Total	1.662

Thus, profits are 6.580M − 2.409M − 1.662M = \$2.590M, which we have attributed wholly to firms in view of the fact that large contractors predominate in this sector.

Government, Education (38)

In the absence of regional data, our basic source was Meerman's (1979) study of public expenditure in Malaysia, from which we derived expenditure on education by the federal government in the states of Kedah and Perlis. On the assumption that expenditures in the various subregions of Kedah were proportional to population, we have derived regional gross output at \$43.221M. Applying the input coefficients from NIOT, we obtain gross value of inputs at producer prices of \$3.587M, which yields value added of \$42.377M. By definition, the latter is comprised exclusively of wages and salaries, which accrue wholly to regional nonfarm households (H_5).

Rural Housing (44)

Although the interindustry coefficients for rural housing are few and weak, this sector accounts for a large proportion of rural incomes.

GROSS VALUE OF OUTPUT. The FAO/World Bank household survey reveals the average age and value of rural houses to be thirteen years and \$1,040, respectively. We have assumed a 15 percent interest rate and a house life of twenty-five years in calculating the replacement cost C, as $C - 13(0.04C) = 1{,}040$, or $C = \$2{,}167$. Hence, annual interest charges are \$156.0 and yearly depreciation is \$86.7, giving a total of \$242.7. When this sum is added to mean household annual outlays on maintenance of \$12.7, the annual imputed rent on an average house is estimated to be \$255.4 in 1973. The ratio of maintenance costs to gross output is therefore 0.05, compared with 0.11 from NIOT. Since the latter covers both urban and rural dwellings, this discrepancy does not seem unreasonable. Assuming that all rural households live in similar dwellings and that there was a notional rise in rents between 1972 and 1973, and adding interest charges on maintenance (see below), we have estimated gross output in 1972 to be \$26.062M.

GROSS VALUE OF INPUTS. The sole inputs are maintenance and the interest charges on borrowings for maintenance. In keeping with our work and assumptions for residential construction (sector 35), deliveries of maintenance from construction are \$1.301M. The FAO/World Bank household survey shows the average household as borrowing \$1.805 annually for repairs and maintenance. Hence, using our calculations for sector 14, we estimate interest payments to traditional sources of credit to have been \$0.044M. Thus, the gross value of inputs is \$1.345M.

VALUE ADDED AND FACTOR COSTS. Value added is \$24.717M, which was allocated to the classes of rural households in proportion to their numbers.

Rest of the World

For the most part, outlays by the Rest of the World have been estimated as residuals. The exception is receipts accruing to state government. These comprise grants and loans from the federal government, which have been taken from FAO (1975, table 54) and Treasury of Malaysia, *Economic Report, 1973–74*, table 4.9, respectively.

References

The word "processed" describes works that are reproduced from typescript by mimeograph, xerography, or similar means; such works may not be cataloged or commonly available through libraries, or may be subject to restricted circulation.

Adelman, Irma, and Sherman Robinson. 1978. *Income Distribution Policy in the Developing Countries*. Oxford: Oxford University Press.

Affifudin, H. O. 1973*a*. "Some Aspects of the Socio-Economic Value System of the Muda Rice Farmers in Developmental Perspectives." Telok Chengai, Alor Setar: General Manager's Office, Muda Agricultural Development Authority.

———. 1973*b*. "The Social, Political and Economic Framework of Muda Rice Farmers: An Historical Perspective." Telok Chengai, Alor Setar: General Manager's Office, Muda Agricultural Development Authority.

Affifudin, H. O., and A. A. Nor. n.d. "A Brief Socio-Economic Study on Rural Farm Housewives in the Muda Scheme." Alor Setar. Processed.

Ali, S. Husin. 1975. *Malay Peasant Society and Leadership*. Kuala Lumpur: Oxford University Press.

Anand, Sudhir. 1977. "A Comparison of the IBRD and Little-Mirrlees Appraisal of a Highway Project in Malaysia." Washington, D.C.: World Bank. Processed.

———. Forthcoming. *Inequality and Poverty in Malaysia: Measurement and Decomposition*. New York: Oxford University Press.

Atkinson, A. B. 1970. "On the Measurement of Inequality." *Journal of Economic Theory* 2 (September):234–36.

Barnum, Howard N., and Lyn Squire. 1979. "An Econometric Application of the Theory of the Farm Household." *Journal of Development Economics* 6:79–102.

———. 1980. *A Model of an Agricultural Household*. Baltimore, Md.: Johns Hopkins University Press.

Bell, Clive. 1980. "The Future of Rice Monocultures in Malaysia." In P. C. Reining and B. Lenkerd, eds. *Village Viability in Contemporary Society*. Boulder, Colo.: Westview Press.

Bell, Clive, and Shantayanan Devarajan. 1980*a*. "Semi-Input-Output and Shadow Prices: A Critical Note." *Oxford Bulletin of Economics and Statistics* 42 (August):251–56.

———. 1980*b*. "Shadow Prices for Project Evaluation under Alternative Macroeconomic Regimes." Washington, D.C.: World Bank. Processed. (*Quarterly Journal of Economics*, forthcoming.)

Bisschop, Johannes, and A. Meeraus. "Matrix Augmentation and Partitioning in the Updating of the Basis Inverse." *Mathematical Programming* 13 (1977): 241–54.

Blitzer, C. R., I. M. D. Little, and Lyn Squire. 1978. "Multiplier Effects and Project Evaluation." Washington, D.C.: World Bank. Processed.

Blitzer, C. R., P. Dasgupta, and J. E. Stiglitz. 1981. "Project Appraisal and Foreign Exchange Constraints." *Economic Journal* 91 (March): 58–74.

Bródy, András, and Anne P. Carter, eds. 1972. *Input-Output Techniques: Proceedings of the Fifth International Conference on Input-Output Techniques*. Amsterdam: North-Holland.

Byron, R. P. 1978. "The Estimation of Large Social Accounts Matrices." *Journal of the Royal Statistical Society*, ser. A, 141:359–67.

Canada, Government of. 1977. *La Structure par entrées-sorties de l'économie canadienne: 1961–1971*. Ottawa: Statistique Canada.

Carter, Anne P., and András Bródy, eds. 1972. *Applications of Input-Output Analysis: Proceedings of the Fifth International Conference on Input-Output Techniques*. Amsterdam: North-Holland.

Chayanov, A. V. 1966. *The Theory of Peasant Economy*. Homewood, Ill.: Richard D. Irwin.

Chenery, Hollis B. 1959. "The Interdependence of Investment Decisions." In Moses Abramovitz and others. *The Allocation of Economic Resources*. Stanford, Calif.: Stanford University Press.

Day, R. H. 1963*a*. "On Aggregating Linear Programming Models of Production." *Journal of Farm Economics* 45:797–813.

———. 1963*b*. *Recursive Programming and Production Response*. Amsterdam: North-Holland.

Day, R. H., and I. J. Singh. 1977. *Economic Development as an Adaptive Process*. London: Cambridge University Press.

Department of Statistics. Various years. *Survey of Manufacturing Industries*. Kuala Lumpur.

———. 1968*a*. *Census of Manufacturing Industries, West Malaysia*. Kuala Lumpur.

———. 1968*b*. *Survey of Distributive Trades*. Kuala Lumpur.

———. 1972*a*. *Survey of Construction Industries*. Kuala Lumpur.

———. 1972*b*. *Census of Wholesale and Retail Establishments 1970*. Kuala Lumpur.

———. 1972*c*. *Survey of Retail Establishments in Western Malaysia*. Kuala Lumpur.

———. 1972*d*. *Yearbook of Transport Statistics*. Kuala Lumpur.

———. 1972*e*. *Principal Statistics of Wholesale Establishments in Alor Setar*. Kuala Lumpur.

———. 1974. *Monthly Bulletin of Statistics*. Kuala Lumpur.

Dobby, E. H. G. 1951. "The North Kedah Plain." *Economic Geography* 28:314–33.

Doering, Otto C. 1972. "Data Report of the 1970 Muda Agro-Economic Survey." Ithaca, N.Y.: Cornell University, Department of Agricultural Economics. Processed.

———. 1973. "Malaysian Rice Policy and the Muda River Irrigation Project." Ph.D. dissertation, Cornell University.

Duloy, J. H., and R. D. Norton. 1973. "Part IV: The Agricultural Sector." In L. M.

Goreux and A. S. Manne, eds. *Multi-Level Planning: Case Studies in Mexico*. Amsterdam: North-Holland.

———. 1975. "Prices and Incomes in Linear Programming Models." *American Journal of Agricultural Economics* 57:591–600.

Egypt, Arab Republic of. 1975. *Statistical Yearbook: 1952–1974*. Cairo.

FAO. 1975. *The Muda Study: A First Report*. 2 vols. Rome.

Forssell, Osmo. "Explaining Changes in Input-Output Coefficients for Finland." In Bródy and Carter (1972).

Goldman, R. N. 1975. "Staple Food Self-Sufficiency and the Distributive Impact of Malaysian Rice Policy." School of Comparative Social Sciences, Universiti Sains Malaysia. Processed.

Goldman, R. N., and Lyn Squire. 1977. "Technical Change, Labor Use and Income Distribution in the Muda Irrigation Project." Washington, D.C.: World Bank. Processed.

———. 1981. "Short-Run Direct Employment, Price and Income Effects of the Muda Irrigation Scheme." *Economic Development and Cultural Change*, forthcoming.

Hazell, P. B. R. 1979. "Endogenous Input Prices in Linear Programming Models." *American Journal of Agricultural Economics* 60 (August):477–81.

Hirschman, Albert O. 1959. *The Strategy of Economic Development*. New Haven and London: Yale University Press.

Hoerr, O. D. 1970. "Education, Income and Equity in Malaysia." Paper presented to the Conference of the Development Advisory Service, Harvard University, held at Dubrovnik.

Jegatheesan, S. 1972. "Contribution of Economic Research to Rice Mechanization in West Malaysia with Specific Reference to the Muda Irrigation Scheme." In H. Southworth, ed. *Farm Mechanisation in East Asia*. New York: Agricultural Development Council.

———. 1977. "The Green Revolution and the Muda Irrigation Scheme." Telok Chengai, Alor Setar: General Manager's Office, Muda Agricultural Development Authority.

Johnston, Bruce F., and Peter Kilby. 1975. *Agricultural and Structural Transformation*. London: Oxford University Press.

King, B. B. 1981. "What Is a SAM? A Layman's Guide to Social Accounting Matrices." Washington, D.C.: World Bank. Processed.

Krugman, Paul, and Lance Taylor. 1978. "Contractionary Effects of Devaluation." *Journal of International Economics* 8:445–56.

Kuchiba, M., and Y. Tsubouchi. 1966. "A Malay Padi Farming Community in the North Western Part of Malaya: A Sociological Analysis of Marriage, Divorce and Family." *Southeast Asian Studies* 1.

Kuyvenhoven, Arie. 1975. "The Integration of Project and Sector Analysis—Some Further Remarks." *Oxford Bulletin of Economics and Statistics* 38 (February):66–77.

———. 1978. *Planning with the Semi-Input-Output Method*. Leiden: Martinus Nyhoff.

Lele, Uma. 1975. *The Design of Rural Development*. Baltimore, Md.: Johns Hopkins University Press.

Little, I. M. D., and J. A. Mirrlees. 1974. *Project Appraisal and Development Planning for the Developing Countries*. London: Heinemann.

Logan, J. R. 1851. "Notes at Pinang and Kidah." *Journal of Indian Archipelago and Eastern Asia* 5.

Lysy, Frank J., and Lance Taylor. 1977. "A Computable General Equilibrium Model for the Functional Income Distribution: Experiments for Brazil, 1959–71." Washington, D.C.: World Bank. Processed.

MADA. 1970*a*. "A Comparative Economic Analysis of Conventional and Pedestrian Tractors in Rotary Cultivation." Telok Chengai, Alor Setar: General Manager's Office, Muda Agricultural Development Authority.

———. 1970*b*. "An Economic Report on Combine Harvesting in the Muda Irrigation Scheme." Telok Chengai, Alor Setar: General Manager's Office, Muda Agricultural Development Authority.

———. 1972. An Exercise in Integrated Agricultural Development." Telok Chengai, Alor Setar: General Manager's Office, Muda Agricultural Development Authority.

———. 1974. "Notes on the Urea Situation and the Prospects of Urea Manufacture." Telok Chengai, Alor Setar: General Manager's Office, Muda Agricultural Development Authority.

Malaysia, Government of. 1971*a*. *Financial Statement for 1970. State of Kedah*. Alor Setar: Audit Department.

———. 1971*b*. *Financial Statement for 1970. State of Perlis*. Kangor: Audit Department.

Masuo, K., and T. Yoshihiro. 1967. "Paddy Farming and Social Structure in a Malay Village." *The Developing Economies* 3 (September):9–34.

Meerman, Jacob P. 1979. *Public Expenditure in Malaysia*. New York: Oxford University Press.

Mellor, John W. 1976. *The New Economics of Growth*. Ithaca, N.Y.: Cornell University Press.

Middelhoek, A. J. "Tests of the Marginal Stability of Input-Output Coefficients." In Carter and Bródy (1972).

Ministry of Agriculture and Cooperatives in West Malaysia. 1967. *Farm Economic Survey of the Muda River Project, 1966/67*. Kuala Lumpur.

Ministry of Agriculture and Fisheries. Various years. *Agricultural Statistical Digest*. Kuala Lumpur.

———. 1971. *Statistical Digest: West Malaysia*. Kuala Lumpur.

Pyatt, Graham, J. Bharier, R. M. Lindley, R. M. Mabro, and Y. Sabolo. 1973. "A Methodology for Macro-Economic Projections." Mission Working Paper no. 12. In *Employment and Income Policies for Iran*. Geneva: International Labour Office.

Pyatt, Graham, Alan R. Roe, and associates. 1977. *Social Accounting for Development Planning: With Special Reference to Sri Lanka*. London: Cambridge University Press.

Pyatt, Graham, and Jeffrey Round. 1977. "Social Accounting Matrices for Development Planning." *Review of Income and Wealth*, ser. 23, pp. 339–64.

———. 1978. "The Distribution of Income and Social Accounts: A Study of Malaysia in 1970." Washington, D.C.: World Bank. Processed.

Rim, S. C. 1969. *Rubber Intercropping*. Kuala Lumpur: Rubber Research Institute.

Roumasset, J. A. 1976. *Rice and Risk*. Amsterdam: North-Holland.

Samy, J., R. H. Geissenhainer, Phang Cheng Chong, and Anna Wong. 1970. "Fertilizer Work with Padi Ria in the Muda Project." Telok Chengai, Alor Setar: Padi Experiment Station.

Scitovsky, Tibor. 1954. "Two Concepts of External Economies." *Journal of Political Economy* 62:143–51.

Scott, M. F-G. 1976. "Shadow Wages in Mauritius." In I. M. D. Little and M. F-G. Scott, eds. *Using Shadow Prices*. London: Heinemann.

Seton, F. 1973. "Shadow Prices for Chile." In R. Eckaus and P. Rosenstein-Rodan, eds. *Analysis of Development Problems: Studies in the Chilean Economy*. Amsterdam: North-Holland.

Sevaldson, Per. "The Stability of Input-Output Coefficients." In Carter and Bródy (1972).

Soo, Swee Weng. 1972. "Semi-detailed Soil Survey of the Kedah-Perlis Coastal Plain." Kuala Lumpur: Department of Agriculture. Processed.

Squire, Lyn, and Herman G. van der Tak. 1975. *Economic Analysis of Projects*. Baltimore, Md.: Johns Hopkins University Press.

Srinivasan, T. N. 1978. "General Equilibrium Theory, Project Evaluation and Economic Development." Walras-Bowley Lecture. Econometric Society, Boulder, Colo.

Stone, Richard. "Foreword." In Pyatt, Roe, and associates (1977).

Swift, M. G. 1967. "Economic Concentration and Malay Peasant Society." In M. Freedman, ed. *Social Organisation: Essays Presented to Raymond Firth*. Chicago: Aldine.

Tinbergen, Jan. 1966. "Some Refinements of the Semi-Input-Output Method." *Pakistan Development Review* 6:243–47.

Treasury of Malaysia. Various years. *Economic Report*. Kuala Lumpur.

University Sains. 1974. "Land Tenure Survey." Pinang: Muda Agricultural Development Authority and University Sains.

UNIDO. 1972. *Guidelines for Project Evaluation*. New York: United Nations.

Veitch, M. D. 1976. "National Parameters for Project Appraisal in Malaysia." Vol. 1. Kuala Lumpur: Economic Planning Unit.

Visaria, Pravin. 1979. "Incidence of Poverty and the Characteristics of the Poor in Peninsular Malaysia." Washington, D.C.: World Bank. Processed.

Wanhill, S. R. C. 1974. "A Note on the Integration of Project and Sector Analysis." *Oxford Bulletin of Economics and Statistics* 36:109–13.

Wincott, Nicholas. 1975*a*. "A Brief Methodological Note on the 'Significant Unit' for Socio Economic Analysis in the Study of Farmers' Behavior in the Muda Irrigation Scheme." Washington, D.C.: World Bank. Processed.

———. 1975*b*. "A Tentative Suggestion as to the Significance of Cyclical Changes in the Structure of the Domestic Group for the Study of the Dynamics of Padi Farming in Kawasan Muda." Washington, D.C.: World Bank. Processed.

Wong, I. F. T. 1966. "The Present Land Use of West Malaysia." Kuala Lumpur: Department of Agriculture.

Yamashita, M., S. Jegatheesan, and C. Y. Wong. 1976. "Agro-Economic Studies in the Muda Region." Telok Chengai, Alor Setar: General Manager's Office, Muda Agricultural Development Authority.

Index

Page numbers in italics indicate tables.

Clive Bell is a senior economist in the Development Research Department, and Roger Slade is an economist in the South Asia Projects Department, at The World Bank; Peter Hazell is a research fellow at the International Food Policy Research Institute.